METHODS IN MOLECULAR BIOLOGY™

Series Editor
John M. Walker
School of Life Sciences
University of Hertfordshire
Hatfield, Hertfordshire, AL10 9AB, UK

For other titles published in this series, go to
www.springer.com/series/7651

Fluorescence in situ Hybridization (FISH)

Protocols and Applications

Edited by

Joanna M. Bridger

Division of Biosciences, School of Health Sciences & Social Care, Brunel University, Uxbridge, UK

Emanuela V. Volpi

Wellcome Trust Centre for Human Genetics, University of Oxford, Oxford, UK

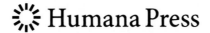 Humana Press

Editors
Joanna M. Bridger, Ph.D.
School of Health Sciences & Social Care
Brunel University
Uxbridge
UK
joanna.bridger@brunel.ac.uk

Emanuela V. Volpi, Ph.D.
Wellcome Trust Centre for Human Genetics
University of Oxford
Oxford
UK
emanuela.volpi@well.ox.ac.uk

ISSN 1064-3745   e-ISSN 1940-6029
ISBN 978-1-60761-788-4   e-ISBN 978-1-60761-789-1
DOI 10.1007/978-1-60761-789-1
Springer New York Dordrecht Heidelberg London

Library of Congress Control Number: 2010935354

Printed on acid-free paper

Humana Press is part of Springer Science+Business Media (www.springer.com)

Preface

Fluorescence in situ Hybridization (FISH) belongs to that special category of well-established molecular biology techniques that, since their inception a few decades ago, have succeeded in keeping a prominent position within the constantly expanding list of laboratory procedures for biomedical research and clinical diagnostics.

The design simplicity and cost-effectiveness of the early FISH protocols, combined with the significant acceleration of discoveries in related technical areas such as fluorescence microscopy, digital imaging, and nucleic acid technology have prompted the diversification of the original technique into an outstanding number of imaginative and useful applications, and thus have not only held back its outmoding but have also promoted its expansion into different areas of basic and applied research in the post-genomic era.

The 34 chapters included in this book aim at portraying the vibrant complexity and diversity of the current FISH protocol landscape, providing cutting-edge examples of various applications for genetic and developmental research, cancer research, reproductive medicine, diagnostic and prognostic purposes, microbial ecology, and evolutionary studies. The book is divided in four parts: (I) Core Techniques, (II) Technical Advancements and Novel Adaptations, (III) Translational FISH: Applications for Human Genetics and Medicine, and (IV) Protocols for Model Organisms.

Part I brings together a comprehensive range of "foundation" protocols from well-established molecular cytogenetics and chromosome biology laboratories. The understanding of the theoretical and practical aspects of these core protocols should provide the beginners with a solid and inspiring introduction to the technique but should also function more widely as an updated and reliable groundwork reference for every person in the field. Protocols for notoriously challenging techniques such as RNA-FISH, Multiplex-FISH, and immunoFISH on chromatin fibers are included and explained in detail in Chaps. 3, 6, and 7, respectively. This section also includes newer applications of basic FISH on tissue microarrays and tissue sections in Chaps. 4 and 5, respectively.

The following three parts show how the original protocols have been aptly and deftly modified and/or applied to specific investigative purposes. More specifically, Part II brings together protocols that are particularly innovative such as COMET-FISH, the use of Quantum dots in FISH, CO-FISH, Cryo-FISH, and multicolor banding. Of particular note is the closing chapter (Chap. 17) describing an exciting new development that pushes the boundaries of FISH so far that FISH no longer needs to be performed on fixed material but can be elicited in living cells.

Part III focuses on the role of FISH and the future prospects in medical and clinical settings for genetic-based diagnostics and prognosis. This part starts with an important chapter (Chap. 18) that sets out how to ensure quality control in the laboratory and how important this is to clinical uses of FISH. The section further develops into the use of FISH in prenatal and pre-implantation diagnosis (Chaps. 19 and 20) and covers techniques for parental origin determination (Chap. 22), high-resolution CGH, and isolated nuclei from biopsy material. The last two chapters (Chaps. 25 and 26) describe a new way of applying FISH in a 3D tissue culture model of tumor biology and a combined flow sorting/FISH protocol for HIV-1 disease monitoring.

Part IV focuses on the development of FISH for different model organisms, and this section transverses the animal kingdom, starting with Chap. 27 and FISH on bacteria, through yeast (Chap. 28) and *Drosophila* (Chap. 30) to mammals, especially 3D preserved in vitro constructed embryos (Chaps. 31 and 34).

The subdivision of chapters into four parts provides a framework for an ordered consultation of the book. However, this "vertical" organization is largely conventional and should not deter the readers from broadening their perspective by also consulting the book in a "horizontal" fashion, cross-referencing between different parts and comparing protocols and troubleshooting notes from a range of laboratories.

Coordinating the efforts of so many distinguished and resourceful investigators has been an honor, and editing this book has been a profoundly invigorating and uplifting experience. Once again, thank you all for your hard work and beautiful contributions.

Uxbridge, UK   *Joanna M. Bridger*
Oxford, UK   *Emanuela V. Volpi*

Contents

Contributors

MARIA CARMELA ACCARDO • *Dipartimento di Genetica e Biologia Molecolare and Istituto Pasteur-Fondazione Cenci-Bolognetti, Università "La Sapienza", Rome, Italy*

TIPHAINE AGUIRRE-LAVIN • *INRA, UMR 1198 Biologie du Développement et Reproduction, Jouy en Josas, France*

PETER F. AMBROS • *St. Anna Kinderkrebsforschung, CCRI, Children's Cancer Research Institute, Vienna, Austria*

RHONA ANDERSON • *Laboratory of Genome Damage, Division of Biosciences, Centre for Cell and Chromosome Biology, Brunel University, Uxbridge, Middlesex, UK*

SUSAN M. BAILEY • *Department of Environmental & Radiological Health Sciences, Colorado State University, Fort Collins, CO, USA*

ISABELLE BATAILLON • *INRA, UMR 1198 Biologie du Développement et Reproduction, Jouy en Josas, France*

NATHALIE BEAUJEAN • *INRA, UMR 1198 Biologie du Développement et Reproduction, Jouy en Josas, France*

LAURENT A. BENTOLILA • *University of Chemistry and Biochemistry, California NanoSystems Institute, University of California at Los Angeles, Los Angeles, CA, USA*

CARMEN BOXLER • *Kirchhoff-Institute of Physics, University of Heidelberg, Heidelberg, Germany*

EVA BOZSAKY • *St. Anna Kinderkrebsforschung, CCRI, Children's Cancer Research Institute, Vienna, Austria*

ALESSANDRO BRERO • *Institute of Human Genetics, Biozentrum (LMU), University of Münich, Planegg-Martinsried, Germany*

JOANNA M. BRIDGER • *Laboratory of Nuclear and Genomic Health, Centre for Cell and Chromosome Biology, Division of Biosciences, School of Health Sciences and Social Care, Brunel University, Uxbridge, Middlesex, UK*

ANNEKE K. BROUWER • *Department of Molecular Cell Biology, Leiden University Medical Centre, Leiden, The Netherlands*

JILL M. BROWN • *MRC Molecular Haematology Unit, John Radcliffe Hospital, Weatherall Institute of Molecular Medicine, Headington, Oxford, UK*

VERONICA J. BUCKLE • *MRC Molecular Haematology Unit, John Radcliffe Hospital, Weatherall Institute of Molecular Medicine, Headington, Oxford, UK*

MAHESH A. CHOOLANI • *Department of Obstetrics and Gynaecology, Yong Loo Lin School of Medicine, National University of Singapore, Singapore*

MICHAEL N. CORNFORTH • *Department of Radiation Oncology, University of Texas Medical Branch, Galveston, TX, USA*

CHRISTOPH CREMER • *Kirchhoff-Institute of Physics, University of Heidelberg, Heidelberg, Germany*

MARION CREMER • *Institute of Human Genetics, Biozentrum (LMU),
University of Münich, Planegg-Martinsried, Germany*

THOMAS CREMER • *Institute of Human Genetics, Biozentrum (LMU),
University of Münich, Planegg-Martinsried, Germany*

PASCALE DEBEY • *INRA, UMR 1198 Biologie du Développement et Reproduction,
Jouy en Josas, France*

PATRIZIO DIMITRI • *Dipartimento di Genetica e Biologia Molecolare and Istituto
Pasteur-Fondazione Cenci-Bolognetti, Università "La Sapienza", Rome, Italy*

ROELAND W. DIRKS • *Department of Molecular Cell Biology, Leiden University
Medical Centre, Leiden, The Netherlands*

LAUREN S. ELCOCK • *Laboratory of Nuclear and Genomic Health,
Division of Biosciences, Centre for Cell and Chromosome Biology,
School of Health Sciences and Social Care, Brunel University, West London, UK*

HELEN A. FOSTER • *Division of Biosciences, Brunel University,
Uxbridge, Middlesex, UK*

ELISA GARIMBERTI • *Laboratory of Leukaemia and Chromosome Research, Division
of Biosciences, Centre for Cell & Chromosome Biology, Brunel Institute for Cancer
Genetics and Pharmacogenomics, Brunel University, Uxbridge, Middlesex, UK*

EDWIN H. GOODWIN • *Kroma TiD Inc., Fort Collins, CO, USA*

DARREN K. GRIFFIN • *Department of Biosciences, University of Kent,
Canterbury, Kent, UK*

MADELEINE GROSS • *Institut für Humangenetik, University of Jena, Jena, Germany*

SAMIR HAMAMAH • *Service de Biologie de la Reproduction, Hôpital Arnaud de
Villeneuve, CHRU Montpellier, Montpellier, France*

ROS HASTINGS • *UK NEQAS for Clinical Cytogenetics, Women's Centre,
John Radcliffe Hospital, Oxford, UK*

MICHAEL HAUSMANN • *Kirchhoff-Institute of Physics, University of Heidelberg,
Heidelberg, Germany*

LAETITIA HERZOG • *INRA, UMR 1198 Biologie du Développement et Reproduction,
Jouy en Josas, France*

SOPHIE HINREINER • *Institut für Humangenetik und Anthropologie,
University of Jena, Jena, Germany*

SHERRY S.Y. HO • *Department of Obstetrics and Gynaecology, Yong Loo Lin School
of Medicine, National University of Singapore, Singapore*

WANNAPORN ITTIPRASERT • *Biomedical Research Institute, Rockville, MD, USA*

ANDREW JEFFERSON • *Department of Cardiovascular Medicine, John Radcliffe
Hospital, Oxford, UK*

NINA KETTERL • *Institute of Human Genetics, Biozentrum (LMU),
University of Münich, Planegg-Martinsried, Germany*

MATTY KNIGHT • *Biomedical Research Institute, Rockville, MD, USA*

DANIELA KOEHLER • *Institute of Human Genetics, Biozentrum (LMU),
University of Münich, Planegg-Martinsried, Germany*

NADEZDA KOSYAKOVA • *Institut für Humangenetik und Anthropologie,
University of Jena, Jena, Germany*

AGATA KOWALSKA • *St. Anna Kinderkrebsforschung, CCRI, Children's
Cancer Research Institute, Vienna, Austria*

ROLAND KRÄMER • *Institute of Inorganic Chemistry, University of Heidelberg, Heidelberg, Germany*

LIRON-MARK LAVITAS • *MRC Clinical Sciences Centre, Imperial College School of Medicine, London, UK*

HENRY J. LEESE • *Biology Department, University of York, York, UK*

THOMAS LIEHR • *Institut für Humangenetik und Anthropologie, University of Jena, Jena, Germany*

JOSEF LOIDL • *Department of Chromosome Biology, University of Vienna, Vienna, Austria*

WALID MAALOUF • *INRA, UMR 1198 Biologie du Développement et Reproduction, Jouy en Josas, France; University of Edinburgh, Edinburgh, UK*

GORAN MATTSSON • *Dako A/S, Glostrop, Denmark*

KAREN J. MEABURN • *National Cancer Institute, NIH, Bethesda, MD, USA*

ANDRE MILLER • *Biomedical Research Institute, Rockville, MD, USA*

HASMIK MKRTCHYAN • *Institut für Humangenetik, University of Jena, Jena, Germany*

ANDRIY MOKHIR • *Institute of Inorganic Chemistry, University of Heidelberg, Heidelberg, Germany*

ZOIA L. MONACO • *Wellcome Trust Centre for Human Genetics, University of Oxford, Oxford, UK*

DANIELA MORALLI • *Wellcome Trust Centre for Human Genetics, University of Oxford, Oxford, UK*

CÉCILE MONZO • *Service de Biologie de la Reproduction, Hôpital Arnaud de Villeneuve, CHRU Montpellier, Montpellier, France*

KRISTIN MRASEK • *Institut für Humangenetik und Anthropologie, University of Jena, Jena, Germany*

PATRICK MÜLLER • *Kirchhoff-Institute of Physics, University of Heidelberg, Heidelberg, Germany*

EDWIN C. ODOEMELAM • *Division of Biosciences, Brunel University, Uxbridge Middlesex, UK*

CAROLINE MACKIE OGILVIE • *Guy's and St Thomas' Centre for Preimplantation Genetic Diagnosis, Guy's Hospital, London, UK*

BRUCE PATTERSON • *Department of Pathology, Stanford University Medical School, Stanford, CA, USA*

FRANCK PELLESTOR • *Service de Biologie de la Reproduction, Hôpital Arnaud de Villeneuve, CHRU Montpellier, Montpellier, France*

ANA POMBO • *MRC Clinical Sciences Centre, Imperial College School of Medicine, London, UK*

NITHYA RAGHAVAN • *Biomedical Research Institute, Rockville, MD, USA*

HARRY SCHERTHAN • *Institut für Radiobiologie der Bundeswehr, Munich, Germany; Max Planck Institute for Molecular Genetics, Berlin, Germany*

EBERHARD SCHMITT • *Kirchhoff-Institute of Physics, University of Heidelberg, Heidelberg, Germany; Leibniz-Institute for Age Research, Jena, Germany*

JUTTA SCHWARZ-FINSTERLE • *Kirchhoff-Institute of Physics, University of Heidelberg, Heidelberg, Germany*

PAUL N. SCRIVEN • *Cytogenetics Department, Guy's Hospital, London, UK*

Janet M. Shipley • *Molecular Cytogenetics, Male Urological Cancer Research Centre, Institute of Cancer Research, Sutton, Surrey, UK*

Asli N. Silahtaroglu • *Wilhelm Johannsen Centre for Functional Genome Research, ICMM, The Panum Institute, Copenhagen N, Denmark*

Irina Solovei • *Institute of Human Genetics, Biozentrum (LMU), University of Münich, Planegg-Martinsried, Germany*

Graciela Spivak • *Department of Biology, Stanford University, Stanford, CA, USA*

Stefan Stein • *Kirchhoff-Institute of Physics, University of Heidelberg, Heidelberg, Germany*

Paula J. Stokes • *Biology Department, University of York, York, UK*

Roger G. Sturmey • *Biology Department, University of York, York, UK*

Beth Sullivan • *Department of Molecular Genetics and Microbiology, Institute for Genome Sciences & Policy, Duke University, Durham, NC, USA*

Brenda M. Summersgill • *Molecular Cytogenetics, Male Urological Cancer Research Centre, Institute of Cancer Research, Sutton, Surrey, UK*

Soo-Yong Tan • *Department of Pathology, Singapore General Hospital, Singapore*

Hans J. Tanke • *Department of Molecular Cell Biology, Leiden University Medical Centre, Leiden, The Netherlands*

Sabrina Tosi • *Laboratory of Leukaemia and Chromosome Research, Division of Biosciences, Centre for Cell & Chromosome Biology, Brunel Institute for Cancer Genetics and Pharmacogenomics, Brunel University, Uxbridge, Middlesex, UK*

Emanuela V. Volpi • *Wellcome Trust Centre for Human Genetics, University of Oxford, Oxford, UK*

Anja Weise • *Institut für Humangenetik und Anthropologie, University of Jena, Jena, Germany*

Joop Wiegant • *Department of Molecular Cell Biology, Leiden University Medical Centre, Leiden, The Netherlands*

Eli S. Williams • *Drug Discovery Program, H. Lee Moffitt Cancer Centre and Research Institute, Tampa, FL, USA*

Vera Witthuhn • *Institut für Humangenetik und Anthropologie, University of Jena, Jena, Germany*

Eckhard Wolf • *Institute of Human Genetics, Biozentrum (LMU), University of Münich, Planegg-Martinsried, Germany*

Sheila Q. Xie • *MRC Clinical Sciences Centre, Imperial College School of Medicine, London, UK*

Valeri Zakhartchenko • *Institute of Human Genetics, Biozentrum (LMU), University of Münich, Planegg-Martinsried, Germany*

Katrin Zwirglmaier • *British Antarctic Survey, Cambridge, UK*

Part I

Core Techniques

Chapter 1

Fluorescence *in situ* Hybridization (FISH), Basic Principles and Methodology

Elisa Garimberti and Sabrina Tosi

Abstract

Fluorescence in situ hybridization (FISH) is widely used for the localization of genes and specific genomic regions on target chromosomes, both in metaphase and interphase cells. The applications of FISH are not limited to gene mapping or the study of genetic rearrangements in human diseases. Indeed, FISH is increasingly used to explore the genome organization in various organisms and extends to the study of animal and plant biology. We have described the principles and basic methodology of FISH to be applied to the study of metaphase and interphase chromosomes.

Key words: Chromosomes, DNA probes, Epifluorescence microscopy, Fluorescence in situ Hybridization, Interphase nuclei

1. Introduction

The technique of fluorescence in situ hybridization (FISH) is based on the same principle as any DNA hybridization method that uses the ability of single-stranded DNA to anneal to complementary DNA. In the case of FISH, the target DNA, which may be metaphase chromosomes, interphase nuclei or tissue sections, is attached to a glass microscope slide. The advent of FISH represented an important step in the field of cytogenetics and since its development (1), it has been used extensively in both research and diagnostics. The advantage of FISH over other in situ hybridization methods (radioactive or immunocytochemical) is mainly due to a markedly improved speed and spatial resolution. Moreover, labeled probes are stable and can be stored over long periods, and statistical analysis due to nonspecific signals on metaphase chromosomes is no longer required. Furthermore, FISH offers the possibility

Joanna M. Bridger and Emanuela V. Volpi (eds.), *Fluorescence in situ Hybridization (FISH): Protocols and Applications*, Methods in Molecular Biology, vol. 659, DOI 10.1007/978-1-60761-789-1_1, © Springer Science+Business Media, LLC 2010

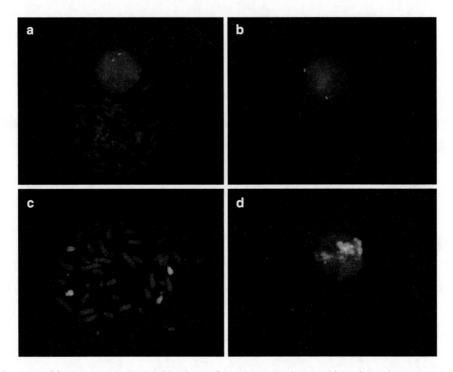

Fig. 1. Examples of fluorescence in situ hybridization performed on metaphase and interphase chromosomes obtained from the peripheral blood lymphocytes of a normal individual. (**a**, **b**) Single color FISH using the BAC clone RP11-299P2 containing the *BCL2* gene (courtesy of Mariano Rocchi, University of Bari, Italy). The BAC clone has been labeled with biotin and detected with Texas Red conjugated to avidin. Red fluorescent signals are visible on nuclei (**a**, **b**) as well as on both chromosomes 18 on the metaphase spread (**a**). (**c**, **d**) Dual color FISH has been performed using a biotin-labeled whole chromosome paint for chromosome 14 (Cambio, Cambridge, UK) detected with Texas Red conjugated avidin and a FITC labeled whole chromosome paint for chromosome 18 (Cambio). Both chromosomes 14 (*in red*) and 18 (*in green*) are visible on the metaphase (**c**) and the respective chromosome territories are visible in the nucleus (**d**). Chromosomal DNA is stained with DAPI (*blue*).

of detection of multiple targets simultaneously in different colors. A range of probes is currently available for FISH, and numerous combinations are available commercially. Specific probe sets have been designed for the detection of chromosome rearrangements in different pathologies; these include constitutional abnormalities and cancer associated aberrations (see Fig. 1).

2. Materials

2.1. Preparation of Target Material

1. Sodium heparin tubes (BD).

2. RPMI 1640 medium.

2.1.1. Peripheral Blood Cell Culture

3. Fetal calf serum.

4. Penicillin.

5. Streptomycin.

6. Glutamine.

7. Phytoemagglutinin (PHA) (Invitrogen, Paisley, UK).

8. Counting chamber (Fisher Scientific).

2.1.2. Harvesting of Cell Cultures and Slide Preparation

1. Colcemid (10 µg/mL) (Invitrogen).

2. Hypotonic solution: 0.075 M Potassium chloride (KCl).

3. Fixative solution: 3:1 Methanol: Glacial acetic acid.

4. Superfrost™ microscope slides (Fisher Scientific).

5. Silica gel (GeeJay Chemicals Ltd).

6. Slide storage boxes (Kartell).

2.2. Preparation of the Probe DNA

2.2.1. DNA Extraction

2.2.1.1. BAC and PAC DNA

1. LB medium: 1% (w/v) Sodium Chloride (NaCl), 1% (w/v) Bactotryptone (Fisher Scientific), and 0.5% (w/v) Yeast extract (Fisher Scientific).

2. Agar LB medium: LB medium, 1.5% (w/v) Agar.

3. Chloramphenicol.

4. Kanamycin.

5. Petri dishes.

6. Sterile plastic loops.

7. Glycerol (Acros Organic).

8. P1 solution: 15 mM Tris-HCl pH 8.0, 10 mM Ethylene-diamine Tetraacetic Acid (EDTA) (Fisher Scientific), 100 µg/mL RNase A (Sigma-Aldrich).

9. P2 solution: 0.2 M Sodium Hydroxide (NaOH) (Fisher Scientific), 1% Sodium Dodecyl Sulphate (SDS) (Fisher Scientific).

10. P3 solution: 3 M Potassium Acetate.

11. Ethanol.

12. Agarose.

2.2.1.2. Cosmid DNA

1. QIAprep® Miniprep Kit (Qiagen).

2. Agarose.

2.2.2. DNA Labeling

2.2.2.1. Nick Translation

1. Nick Translation System (Invitrogen).

2. Biotin-16-dUTP (Roche).

3. Digoxigenin-11-dUTP (Roche).

4. Agarose.

5. Microspin G50 Columns (Amersham).

6. Salmon Sperm DNA (Invitrogen).

7. Sodium Acetate 3 M.

8. Ethanol.

9. Vacuum desiccator.

Postlabeling Treatment of the Probe

1. C_0t–1 human DNA (Roche).

2. 20× Saline Sodium Citrate (SSC): 175.3 g of Sodium Chloride, 88.2 g of Sodium Citrate. Add 1 L of dd H_2O, pH 7.0, autoclave before storage at room temperature.

3. Hybridization buffer: 50% (v/v) Formamide (purity: 99.5%+), 10% (w/v) dextran sulphate, 1% (v/v) Tween®20, 2× SSC, pH 7.0.

2.2.2.2. Degenerated Oligonucleotide Primer (DOP)-PCR

1. 10× Oligonucleotide primer (DOP)-PCR buffer (Helena Biosciences).

2. Nucleotides (dATP, dCTP, dGTP, and dTTP) (Eppendorf, Fisher Scientific).

3. DOP-PCR Primers (Sigma-Aldrich).

4. Taq Supreme Polymerase (Kapa Biosystem).

5. DNase-free dd H_2O (VWR International).

6. Agarose.

7. Biotin-16-dUTP (Roche).

8. Digoxigenin-11-dUTP (Roche).

9. Vacuum desiccator.

Postlabeling Treatment of the Probe

1. C_0t–1 Human DNA (Roche).

2. Salmon Sperm DNA (Invitrogen).

3. Sodium Acetate.

4. Ethanol.

5. 20× SSC: 175.3 g of Sodium Chloride, 88.2 g of Sodium Citrate. Add 1 L of dd H_2O, pH 7.0, autoclave before storage at room temperature.

6. Hybridization buffer: 50% (v/v) Formamide (purity: 99.5%+), 10% (w/v) Dextran Sulphate, 1% (v/v) Tween®20, 2× SSC, pH 7.0.

2.3. Fluorescence In Situ Hybridization

2.3.1. Pretreatment of the Slides

1. 20× SSC: 175.3 g of Sodium Chloride, 88.2 g of Sodium Citrate. Add 1 L of dd H_2O, pH 7.0, autoclave before storage at room temperature.

2. Parafilm M.

3. Ethanol.

4. Denaturing solution (70% Formamide in 2× SSC): 10 mL of dd H_2O + 5 mL of 20× SSC, 25 µL of 250 mM EDTA, 35 mL of formamide (purity: 99.5%+), pH 7.0, make up fresh.

5. Coplin jars.

2.3.2. Hybridization and Posthybridization Washes	1. Plastic slide storage boxes (Kartell). 2. Rubber solution (Weldtite Product). 3. Microscope coverslips (22 × 22 mm). 4. Parafilm M. 5. 20× SSC: 175.3 g of Sodium Chloride, 88.2 g of Sodium Citrate. Add 1 L of dd H_2O, pH 7.0, autoclave before storage at room temperature. 6. 4× SSC/Tween®20: Dilute the stock solution 20× SSC to a final dilution of 4× SSC and add 500 µL of Tween®20 to reach a final concentration of 0.05%. 7. Coplin jars.

2.3.2. Hybridization and Posthybridization Washes

1. Plastic slide storage boxes (Kartell).

2. Rubber solution (Weldtite Product).

3. Microscope coverslips (22 × 22 mm).

4. Parafilm M.

5. 20× SSC: 175.3 g of Sodium Chloride, 88.2 g of Sodium Citrate. Add 1 L of dd H_2O, pH 7.0, autoclave before storage at room temperature.

6. 4× SSC/Tween®20: Dilute the stock solution 20× SSC to a final dilution of 4× SSC and add 500 µL of Tween®20 to reach a final concentration of 0.05%.

7. Coplin jars.

2.3.3. Detection

1. Slide storage boxes (Kartell).

2. 20× SSC: 175.3 g of Sodium Chloride, 88.2 g of Sodium Citrate. Add 1 L of dd H_2O, pH 7.0, autoclave before storage at room temperature.

3. 4× SSC/Tween®20: Dilute the stock solution 20× SSC to a final dilution of 4× SSC and add 500 µL of Tween®20.

4. Blocking solution: 1.8 g (3% w/v) of bovine serum albumin (BSA) (Acros Organic) in 60 mL of 4× SSC/Tween®20. Filter through a 0.20 µm syringe filter.

5. 1× Phosphate Buffered Saline (PBS) (Sigma-Aldrich).

6. Coplin jars.

7. DAPI (4′,6-diamidino-2-phenylindole dihydrochloride)/Antifade (0.1 µg/mL) (MP Biochemicals).

8. Glass coverslips (22 × 40 mm).

9. Rubber solution (Weldtite Product).

2.3.3.1. Biotin-Labeled Probes

1. Avidin D-Texas Red (Vector Laboratories).

2. Biotinylated anti-Avidin D (Vector Laboratories).

2.3.3.2. Digoxigenin-Labeled Probes

1. Mouse anti-Digoxigenin (Sigma-Aldrich).

2. Rabbit anti-Mouse-FITC (Sigma-Aldrich).

3. Monoclonal anti-Rabbit-FITC (Sigma-Aldrich).

3. Methods

3.1. Preparation of Target Material

3.1.1. Peripheral Blood Cell Culture

Cell cultures can be set up using a variety of sources, depending on the particular aim of the study. Briefly, we describe here the cell culture and harvesting methods for lymphocytes derived from human peripheral blood. For different cell types and specific methods, see ref. 2.

1. Fresh peripheral blood samples are collected in a 10 mL centrifuge tube containing Sodium Heparin (see Note 1), and the number of cells is assessed using a counting chamber.

2. Set up one to four cultures using RPMI 1640 medium supplemented with 20% (v/v) of fetal calf serum, 50 IU/mL of Penicillin, 50 mg/mL of Streptomycin, 2 mM L-glutamine, and 0.225 mg/mL of Phytoemagglutinin. Add blood to a final concentration of 1×10^6 cells/mL.

3. Incubate at 37°C for 72 h (see Note 2).

3.1.2. Harvesting of Cell Cultures and Slide Preparation

1. At the end of the culture time, add colcemid (0.1 µg/mL) to the cell cultures and incubate at 37°C for 1 h.

2. Harvest the cells by centrifuging at $250 \times g$ for 5 min.

3. Discard the supernatant, add drop wise 10 mL of hypotonic solution, and incubate for 15 min at 37°C.

4. Add 0.5 mL of fixative solution and centrifuge at $250 \times g$ for 5 min.

5. Discard the supernatant, add 1 mL of fixative solution, and resuspend the pellet by vortexing.

6. Add 9 mL of fixative solution slowly and incubate at room temperature for 15 min.

7. Centrifuge at $250 \times g$ for 5 min.

8. Discard the supernatant, add 9 mL of fixative solution, and centrifuge at $250 \times g$ for 5 min.

9. Repeat point eight until the cell suspension is clear.

10. Resuspend the cell pellet in 2 mL of fixative solution.

11. Drop 8 µL of cell suspension onto a clean glass slide, and let to air-dry (see Note 3).

12. Check the concentration and quality (spread and humidity) of nuclei and chromosomes at the microscope (see Note 4).

13. Slides can be left at room temperature and used for FISH after 12 h. If FISH is planned after more than 5 days, storage of slides at –20°C is advised. In this case, slides should be placed in appropriate storage boxes in the presence of silica gel. Storage boxes should be sealed with tape in order to avoid penetration of moisture (see Note 5).

3.2. Preparation of Probe DNA

Different cloning vectors can be used for FISH (see Table 1); however, the ones most commonly employed in conventional two-dimensional (2D) FISH are BAC, PAC, and cosmid clones. The FISH protocol described in this chapter is based on the use of these clones as well as the use of painting probes generated by DOP-PCR.

Table 1
Examples of cloning vectors used for FISH

Vector	Host	Structure	Insert size (kb)
Cosmids	*Escherichia coli*	Circular plasmid	35/45
P1 clones	*E. coli*	Circular plasmid	70/100
BACs	*E. coli*	Circular plasmid	Up to 300
PACs	*E. coli*	Circular plasmid	100/300
YACs	*Saccharomyces cerevisiae*	Linear chromosomes	100/2,000

BACs bacterial artificial chromosomes; *PACs* P1-derived chromosomes; *YACs* yeast artificial chromosomes (3)

3.2.1. DNA Extraction

3.2.1.1. BAC and PAC DNA

The bacterial cells containing the probe can be delivered either as agar stabs or in vials containing glycerol; the latter can be stored at −80°C for long periods of time while the former can be stored at 4°C only for few days. The bacterial cells are first grown as streaks on an agar plate; colonies are then expanded in a liquid culture.

1. Solid culture: Spread the bacterial cells with a sterile loop onto the surface of LB agar in a petri dish, supplemented with the required antibiotic (usually 12.5 µg/mL of chloramphenicol or kanamycin) (see Note 6).

2. Incubate the Petri dish at 37°C for at least 24 h, and check the bacterial growth.

3. Liquid culture: With a sterile loop, collect several colonies and inoculate 5 mL of LB medium supplemented with the required antibiotic (see Note 7).

4. Grow the cells at 37°C with shaking overnight or longer (see Note 8).

5. Harvest the cells by centrifugation at $1,300 \times g$ for 10 min.

6. Add 300 µL of P1 solution and vigorously resuspend the cells (no cell clumps should be visible after resuspension of the pellet) and transfer the solution to a 1.5 mL microcentrifuge tube.

7. Add 300 µL of P2 solution and mix gently by inverting the tube 4–6 times. From this step onwards, the DNA is already suspended in the solution, hence vortexing should be avoided as it can lead to DNA shearing.

8. Incubate the DNA suspension at room temperature for 5 min.

9. Add 300 µL of P3 solution drop by drop and mix gently by inverting the tube four to six times (do not vortex).

10. Incubate the suspension on ice for 10 min and centrifuge at 9,500 × g for 10 min at 4°C.

11. Transfer the supernatant (discard the pellet) into a 1.5 mL microcentrifuge tube containing 800 μL of ice-cold isopropanol and mix by inverting the tube several times.

12. Place the tube at –20°C overnight to allow DNA precipitation.

13. Centrifuge at 9,500 × g for 15 min at 4°C.

14. Discard the supernatant, add 500 μL of ice-cold 70% ethanol, and invert the tube several times in order to wash the DNA pellet.

15. Centrifuge at 9,500 × g for 5 min at 4°C, discard the supernatant, and allow the DNA pellet to dry at room temperature.

16. Resuspend the DNA pellet in 40 μL of sterile dd H_2O by gentle occasional tapping of the tube. The DNA can be stored at –20°C until required.

It is good practice to check the quality of the DNA by electrophoresis (using a 1% agarose gel). This method can also be used for an approximate estimation of DNA concentration. The average yield of the DNA should be approximately 2–4 μg.

3.2.1.2. Cosmid DNA

The bacterial cells containing the cosmid clone can be delivered either as agar stabs or in vials of glycerol as for BAC and PAC containing cells. For colony growth, follow the same protocol as for BAC and PAC clones. For the extraction of cosmid DNA, we use a commercial extraction kit (Qiagen) and we follow the manufacturer instructions. Cosmid DNA quality and concentration are checked by DNA electrophoresis using a 1% agarose gel.

3.2.2. DNA Labeling

3.2.2.1. Nick Translation

Nick translation is the labeling technique most widely used for labeling of DNA FISH probes. This allows the incorporation of modified nucleotides conjugated with either biotin or digoxigenin (e.g., Biotin-16-dUTP, Digoxigenin-14-dUTP). Labeling can be performed also using dUTPs conjugated to specific fluorochromes such as fluorescein isothiocyanate (FITC) or Cyanine dyes such as Cy3 and Cy5.

Several commercial kits are now available for nick translation. We use the Nick Translation System from Invitrogen. Our labeling protocol is a modified version of the one provided by the manufacturer.

1. Mix 5 μL of dNTP Mix (minus dTTP), 1 μg of probe DNA, 1 μL of dUTP conjugated with biotin or digoxigenin, and sterile dd H_2O up to a volume of 45 μL.

2. Add 5 μL of PolI/DNaseI Mix, mix gently, and centrifuge for few seconds.

3. Incubate at 15°C overnight (see Note 9).

4. Once the reaction is complete, check that the probe DNA has reached a suitable size; load 5 µL of labeled probe onto a 2% agarose gel and proceed with electrophoresis.

5. In order to remove the unincorporated nucleotides, the labeled probes are run through Amersham Microspin G50 columns (following the manufacturer's instructions).

6. After purification, the DNA suspension is collected in a new 1.5 mL microcentrifuge tube.

7. Add 50 µg of salmon sperm DNA, 10 µL of Sodium Acetate 3 M, 2.25 V of ice-cold 100% ethanol.

8. Leave the DNA to precipitate at –80°C for 1 h.

9. Centrifuge at $16,000 \times g$ for 30 min at 4°C.

10. Discard the supernatant (a white pellet should be seen).

11. Wash the pellet by adding 200 µL of ice-cold 70% ethanol and inverting the tube several times.

12. Centrifuge at $16,000 \times g$ for 15 min at 4°C, discard the supernatant, and dry the pellet at room temperature for several hours or on heat block at 45°C for 30 min. (see Note 10).

13. Add 20 µL of water and resuspend the pellet at room temperature by gentle occasional tapping of the tube.

14. The DNA is stable in water for long periods if stored at –20°C until use.

Postlabeling Treatment of the Probe

Before hybridization, competitor DNA is added to the probe DNA (for amounts of C_0t–1 human DNA, see Table 2) (see Note 11).

Table 2
Amounts of DNA probe and COT human DNA suggested per hybridization

Probe type	Amount of probe per slide (ng)	Amount of C_0t human DNA per slide (mg)
Cosmid	100	2.5
PAC/BAC	250	5
Centromeric probes	100	1.5
PCR amplified chromosomes/ chromosome fragments	100	6

1. Dry the pellet in a vacuum desiccator for 15 min (see Note 10).

2. Resuspend the pellet in 12 µL of hybridization buffer (per slide) for at least 1 h at room temperature (see Note 12). The DNA is stable at this stage and can be left at room temperature for use on the same day. Alternatively, it can be kept at −20°C for long periods of time.

3.2.2.2. Degenerated Oligonucleotide Primer (DOP)-PCR

This method is used mostly for amplification and labeling of large genomic regions to produce whole chromosome paints (WCP) or partial chromosome paints (PCP). The templates are usually flow-sorted chromosomes or chromosomal microdissected regions. The technique involves two different rounds of PCR: the first one allows the amplification of chromosomal DNA while the second one allows the incorporation of labeled nucleotides (see Note 13).

1. First PCR round (amplification): In a sterile 0.5 mL centrifuge tube, mix 10 µL of 10× DOP-PCR Buffer, 10 µL of Nucleotide Mix (2 mM of each dATP, dCTP and dGTP), 10 µL of dTTP (2 mM), 10 µL of DOP PCR Primers (20 µM), 1 µL of Taq Supreme Polymerase (2.5 units), 500 chromosomes or chromosomal fragments (usually resuspended in 2 µL), and DNase-Free dd H_2O up to a final volume of 100 µL (see Note 14).

2. The PCR reaction is then performed in a thermocycler with the following settings: an initial denaturing step at 94°C for 9 min followed by 30 cycles (1 min at 94°C, 1 min at 62°C, and 1.5 min at 68°C) and a last step at 68°C for 8 min. After the last step, the temperature is kept at 4°C until the sample is removed.

3. Gel electrophoresis (1% agarose gel) has to be performed to check that DNA amplification has occurred.

4. Second PCR round (labeling): In a sterile 5 mL centrifuge tube, mix 10 µL of 10× DOP-PCR Buffer, 10 µL of Nucleotide Mix (2 mM of each dATP, dCTP and dGTP), 4 µL of dTTP (2 mM), 20 µL of Biotin-16-dUTP (1 mM), 10 µL of DOP-PCR Primers (20 µM), 1 µL of Taq Supreme Polymerase (2.5 units), 10 µL of PCR Amplified Template DNA, and DNase-free dd H_2O up to a final volume of 100 µL (see Notes 13, 14 and 15).

5. The second PCR round is performed using the same settings on the PCR machine as for the first PCR round (see step 2).

6. The size of labeled PCR products should be checked by electrophoresis on a 1% agarose gel (expected size of DNA fragments should be 200–600 bp). Ideally, the average yield of the labeled DNA obtained should be approximately 20–50 ng/µL.

7. The labeled PCR products are stored at −20°C until use.

Postlabeling Treatment
of the Probe

1. Before use for FISH, the labeled PCR products should be prepared as follows (quantities refer to the hybridization of one slide): In a 1.5 mL microcentrifuge tube, mix 8 µL of labeled probe DNA solution (which should contain approximately 200–400 ng of DNA), 7 µL of human C_0t–1 DNA (1 µg/µL), 3 µL of salmon sperm DNA (1 µg/µL), 1/10th volume of 3 M Sodium Acetate (NaAc), and two volumes of ice-cold 100% ethanol (see Note 11).

2. Place the microcentrifuge tube at –80°C for at least 30 min.

3. Centrifuge at $16,000 \times g$ at 4°C for 30 min.

4. After centrifugation a white pellet should be seen, discard the supernatant and wash the pellet with 200 µL of ice-cold 70% ethanol.

5. Centrifuge at $16,000 \times g$ at 4°C for 15 min and discard the supernatant without removing the pellet.

6. Allow to air-dry at 37°C in a heat block or vacuum desiccator (see Note 10).

7. Add 12 µL of Hybridization Buffer (per slide) and resuspend the pellet at room temperature for several hours or in a water bath at 50°C for at least 2 h.

3.3. Fluorescence In Situ Hybridization

3.3.1. Pretreatment of the Slides

1. Wash the slides in 2× SSC in a Coplin jar for 5 min on a shaking platform.

2. Dehydrate the slides through an alcohol series (make up fresh): 3 min in 70% ethanol, 3 min in 90% ethanol, and 3 min in 100% ethanol.

3. Allow to air-dry for at least 5 min.

4. Denature the slides for 5 min in freshly made prewarmed Denaturing Solution in a Coplin jar placed in a water bath at 70°C.

5. Wash the slides in ice-cold 2× SSC for 5 min on a shaking platform.

6. Dehydrate the slides through an alcohol series as in point three.

7. Allow to air-dry for at least 5 min.

8. Leave the slides at room temperature if to be used on the same day for FISH.

3.3.2. Pretreatment of Hybridization Mixture

As described earlier (see Subheading 3.2), the labeled probe DNA has been mixed with the appropriate amount of salmon sperm DNA and C_0t–1 human DNA. Subsequently, the probe mixture has been resuspended in hybridization buffer.

Before hybridization:

1. Place the 1.5 mL microcentrifuge tube containing the in hybridization mixture (probe mixed with hybridization buffer) at 75°C for 10 min to allow denaturation.

2. Transfer the tube to ice and leave for 2 min (optional).

3. Place the tube at 37°C for 15 min to 2 h in a water bath to allow preannealing of competitor DNA to probe DNA (optional) (see Note 16).

If steps 2 and 3 are not required, the denatured probe in hybridization mixture can be placed directly onto the slide (see Subheading 3.3.3). Preannealing will occur during hybridization.

3.3.3. Hybridization and Posthybridization Washes

1. Drop 12 µL of the denatured hybridization mixture (containing the probe mixed with hybridization buffer) on each slide.

2. Cover with a 22 × 22 mm coverslip (avoiding bubbles) and seal the coverslip with rubber solution.

3. Place the slides in a moist chamber (see Note 17) and let it float in a water bath at 42°C overnight or place in a 42°C incubator (see Note 18).

4. Gently remove the rubber solution avoiding movements of the coverslips that can result in damage to chromosomes and nuclei.

5. Wash the slides in a Coplin jar containing 2× SSC for 5 min on a shaking platform in order to remove the coverslips.

6. To remove the unbound or nonspecifically – bound probe fragments, the slides will be washed under stringent conditions.

7. Place the slides for 5 min in a prewarmed Coplin jar containing 0.4× SSC for 5 min at 70°C.

8. Transfer the slides in a Coplin jar containing 2× SSC for 5 min at room temperature on a shaking platform.

3.4. Detection

The following detection steps involve the use of fluorochrome conjugated antibodies or, in the case of biotin labeled probes, the use of fluorochrome conjugated avidin and antiavidin antibodies. These detection steps are not required when probes labeled directly with fluorochromes are used. In both cases (directly and indirectly labeled probes), a wash in 1× PBS is required before mounting the slides with DAPI/Antifade solution.

3.4.1. Biotin-Labeled Probes

1. Place 150 µL of blocking solution onto each slide, cover with parafilm (see Note 19), and incubate in moist chamber (see Note 17) at 37°C for at least 1 h (see Note 20).

2. Dilute 1.5 µL of Avidin D-Texas Red in 1 mL of blocking solution (final concentration of 2 µg/mL). Place 75 µL of this solution onto each slide, cover with parafilm, and incubate in moist chamber at 37°C for 20 min.

3. Remove the parafilm and wash the slides in 4× SSC/Tween®20 three times for 3 min each on a shaking platform, in the dark.

4. Dilute 10 μL of Biotinylated anti-Avidin D in 1 mL of blocking solution (final concentration of 5 μg/mL). Place 75 μL of this solution onto each slide, cover with parafilm, and incubate in moist chamber at 37°C for 20 min.

5. Remove the parafilm and wash the slides in 4× SSC/Tween®20 three times for 3 min each on a shaking platform, in the dark.

6. Repeat step 2 (this amplification step is not required when using WCP, PCP, or centromeric probes) (see Note 21).

7. Remove the parafilm and wash the slides in 4×SSC/Tween®20 twice for 3 min on a shaking platform, in the dark.

8. Wash the slides in 1× PBS for at least 5 min on a shaking platform, in the dark.

9. Place 15 μL of DAPI containing mountant on each slide, cover with a 22×40 mm coverslip, and seal with rubber solution.

3.4.2. Digoxigenin-Labeled Probes

1. Place 150 μL of blocking solution onto each slide, cover with parafilm (see Note 19), and incubate in moist chamber (see Note 17) at 37°C for 1 h (see Note 20).

2. Dilute 1.5 μL of mouse anti-digoxigenin in 1 mL of blocking solution (final dilution of 1:666). Place 75 μL of this solution onto each slide, cover with parafilm, and incubate in moist chamber at 37°C for 20 min.

3. Remove the parafilm and wash the slides in 4× SSC/Tween®20 three times for 3 min each on a shaking platform, in the dark.

4. Dilute 1 μL of Rabbit anti-mouse-FITC in 1 mL of blocking solution (final dilution of 1:1,000). Place 75 μL of this solution on each slide, cover with parafilm, and incubate in moist chamber at 37°C for 20 min.

5. Remove the parafilm and wash the slides in 4× SSC/Tween®20 three times for 3 min on a shaking platform, in the dark.

6. Dilute 10 μL of monoclonal anti-Rabbit-FITC in 1 mL of blocking solution (final dilution of 1:100). Place 75 μL of this solution on each slide, cover with parafilm, and incubate in moist chamber at 37°C for 20 min.

7. Remove the parafilm and wash the slides in 4× SSC/Tween®20 twice for 3 min on a shaking platform, in the dark.

8. Wash the slides in 1× PBS for at least 5 min on a shaking platform and in the dark.

9. Place 15 μL of DAPI containing mountant on each slide, cover with a 22×40 mm coverslip, and seal with rubber solution.

*3.4.3. Dual Color Detection
(See Note 22)*

1. Place 150 μL of blocking solution onto each slide, cover with parafilm (see Note 18), and incubate in moist chamber (see Note 16) at 37°C for 1 h (see Note 19).

2. Dilute in 1 mL of blocking solution 1.5 μL of Avidin D-Texas Red (final concentration of 2 μg/mL) and 1.5 μL of mouse anti-digoxigenin (final dilution of 1:666). Place 75 μL of this solution onto each slide, cover with parafilm, and incubate in moist chamber at 37°C for 20 min.

3. Remove the parafilm and wash the slides in 4× SSC/Tween®20 three times for 3 min each on a shaking platform, in the dark.

4. Dilute in 1 mL of blocking solution 10 μL of biotinylated anti-Avidin D (final concentration of 5 μg/mL) and 1 μL of rabbit anti-mouse-FITC (final dilution of 1:1,000). Place 75 μL of this solution onto each slide, cover with parafilm, and incubate in moist chamber at 37°C for 20 min.

5. Remove the parafilm and wash the slides in 4× SSC/Tween®20 three times for 3 min on a shaking platform, in the dark.

6. Dilute in 1 mL of blocking solution 1.5 μL of Avidin D-Texas Red (final concentration of 2 μg/mL) and 10 μL of monoclonal anti-rabbit-FITC (final dilution of 1:100). Place 75 μL of this solution onto each slide, cover with parafilm, and incubate in moist chamber at 37°C for 20 min.

7. Remove the parafilm and wash the slides in 4× SSC/Tween®20 twice for 3 min on a shaking platform, in the dark.

8. Wash the slides in 1×PBS for at least 5 min on a shaking platform and in the dark.

9. Place 15 μL of DAPI containing mountant on each slide, cover with a 22×40 mm coverslip, and seal with rubber solution.

3.5. Microscopy and Image Analysis

Visualization of fluorescent hybridization signals requires the use of an epifluorescence microscope equipped with appropriate filter set specific for the fluorochromes to be viewed (see Table 3). In the last decade or so, a range of imaging systems have been developed in order to improve the analysis and acquisition of FISH images. Charge-coupled device (CCD) cameras are now widely used in conjunction with appropriate software for image analysis.

In our laboratory, the system in use for 2D-FISH analysis is a Zeiss microscope (Axioplan2 Imaging) equipped with a Sensys cooled CCD camera. Appropriate filters are mounted in an eight position computerized filter wheel connected with the microscope. Software used for acquisition and storage of FISH analysis is Smart Capture 2 or 3 (Digital Scientific).

Table 3
Fluorescence dyes most commonly used in FISH

Fluorochrome	Absorbance (nm)	Emission (nm)	Color
DAPI	350	456	Blue
Spectrum aqua[a]	433	480	Blue
FITC	490	520	Yellow/green
Spectrum green[a]	497	524	Green
Rhodamine	550	575	Red
Cy3[b]	554	568	Red
Spectrum orange[a]	559	588	Orange
Cy3.5[b]	581	588	Red
Spectrum red[a]	587	612	Red
Texas Red	595	615	Deep red
Cy5[b]	652	672	Far red
Cy5.5[b]	682	703	Near infrared
Cy7[b]	755	778	Near infrared

[a]Vysis
[b]Amersham

4. Notes

1. The anticoagulant of choice is Sodium or Lithium Heparin. Other anticoagulants such as EDTA are toxic to the cells and will hamper cell culture.

2. If using cells other than peripheral blood, the timing of cell culture varies according to the cell cycle. For instance, for bone marrow cultures, harvesting is recommended after 24/48 h.

3. Cells should be dropped on the center of the slide, which will be covered by a 22×22 mm coverslip during hybridization (see Subheading 3.3.3).

4. In order to get good FISH results, it is essential to achieve good quality chromosome spreads. Each slide should be carefully checked under a phase-contrast microscope immediately after the spread. The quality of the spread depends on the rate of evaporation of the fixative; this is directly related to atmospheric conditions such as humidity and temperature (4). Our aim is to obtain pale grey chromosomes and nuclei

free of cytoplasmic material, when viewed by a brightfield microscope. The metaphases should, moreover, cover a suitable area without any overlapping chromosomes or chromosome loss. A quick evaporation of the fixative solution due to high temperature or low humidity will produce overspread metaphases with pale chromosomes and nuclei. High humidity or low temperature will cause metaphases with overlapping dark black or glossy chromosomes and shiny nuclei. All these events might result in low efficiency during the hybridization protocol. The amount of overlapping chromosomes and chromosome loss is also related to the concentration of the sample, in fact, a too dense cell suspension usually results in poorly spread metaphases.

5. When defrosting the slides, leave the box unopened on the bench until room temperature is reached. This is to avoid condensation that would impact on the quality of target material.

6. The vector in the bacterial cells, in addition to the probe DNA, contains a gene for resistance to a specific antibiotic. The use of this antibiotic in the culture medium allows selective growth of those bacteria containing the clone of interest.

7. Selection of single clone colonies is not necessary when the DNA is used for FISH. It is advisable to use a loop of bacterial growth from the agar plate for inoculation into liquid culture.

8. After the liquid culture and before harvesting the bacterial cells, it is recommended to prepare one or more aliquots of 0.5 mL of bacteria culture mixed with 0.5 mL of glycerol (Acros organic) in a freezable vial. This can be stored at –80°C for long periods of time.

9. In our laboratory, the nick translation reaction is performed at 15°C overnight rather than at 15°C for 1 h (as instructed by the manufacturer). A 1 h incubation results in larger DNA fragments, whereas an overnight incubation is more likely to produce DNA fragments in the desired range (200–600 bp). Larger DNA fragments would result in bright fluorescent signals all over the slide. If the size of labeled DNA is larger than 600 bp, it should be recut using DNase I. When labeled DNA fragments are too small (<200 bp), they might not hybridize efficiently to chromosomal DNA; hence, FISH signals would not be visible.

10. The DNA usually becomes completely transparent when dry. However, over-drying the DNA pellet might make it difficult to resuspend it. DNA can be dried at room temperature for several hours, in a heating block at maximum 40°C

for approximately 30 min or in vacuum dessicator for approximately 15 min. The use of high temperatures should be avoided in order to prevent the denaturation of the probe at this stage.

11. The human genome presents a high content of interspersed reiterated elements that may interfere with the hybridization. In order to suppress the signals due to the presence of these repetitive sequences, competitive in situ suppression is widely applied (5). The most common method to achieve this uses a large excess of unlabeled total genomic C_0t-1 human DNA. The amount of probe and C_0t-1 human DNA for each hybridization is calculated on the basis of size of cloned DNA (see Table 2).

12. The resuspension of the DNA pellet in water or hybridization buffer is possible at room temperature for several hours, with occasional gentle tapping on the bottom of the tube, or overnight at 4°C.

13. In the DOP-PCR protocol, DNA contamination can easily occur and will result in unwanted DNA amplification that would impact on the FISH results; hence, this type of protocol requires sterile working conditions. This means a sterile working place and autoclaved tubes and tips. We set out a working area under a laminar flow hood to ensure the lowest contamination possible. A negative control in which the template DNA is replaced with DNase-free dd H_2O should always be included in DOP-PCR experiments since signs of contamination would be detected easily.

14. All the reagents used in the DOP-PCR protocols should be kept on ice during use. In particular, the Taq Polymerase should be kept at –20°C until use and replaced immediately after use.

15. Digoxigenin-11-dUTP can be used instead of Biotin-16-dUTP during the second round of DOP-PCR.

16. An incubation of 2 h is usually sufficient for the annealing of all repetitive sequences present in the probe. Incubation longer than 2 h might result in reannealing of specific sequences within the probe and therefore, in a less effective FISH hybridization.

17. A plastic slide storage box containing moist tissue paper can be used as moist chamber.

18. The hybridization temperature is decided according to the percentage of formamide contained in the hybridization buffer. Our hybridization buffer contains 50% formamide, and hybridization occurs at 42°C. The majority of commercial probes use a hybridization buffer with a higher content of

formamide (usually 60 or 70%) and lower hybridization temperature (usually 37°C).

19. In the blocking step and all the antibody incubations, we use a piece of parafilm cut in the shape and size of the microscope slide in place of a glass coverslip. The parafilm can be easily peeled off the slide avoiding damage to nuclei and chromosomes.

20. Blocking solution containing BSA can be replaced with a solution containing 4% (w/v) powdered milk in 4× SSC/ Tween®20 for the blocking step (incubation at 37°C). This step should be performed for at least 1 h, as incubations shorter than 1 h usually result in higher background. However, antibodies should always be diluted in 3% BSA/4× SSC/ Tween®20 and not in a solution of dried milk.

21. The protocol described here, with three antibody layers, is normally used for single-copy probes. The number of layers can be increased by repeating some steps. However, increasing the number of amplification steps can result in a higher background. For repetitive centromeric probes and painting probes one single layer of antibody is usually sufficient to achieve good signal intensity.

22. In dual-color FISH, two differentially labeled probes are added to the hybridization buffer. An appropriate amount of C_0t-1 human DNA should be added accordingly to suppress the repetitive sequences present in both clones. This means that, for example, for hybridizations employing two BAC probes, a total of 10 µL of C_0t-1 human DNA should be added per slide.

References

1. Pinkel, D., Straume, T., and Gray, J.W. (1986) Cytogenetic analysis using quantitative, high-sensitivity, fluorescence hybridization. *Proc Natl Acad Sci USA* **83**, 2834–2938.

2. Saunders, K., Czepulkowski, B. (2001) Culture of human cells for chromosomal analysis, in *Analyzing Chromosomes* (Czepulkowski, B., ed.), BIOS Scientific Publishers, Oxford, UK, pp 19–68.

3. Kearney, L., Tosi, S., and Jafu, R.J. (2002) Detection of chromosome abnormalities in leukemia using fluorescence in situ hybridization, in *Methods in molecular medicine, vol. 68:* *molecular analysis of cancer* (Boultwood, J., and Fidler, C., eds.), Humana, Totowa, NJ, pp 14–15.

4. Spurbeck, J.L., Zinsmeister, A.R., Meyer, K.J., and Jalal, S.M. (1996) Dynamics of chromosomes spreading. *Am J Med Genet* **61**, 387–393.

5. Jauch, A., Daumer, C., Lichter, P., Murken, J., Schroeder-Kurt, T., and Cremer, T. (1990) Chromosomal in situ suppression hybridization of human gonosomes and autosomes and its use in clinical cytogenetics. *Hum Genet* **85**, 145–150.

Chapter 2

Fluorescence In Situ Hybridization on DNA Halo Preparations and Extended Chromatin Fibres

Lauren S. Elcock and Joanna M. Bridger

Abstract

Although many fluorescence in situ hybridisation (FISH) protocols involve the use of intact, fixed nuclei, the resolution achieved is not always sufficient, especially for physical mapping. In light of this, several techniques are commonly used to create extended chromatin fibres or extruded loops of DNA. As a result, it is possible to visualise and distinguish regions of the genome at a resolution higher than that attained with conventional preparations for FISH. Such methodologies include fibre-FISH and the DNA halo preparation. While fibre-FISH involves the stretching of chromatin fibres across a glass slide, the DNA halo preparation is somewhat more complex; whereby DNA loops instead of chromatin fibres are generated from interphase nuclei. Furthermore, the DNA halo preparation coupled with FISH is a useful tool for examining interactions between the inextractable nuclear matrix and the cell's genome.

In this chapter, we describe how to successfully generate extended chromatin fibres and extruded DNA loops. We will also provide detailed methodologies for coupling either procedure with two distinct FISH procedures; 2D-FISH, which allows for the visualisation of specific chromosomal regions, while telomere peptide nucleic acid (PNA) FISH, enables the detection of all telomeres present within human nuclei.

Key words: 2D FISH, DNA halo preparation, Extended chromatin fibres, Extruded DNA loops, Fibre-FISH, Nuclear matrix, Telomere PNA FISH, Telomeres

1. Introduction

Examining extended chromatin/DNA fibres can be achieved by several procedures: fibre-FISH, the DNA halo preparation, and molecular combing. In this chapter, the latter will not be discussed due to its complexity and its many methodological variations. The release of extended chromatin from interphase nuclei, was first described in 1992 by Heng and co-authors (1), with the possibility of resolving sequences as little as 10 kb apart (2).

Joanna M. Bridger and Emanuela V. Volpi (eds.), *Fluorescence in situ Hybridization (FISH):*
Protocols and Applications, Methods in Molecular Biology, vol. 659,
DOI 10.1007/978-1-60761-789-1_2, © Springer Science+Business Media, LLC 2010

Although this group was the first to publish, other laboratories were also utilising similar methodologies at the time (3–5).

In 1993, Parra and Windle described a DNA mapping procedure that involved performing fluorescence in situ hybridisation (FISH) on DNA stretched along a glass microscope slide. This methodology improved resolution by producing DNA that was extended beyond relaxed linear DNA (6). This was achieved by lysing cells with a detergent and allowing the resulting DNA in solution to be stretched by travelling down the microscope slide under gravitational pull. Although this is a relatively straightforward way of producing extended chromatin fibres, the DNA halo preparation can also be used in high resolution mapping (7, 8); however, it is a more time-consuming and less physiological method. The preparation of DNA halos generates elongated extruded strands of linearised DNA as nucleosomes in a halo around the extracted residual interphase nucleus (Fig. 1). As a result, it can be used for physical mapping. However, one of the main problems with employing DNA halo preparations for mapping is that certain regions of DNA remain condensed within the residual nuclei and are thus, inaccessible (9). This DNA is attached to the insoluble nuclear architecture remaining within the residual nucleus, namely the nuclear matrix. These regions of inaccessible DNA may be the regions of interest for the researcher and so preclude DNA halo preparations from being used in mapping. However, this situation has led serendipitously to DNA halos being a platform to study genome interactions with the

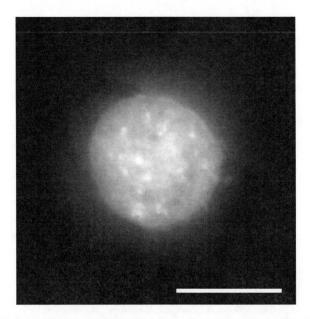

Fig. 1. Nucleus following the DNA Halo preparation and counterstained with DAPI. Scale bar = 10 μm.

nuclear matrix (10). The nuclear matrix is a permanent network of core filaments underlying thicker fibres, which run throughout the nucleus (11). The structure is found to interact with the genome via sequences termed matrix attachment regions (MARs) (12). Research shows that whole chromosome and more specifically, telomere, organization is mediated by the nuclear matrix (e.g. (9, 13–15)). The DNA halo procedure, a derivative of a method described by Berezney and Coffey in 1974, involves the removal of all soluble proteins so that only the nuclear matrix, nuclear matrix-associated proteins and chromosomes, remain intact. Since the methodology contains pre-fixation permeabilization, followed by high-salt extraction steps, any DNA not bound to this nuclear structure creates a "halo" of DNA surrounding the residual nucleus. As a result, it can be assumed that regions of DNA found within the residual nucleus have tight associations with the nuclear matrix, while those in the DNA halo have looser attachments.

Both DNA halo preparation and fibre FISH protocols can be coupled with either two-dimensional fluorescence in situ hybridisation (2D-FISH) or telomere PNA (peptide nucleic acid) FISH, in order to visualise specific regions of the genome. While 2D-FISH allows the detection of specific chromosomes, telomere PNA FISH theoretically paints all telomeres present in the human nucleus.

2. Materials

2.1. DNA Halo Preparation

2.1.1. Cell Culture

1. Dulbecco's Modified Eagle's Medium (DMEM) (Gibco/BRL) supplemented with either 15% (v/v) foetal bovine serum (FBS) or 10% (v/v) newborn calf serum (NCS), depending on the cell line. The medium also needs to contain 2% (v/v) penicillin/streptomycin and 1% (v/v) L-glutamine.

2. Versene (phosphate buffered saline (137 mM NaCl, 2.7 mM KCl, 10 mM Na_2HPO_4, 1.8 mM KH_2PO_4; pH 7.4) with 0.2% (w/v) ethylenediaminetetraacetic acid; EDTA) kept refrigerated at 4°C, pre-warmed to 37°C.

3. 0.25% trypsin produced from a 2.5% stock solution (Gibco/BRL), which has been diluted (1:10 v/v) in Versene.

4. Haemocytometer.

5. SuperFrost™ microscope slides (MENZEL-GLÄSER; available from Thermo Fisher Scientific).

6. 10% (v/v) HCl.

7. QuadriPERM chambers (Greiner Bio One) for growing cells directly upon glass microscope slides.

2.1.2. DNA Halo
Preparation

1. CSK buffer (made up in ddH$_2$0): 100 mM NaCl, 3 mM MgCl$_2$, 0.3 M Sucrose, 10 mM 1,4-Piperazinediethanesulfonic acid (PIPES; pH 7.8), 0.5% (v/v) Triton X-100. Store at 4°C or on ice.

2. 10× phosphate buffered saline (10× PBS): 1.37 M NaCl, 27 mM KCl, 100 mM Na$_2$HPO$_4$, 18 mM KH$_2$PO$_4$; pH 7.4. Dilute accordingly for 1×, 2× and 5× PBS.

3. Extraction buffer (made up in ddH$_2$0): 2 M NaCl, 10 mM PIPES (pH 6.8), 10 mM EDTA, 0.05 mM (v/v) spermine (Sigma Aldrich), 0.125 mM (v/v) spermidine (Sigma Aldrich), 0.1% (w/v) digitonin (Sigma Aldrich; see Notes 1 and 2).

4. Ethanol series comprising of 10, 30, 70 and 95% (v/v) ethanol. Store at room temperature.

5. Glass Coplin jars.

2.2. Slide Preparation
for Fibre-FISH

2.2.1. Cell Culture

1. As per Subheadings 2.1.1 and 2.1.2.

2. Plastic cell culture flasks/dishes.

3. Glass haemocytometer.

4. Phosphate buffered saline (PBS, 1×).

2.2.2. Slide Preparation

1. SuperFrost™ microscope slides.

2. Lysing buffer: 0.5% (w/v) sodium dodecyl sulphate (SDS), 5 mM EDTA, 100 mM Tris–HCl (pH 7).

3. Fixative: methanol:acetic acid (3:1 v/v). Cool on ice.

4. Ethanol row comprising of 70, 90 and 100% (v/v) ethanol at room temperature.

2.3. Two-Dimensional
Fluorescence In Situ
Hybridisation

2.3.1. Slide Denaturation

1. 20× saline sodium citrate (SSC): 3 M NaCl, 0.3 M tri-sodium citrate; pH 7.0. This buffer can then be diluted as necessary.

2. 70% formamide: 70% (v/v) formamide, 2× SSC; pH 7.0. See Note 3 for safety instructions.

3. Ethanol series comprising of 70, 90 and 100% (v/v) ethanol at room temperature.

4. Glass Coplin jars.

5. Warming plate.

6. Oven that can bake slides at 70°C.

2.3.2. Probe Denaturation
and Hybridisation

1. Directly labelled total human chromosome DNA probes (Cambio or see Chapter 1 for how to make DOP-PCR chromosome painting probes).

2. Rubber cement (Halfords).

3. Humid hybridization chamber; ensure that this is moistened so that the slides hybridizing do not dry out (see Note 4).

4. Buffer A: 50% (v/v) formamide, 2× SSC; pH 7.0.

5. Buffer B: 0.1× SSC; pH 7.0.

6. 4× SSC.

7. Vectashield self-sealing mountant containing 4',6-diamidino-2-phenylindole (DAPI), (Vectalaboratories; see Note 5).

8. Glass Coplin jars.

9. Glass Coverslips: 22 × 22 mm and 50 × 22 mm.

2.4. Telomere PNA FISH

1. Telomere PNA FISH kit/FITC, available from Dako (see Note 6). Included in this kit is a PNA telomere probe which contains formamide. See Note 3 for advice regarding work with formamide.

2. Glass Coplin jars.

3. 3.7% (v/v) formaldehyde (see Note 7).

4. Ethanol series comprising of 70, 85 and 96% ethanol. Store at 4°C.

5. Glass Coverslips: 22 × 22 mm and 50 × 22 mm.

6. Vectashield mountant containing 4',6-diamidino-2-phenylindole dihydrochloride (DAPI).

7. Oven for baking slides.

3. Methods

As mentioned in the introduction, both the DNA halo preparation and fibre FISH can be combined with either 2D FISH or telomere PNA FISH. Therefore, this section has been divided into four parts; with the first two, Subheadings 3.1 and 3.2, detailing distinct methodologies for producing extended DNA loops and chromatin fibres (DNA halo preparation and fibre FISH, respectively), while Subheadings 3.3 and 3.4 explain the methodology involved in performing two different FISH procedures (2D FISH and telomere PNA-FISH).

3.1. DNA Halo Preparation

3.1.1. Cell Culture

1. SuperFrost slides should be thoroughly cleaned in 10% (v/v) HCl for an hour; slides are dropped individually into the acid. Following this, the slides are washed in tap water ten times and then in double distilled water, ten times. Next, the slides are rinsed in methanol twice and then, remain in methanol until flaming. Once the flaming is complete, the slide is placed in the QuadriPERM chamber (see Note 8).

2. The respective cells are harvested and then, after using the haemocytometer to determine cell density, 1×10^5 cells are seeded per slide.

3. Cells are grown for at least 48 h at 37°C in 5% CO_2 (see Note 9).

*3.1.2. DNA Halo
Preparation*

1. Once the CSK buffer is made, it is placed on ice or refrigerated at 4°C (see Note 10). Following this, the QuadriPERM chambers, containing the cells are removed from the incubator; the media is then discarded and the slides are labelled accordingly, using a pencil. The slides are placed in an ice-cold Coplin jar, containing CSK buffer and are left on ice for a period of 15 min. This whole process is performed as quickly as possible.

2. After this incubation, the CSK buffer is thrown away and the slides are rinsed in 1× PBS, three times (straight-in, straight-out).

3. The slides are then placed in a fresh Coplin jar containing extraction buffer for an incubation time of 4 min at room temperature (see Note 11).

4. Using the same Coplin jar, slides are rinsed consecutively in 10×, 5×, 2× and 1× PBS (for 1 min each).

5. Slides are taken through an ethanol series comprising of 10, 30, 70 and 95% ethanol (consecutively; straight-in, straight-out).

**3.2. Slide Preparation
for Fibre-FISH**

3.2.1. Cell Culture

Cells are placed in the same medium described in item 1, Subheading 2.1.1 However, the cells are grown in flasks or dishes (not in QuadriPERM chambers, as per Subheading 3.1), at 37°C in 5% CO_2, for 2–4 days (depending on the cell line used). At this point, the cells are harvested (using solutions detailed in Subheadings 2.1.1 and 2.1.2).

3.2.2. Slide Preparation

1. Harvest and then calculate the number of cells using a haemocytometer.

2. Resuspend the cell pellet in 1× PBS (40,000 cells/mL).

3. Using a pipette, place 2 µL from the cell suspension onto one end of the microscope slide. Then, allow the cell suspension to air-dry.

4. Following this, add 30 µL of lysing buffer to the cell area and incubate for 7 min at room temperature.

5. The slide is tilted to a 30° angle; at this point, the chromatin released by cell lysis, will slowly move down the slide.

6. The slide is then air-dried and fixed in ice-cold fixative (methanol: acetic acid; 3:1 v/v) for 2 min.

7. After allowing the slide to air-dry once again, it is taken through an ethanol row (70, 80 and 100%), 5 min in each solution.

8. The slide is then air-dried before it is baked for 1 h at 70°C. The slide is now ready for hybridisation.

3.3. Two-Dimensional Fluorescence In Situ Hybridization

3.3.1. Slide Denaturation

1. Pre-warm waterbaths to 70°C and 37°C. Previously prepared formamide solution is heated to 70°C.

2. Take slides through a 70, 90, 100% ethanol series; 5 min in each solution. Following this, the slides are air-dried using a warming plate and then, baked in an oven at 70°C for 5 min.

3. For denaturation, the slides are placed in 70% formamide for 2 min at 70°C.

4. Next, the slides are taken through the same ethanol series as before, however, with one exception; the 70% ethanol is ice cold. They are then dried on the warming plate.

3.3.2. Probe Denaturation and Hybridisation

1. Probes are denatured at 75°C for 10 min in a waterbath or on a heat-block.

2. Probes are heated at 37°C for 30 min; following this, 10 µL of whole chromosome probe is added to the appropriate slide, covered by a coverslip secured using rubber glue.

3. Next, the slides are left at 37°C, in a humid hybridization chamber for at least 18 h.

4. After this incubation period, slides are removed from the hybridization chamber and the rubber glue is carefully removed using forceps.

5. The slides are then washed in buffer A, pre-warmed to 45°C, for 15 min, with three changes of buffer.

6. Subsequently, the same procedure is performed with buffer B but instead the buffer is incubated at 60°C.

7. The slides are then plunged into 4× SSC at room temperature.

8. At this point, slides can be mounted in Vectashield mountant containing DAPI. They are now ready for visualisation.

3.4. Telomere PNA FISH

1. Pre-heat the oven to 80°C. Also, place a Coplin jar containing Wash solution in a waterbath and allow it to heat to 65°C. Ideally, this should be done a couple of hours before step 3.

2. Fill four Coplin jars with TBS and label TBS1–TBS4 accordingly. In addition, pour 100 mL of Rinse solution into a Coplin jar. These steps should be performed an hour before step 3 in order to allow the solutions to warm to room temperature.

3. Place the slides in TBS1 for approximately 2 min.

4. Following this, the slides are removed and placed in a Coplin jar containing 3.7% formaldehyde for 2 min exactly. This is performed in the fume hood.

5. The slides are washed twice; once in TBS1 for 5 min and then in TBS2 for 5 min.

6. The slides are incubated for 10 min in a Coplin jar containing Pre-Treatment solution.

7. Again, the slides are washed in TBS3 and TBS4, for 5 min each. They are then taken through an ice cold ethanol series (70, 85 and 96%) for 2 min per concentration.

8. Air-dry the slides.

9. When dry, add 10 μL of PNA probe onto each slide and then place a coverslip on top. Incubate at 80°C for 5 min (see Note 12).

10. Remove the slide(s) from the oven and incubate in the dark at room temperature for 40 min at least (see Note 13).

11. Following this incubation period, the slides are placed in Rinse solution for 1 min at room temperature.

12. Slides are then incubated in the Wash solution for 5 min at a temperature of 65°C (see Note 14).

13. Next, the slides go through an ice cold ethanol series (70, 85 and 96%) for 2 min per concentration. After this, they are air-dried vertically and then mounted in Vectashield DAPI mountant.

14. The slides are now ready for visualisation (Fig. 2) and are routinely stored at 4°C (see Note 15).

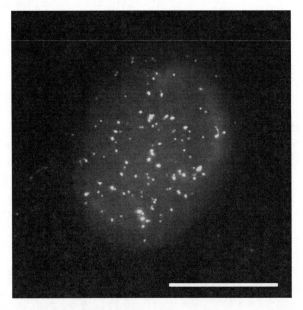

Fig. 2. Nucleus following the DNA halo preparation and Telomere-PNA FISH. Nucleus is counterstained with DAPI (*light grey*) while telomeres are detected in FITC (*bright punctate foci*). Scale bar = 10 μm.

4. Notes

1. Spermine, spermidine, and digitonin are added to the extraction buffer at the end of preparation; by doing this, their biological activity is preserved.

2. When weighing-out digitonin, great care must be taken since it is extremely dangerous; a lab coat, mask, doubled gloves, and safety glasses should be worn. The powder is dissolved separately using heat (60–70°C) and then added to the extraction buffer once cooled. Store at room temperature.

3. Care must be taken when using formamide; if women suspect that they might be or could be pregnant, they should avoid using formamide since it is a teratogen. All applications using formamide should be performed in a fume hood.

4. We construct humid chambers out of sandwich boxes that are covered in foil to exclude light. The floor of the box has three to four layers of tissue that is moistened during a hybridization reaction. We cut plastic pipettes to size and create a platform for the hybridizing slides to rest upon. The chambers will float in a waterbath or can be placed in an oven.

5. DAPI is a DNA intercalator dye and so care should be taken when using – wear gloves when mounting slides.

6. Telomere PNA FISH kit/FITC, available from Dako. This kit is also available in Cy3. These kits provide the majority of reagents required for the procedure. Included is a PNA telomere probe which hybridises to repetitive sequences within human telomeres. Theoretically, all 184 human telomeres should be detected with this assay. Furthermore, signal intensity correlates with telomere length. In addition, each kit provides five foil packages, each of which contains the required components to make 1 L of Tris-Buffered Saline (TBS) pH 7.5 in distilled water. This solution needs to be stored at 4°C. Also included is a Pre-Treatment solution (composed of proteinase K), which needs to be diluted 1:2,000 in TBS. The manufacturers suggest adding 40 µL of Pre-Treatment solution to 80 mL of TBS. This needs to be diluted freshly before each experiment. Concentrated Rinse and Wash solutions are also provided in the kit; the 20 mL bottles for each need to be diluted (1:50) in 980 mL of distilled water. Once diluted, these are kept at 4°C until use and are stable for 1.5 years.

7. Use gloves, safety glasses and a fume hood when using formaldehyde. Do not ingest, inhale, or spill on skin. If this happens, rinse off with copious amounts of water.

8. Flaming slides is a great way to obtain sterile slides immediately prior to use. However, there are hazards associated with this method. Have the QuadriPERM dish close to the beaker containing the slides in methanol; the Bunsen burner should be positioned slightly further away. Use long forceps but make sure that they grip the slides well. Do not have the methanol too deep, just covering the slides is best and then the methanol is not a long way down the forceps when they are dipped into the beaker. Use a large beaker. Always ensure the methanol has evaporated from the forceps and that the forceps have cooled slightly before placing them back into the beaker and the methanol. The beaker should be covered by a piece of aluminium foil; this will starve oxygen from any methanol that catches fire. DO NOT flame slides in any kind of hood where the air is circulated.

9. Ensure that the cell density on the slides is not too high; if the cells are too confluent, then loops of DNA from different nuclei will cross over each other, making the analysis difficult and/or inaccurate. You should aim for 60–70% confluency of cells. If early passage human dermal fibroblasts are to be used and have been set up at a density of $1 \times 10^3/cm^2$, then 1–2 days growth should be sufficient to obtain 60–70% confluency. Other cells, such as transformed cells may achieve this level of confluency earlier and primary cells with increased numbers of senescent cells may take longer.

10. Both the CSK and extraction buffers are made freshly on the day of use.

11. Ensure that the extraction buffer is at room temperature before use with an accurate thermometer.

12. This temperature should be a minimum of 80°C and a maximum of 90°C. Temperatures below 75°C will significantly impair the FISH signals.

13. The kit's manufacturers, DakoCytomation, suggest that slides can be left for 30 min – 4 h.

14. The manufacturers state that the washing conditions must not deviate from this temperature by more than ±3°C. The temperature must be monitored throughout using an accurate thermometer.

15. Once the slides have been mounted in DAPI, the nuclei can be visualised using a fluorescent microscope. The telomeres can be seen as small fluorescent spots within the interphase nucleus. Theoretically, 184 telomere signals should be present; however, in practice, it is extremely rare to record this number. Furthermore, the signal intensity is correlated with the length of the telomere; as a result, the kit can be used to

measure and compare the telomere lengths of different cell lines. When coupled with the DNA halo preparation, telomere signals should be detected both within the residual nucleus as well as in the surrounding DNA halo.

References

1. Heng, H. H., Squire, J., and Tsui, L. C. (1992) High-resolution mapping of mammalian genes by in situ hybridization to free chromatin. *Proc Natl Acad Sci U S A* **89**, 9509–9513.

2. Weier, H. U. (2001) DNA fiber mapping techniques for the assembly of high-resolution physical maps. *J Histochem Cytochem* **49**, 939–948.

3. Haaf, T., and Ward, D. C. (1994) Structural analysis of alpha-satellite DNA and centromere proteins using extended chromatin and chromosomes. *Hum Mol Genet* **3**, 697–709.

4. Senger, G., Jones, T. A., Fidlerova, H., Sanseau, P., Trowsdale, J., Duff, M., and Sheer, D. (1994) Released chromatin: linearized DNA for high resolution fluorescence in situ hybridization. *Hum Mol Genet* **3**, 1275–1280.

5. Fidlerova, H., Senger, G., Kost, M., Sanseau, P., and Sheer, D. (1994) Two simple procedures for releasing chromatin from routinely fixed cells for fluorescence in situ hybridization. *Cytogenet Cell Genet* **65**, 203–205.

6. Parra, I., and Windle, B. (1993) High resolution visual mapping of stretched DNA by fluorescent hybridization. *Nat Genet* **5**, 17–21.

7. Wiegant, J., Kalle, W., Mullenders, L., Brookes, S., Hoovers, J. M., Dauwerse, J. G., van Ommen, G. J., and Raap, A. K. (1992) High-resolution in situ hybridization using DNA halo preparations. *Hum Mol Genet* **1**, 587–591.

8. Lawrence, J. B., Carter, K. C., and Gerdes, M. J. (1992) Extending the capabilities of interphase chromatin mapping. *Nat Genet* **2**, 171–172.

9. Gerdes, M. G., Carter, K. C., Moen, P. T., Jr., and Lawrence, J. B. (1994) Dynamic changes in the higher-level chromatin organization of specific sequences revealed by in situ hybridization to nuclear halos. *J Cell Biol* **126**, 289–304.

10. Elcock, L. S., and Bridger, J. M. (2008) Exploring the effects of a dysfunctional nuclear matrix. *Biochem Soc Trans* **36**, 1378–1383.

11. He, D. C., Nickerson, J. A., and Penman, S. (1990) Core filaments of the nuclear matrix. *J Cell Biol* **110**, 569–580.

12. Gasser, S. M., and Laemmli, U. K. (1986) The organisation of chromatin loops: characterization of a scaffold attachment site. *EMBO J* **5**, 511–518.

13. de Lange, T. (1992) Human telomeres are attached to the nuclear matrix. *EMBO J* **11**, 717–724.

14. Croft, J. A., Bridger, J. M., Boyle, S., Perry, P., Teague, P., and Bickmore, W. A. (1999) Differences in the localization and morphology of chromosomes in the human nucleus. *J Cell Biol* **145**, 1119–1131.

15. Ma, H., Siegel, A. J., and Berezney, R. (1999) Association of chromosome territories with the nuclear matrix. Disruption of human chromosome territories correlates with the release of a subset of nuclear matrix proteins. *J Cell Biol* **146**, 531–542.

Chapter 3

Detection of Nascent RNA Transcripts by Fluorescence In Situ Hybridization

Jill M. Brown and Veronica J. Buckle

Abstract

The development of cellular diversity within any organism depends on the timely and correct expression of differing subsets of genes within each tissue type. Many techniques exist which allow a global, average analysis of RNA expression; however, RNA-FISH permits the sensitive detection of specific transcripts within individual cells while preserving the cellular morphology. The technique can provide insight into the spatial and temporal organization of gene transcription as well the relationship of gene expression and mature RNA distribution to nuclear and cellular compartments. It can also reveal the intercellular variation of gene expression within a given tissue. Here, we describe RNA-FISH methodologies that allow the detection of nascent transcripts within the cell nucleus as well as protocols that allow the detection of RNA alongside DNA or proteins. Such techniques allow the placing of gene transcription within a functional context of the whole cell.

Key words: Transcription, RNA-FISH, RNA-immunoFISH, RNA-DNA-FISH, RNA, Methods, Nascent transcript, RNA probes, Gene expression

1. Introduction

The ability to relate the expression of specific genes to a cellular phenotype is an extremely useful tool in both research and diagnostic settings. Current RNA detection techniques mainly look at gene expression using RNA obtained from whole cell populations; however, it is possible to detect RNA directly in single cells where the structural context is preserved. Mature mRNA for specific proteins can be visualized in frozen sections so that expression patterns across different tissues can be investigated (1, 2). Total levels of transcription can be investigated with the incorporation of bromouridine (BrUTP) into mRNA in living cells (3). In this chapter however, we will concentrate

Joanna M. Bridger and Emanuela V. Volpi (eds.), *Fluorescence in situ Hybridization (FISH): Protocols and Applications*, Methods in Molecular Biology, vol. 659,
DOI 10.1007/978-1-60761-789-1_3, © Springer Science+Business Media, LLC 2010

on the analysis of single gene expression at the level of the individual cell by detection of primary nascent transcripts within the nucleus.

The procedure of RNA-FISH requires cells to be fixed rapidly to prevent deterioration of the RNA, which is then hybridized to a hapten-labelled probe for the gene of interest. The probe could be a denatured genomic clone, a pool of oligonucleotides, or a riboprobe. The type of probe will determine whether only nascent transcript is detected (intronic probe) or whether later forms of the message are also visualized (genomic or exonic probes). Sense and antisense transcripts can only be discriminated when using either oligonucleotide or riboprobes; genomic probes will of course detect both transcript forms. Since splicing is thought to be mainly co-transcriptional (4, 5), nascent transcripts are usually visualized as a punctate dot located at the site of transcription as shown in Fig. 1. Partially spliced RNA transcripts can, however, move away from the site of transcription appearing as elongated tracks, and this is often the case when a gene is either heavily transcribed or highly spliced (6–9). Mature transcripts can be seen throughout the nucleus and, depending on their nature and abundance, may also be cytoplasmic (10, 11). With standard DNA-FISH, both probe and target DNA are denatured and slides are treated with RNase. With RNA-FISH by contrast, the nuclear DNA of the target cells is not denatured so that the single-stranded probe will only hybridize to single-stranded RNA. Such hybridizations require careful controls to exclude background signal, non-specific hybridization of the probe, and to ensure that RNA not DNA is the hybridization target.

RNA-FISH becomes an extremely powerful tool when used in combination with the detection of specific regions of DNA (RNA-DNA-FISH) or cellular proteins (RNA-immunoFISH), examples of which are shown in Fig. 1. These combinations enable the interrogation of many nuclear processes in both space and time. RNA-DNA-FISH can be used for example to investigate nuclear organization and gene activity, or indeed to assess monoallelic expression (12–14). RNA-immunoFISH allows gene activity to be placed in the context of cell type, stage of differentiation or cell cycle (15), or in spatial relation to specific nuclear proteins or bodies (16, 17). We provide combination protocols here and have also included a "gentle" protocol that helps maintain the cytoplasm around nuclei so that cytoplasmic mRNA can also be visualized when using genomic or exonic probes (10, 11). It is also possible to extend the RNA-FISH technique to permit the simultaneous detection of many genes (single gene expression profiling) (18), even in paraffin-embedded tissue (19). The most exciting recent developments involve the visualization of gene expression in live cells (20) and the correlation of such transcription with aspects of nuclear organization (21–23).

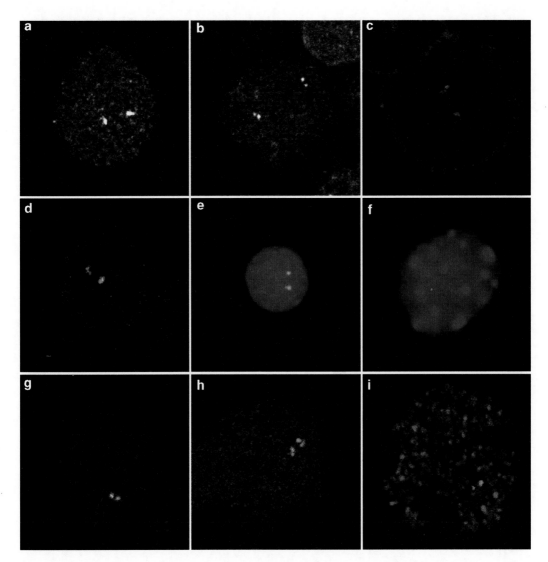

Fig. 1. Detection of nascent transcripts by RNA-FISH. (**a**) Alpha globin transcripts detected in a flow-sorted human intermediate erythroblast using DNP-labelled oligonucleotide probes. (**b**) Transcripts from the c16orf35 gene, located just upstream from the alpha globin gene cluster, detected in a human intermediate erythroblast using a DIG-labelled human genomic probe. Here, the gene has clearly replicated. (**c**) Alpha (*red*) and beta (*green*) globin transcripts detected together in a human intermediate erythroblast, using DNP-labelled oligonucleotide probes for alpha globin and biotinylated genomic plasmid for beta globin. Only one beta globin gene is actively transcribing. (**d**) Alpha globin (*green*) and c16orf35 (*red*) transcripts detected in a mouse foetal blood cell using a DIG-labelled oligonucleotide probe for mouse alpha globin and a biotin-labelled pool of genomic plasmids for c16orf35. Regulatory elements controlling transcription from the alpha globin gene cluster are located within the c16orf35 gene. Here, we show that the two genes can be transcribed simultaneously. (**e**) The same probes detecting transcripts in a cell obtained from mouse foetal liver. Here, only one c16orf35 gene is active. (**f**) The same probes hybridized to dabs of mouse foetal brain. In this cell, no alpha globin locus is active and only one c16orf35 gene is being transcribed. (**g**, **h**) An RNA-DNA-FISH hybridization to the alpha globin genes in a human intermediate erythroblast. (**g**) Both loci (using a DIG-labelled genomic BAC probe detected in *green*) are actively transcribing (using DNP-labelled oligonucleotide probes detected in *red*). (**h**) In this erythroblast, only one of the alpha globin gene clusters is active. (**i**) RNA-immunoFISH detection of alpha globin transcripts (*green*) with H5 antibody to the active elongated form of RNA polymerase II (*red*) in a human intermediate erythroblast. This antibody detects foci of transcription spread throughout the nucleus.

2. Materials

It is vital that diethylpyrocarbonate (DEPC)-treated Milli-Q™ (Millipore) water (DEPC MQW) (see Note 1) is used for making all solutions (see Note 2). Where possible a dedicated work area and items needed e.g. pipettes, tips, glassware, forceps, etc., should be set aside for RNA work alone. Always wear gloves to protect the experiment from RNases present on the skin.

2.1. Glassware, General Equipment, and Working Area Preparation

1. RNaseZap (Ambion) (see Note 3).
2. DEPC MQW: Add 2×5 mL bottles DEPC, approximately 97% (Sigma) (see Note 4) to 10-L MQW in an autoclavable 10-L carboy (Nalgene) in fume hood. Leave to stand 24 h then autoclave to inactivate remaining DEPC. Store at room temperature (RT).

2.2. Slide Cleaning and Preparation

1. RNaseZap.
2. DEPC MQW.
3. Ethanol (AnalaR grade).
4. 0.01% poly-L-lysine solution: Dilute 0.1% poly-L-lysine solution (Sigma) one-tenth with DEPC MQW. Prepare fresh.
5. Superfrost™ slides.
6. Metal slide racks (Solmedia).
7. Large glass trough with lid (RA Lamb).

2.3. Preparation of Cells for Fixation

1. Phosphate buffered saline (PBS) solution: For human cells, make up 1× PBS solution and for mouse cells 0.9× PBS solution (see Note 5) using PBS tablets and DEPC MQW. Autoclave and store at RT.

2.4. Cell Fixation

For all fixation procedures, the following items will be required:

1. RNaseZap treated glass Coplin or Hellendahl and Stretton Young jars (RA Lamb).
2. Square 245 mm×245 mm bioassay dishes (Corning).

2.4.1. For Standard RNA-FISH and DNA-RNA-FISH

1. Saline solution: For stock 10× saline solution (9% w/v), make up 90-g NaCl to 1 L in DEPC MQW. Autoclave and store at RT.
2. Fixative: 4% formaldehyde (FA) and 5% glacial acetic acid in 1× saline. For 100 mL add 11 mL 37% FA (see Note 6) (VWR; store at RT, stable for 2–3 months once opened), 5-mL glacial acetic acid (VWR; AnalaR grade), 10 mL 10× saline, and 74-mL DEPC MQW. Make fresh.

2.4.2. For Gentle RNA-FISH and RNA-ImmunoFISH

1. 3.7% FA fixative solution: 37% FA (VWR) diluted 1:10 (v/v) with PBS. Make fresh.

2.5. Probe Labelling

2.5.1. Oligonucleotide Probes

Oligonucleotide probes (oligos) are useful for detecting nascent transcripts that are in reasonable abundance. Ready-labelled oligos are available from several commercial companies, for example Eurogentec, GeneDetect (see Note 7), or alternatively oligos can be labelled "in-house" using a terminal deoxynucleotidyl transferase (TdT) procedure (24–26). To detect nascent transcripts, use a pool of 3–4 spaced oligos (each ~35–50 bp in length) (27) (see Note 8) specific for intronic sequences from the gene of interest. Oligos specific for exonic sequences will detect both nascent and mature transcripts.

1. TdT labelling kit: Available from many commercial suppliers e.g. Promega, Roche and includes required buffers and TdT enzyme.
2. 10-mM dNTP mix: Mix 10 µL each 100 mM dATP, dGTP, dCTP, and dTTP (GE Healthcare) with 60-µL double distilled H_2O (ddH_2O). Store in aliquots at –20°C.
3. 100 pmol each oligo (see Note 9).
4. Hapten: Biotin-16-dUTP (Roche), digoxigenin-11-dUTP (Roche).
5. Qiaquick nucleotide removal kit (Qiagen).
6. ddH_2O.

2.5.2. Nick-Translated Genomic DNA

1. 1-µg genomic DNA, usually plasmid or cosmid in origin (see Note 10).
2. 10× Nick-translation buffer: 0.5 M Tris–HCl pH 8.0, 50 mM $MgCl_2$, 0.5 mg/mL BSA (Sigma, Fraction V). Store in aliquots at –20°C.
3. 0.1 M β-mercaptoethanol: 0.1-mL β-mercaptoethanol (Sigma, pure) and 14.4-mL ddH_2O. Store in aliquots at –20°C.
4. 0.5 mM dAGC mix: 1 µL of each 100 mM dA/G/CTP (GE Healthcare) in 200 µL of ddH_2O. Store in aliquots at –20°C.
5. DNAse I (Grade 1) stock solution: Make up DNAse I (Roche; 20,000 U) in 1 mL 50 mM NaCl/50% glycerol solution. Store at –20°C.
6. Hapten: Biotin-16-dUTP (Roche) or digoxigenin-11-dUTP (Roche).
7. DNA polymerase I (10 U/µL) (Invitrogen).
8. ddH_2O.

9. 2% TBE agarose gel: Melt 2-g agarose in 100 mL 1× TBE buffer (5× TBE stock buffer; 54-g Tris base, 27.5-g boric acid, 10 mM EDTA pH 8.0 in 1-L MQW).

10. illustra™ G-50 MicroSpin™ columns (GE Healthcare).

2.6. Standard RNA-FISH

2.6.1. Pre-treatment and Hybridization

1. Tris/saline solution (TS): 0.15 M NaCl and 0.1 M Tris–HCl. For 100 mL, add 10 mL of 10× saline (see Subheading 2.4.1, item 1), 5 mL of 2 M Tris–HCl pH 7.5, and 85 mL DEPC MQW. Store at RT.

2. Pepsin solution: 0.02% pepsin and 0.01 M HCl (pH 2.0). For 100 mL, use 2 mL of a 1% stock solution of porcine pepsin A (Sigma; add 100 mg to 10 mL DEPC MQW), 90 μL of concentrated HCl (VWR; sp. gr. 1.18), and 98 mL of DEPC MQW (see Note 11). Prepare fresh.

3. 3.7% FA fixative solution: See Subheading 2.4.2, item 1.

4. RNA hybridization mixture (RNA HM): 25% (v/v) formamide (Ambion; deionized) (see Note 12) with 200 ng/μL salmon sperm DNA (Sigma), 5× Denhardt's solution (50× Denhardt's: 500-mg Ficoll, 500-mg polyvinylpyrrolidone, 500-mg BSA in 50-mL DEPC MQW. Filter sterilize and store at –20°C in aliquots), 50-mM sodium phosphate pH 7, 1 mM EDTA in 2× SSC (see Subheading 2.6.2, item 1 for SSC details). Store at –20°C in aliquots. No more than an additional one-tenth (v/v) of oligos should be added to give the final hybridization solution. If larger volumes of probe are needed, then the concentration of the hybridization mixture should be adjusted accordingly.

5. Ethanol series: 70, 90, and 100% (v/v) ethanol in DEPC MQW. Prepare fresh.

6. RNase-free coverslips: 22 mm × 22 mm × 1.5 mm thickness (VWR). Immerse coverslips in RNaseZap for 2 min, then rinse in DEPC MQW, and store in 70% ethanol (made with DEPC MQW) until required. Air-dry when required.

7. RNaseZap treated Coplin or Hellendahl and Columbia jars (RA Lamb). See Subheading 3.1 for preparation details.

8. Probes: See Subheadings 2.5 and 3.5.

9. Human or mouse C_0t-1 DNA (Roche or Invitrogen, respectively).

10. Single-stranded salmon testes DNA (10 mg/mL) (Sigma).

11. 3 M sodium acetate solution: Make up 24.6-g anhydrous sodium acetate (Sigma) to 100 mL with MQW. Store at RT.

12. 70 and 100% (v/v) ethanol: Make up with DEPC MQW. Store at –20°C.

13. Vulcanizing rubber solution (Weldtite Products).

14. Hybridization chamber: Any box that holds slides that are placed flat is suitable.

2.6.2. Detection

1. 20× SSC in DEPC MQW: For 1 L of 20× SSC, add 175.3-g NaCl and 88.2-g tri-sodium citrate. Dilute to 2× SSC for washing purposes. Store at RT.

2. Tris/saline/Tween solution (TST): 0.15 M NaCl, 0.1 M Tris–HCl, and 0.05% (v/v) Tween 20 (BDH). For 1 L, add 100 mL of 10× saline, 50 mL of 2 M Tris–HCl pH 7.5, 475 μL of Tween-20 (BDH), and 850-mL DEPC MQW. Store at RT.

3. 1.35% blocking solution: 1 in 20 dilution of stock (27% (w/v)) blocking solution (ELISA blocking reagent (Roche) made up in DEPC MQW) in TS. Store stock blocking solution at –20°C; make working solution fresh.

4. Antibodies: Dilute in blocking solution. See Table 1 for antibody purchase and dilution information.

5. 3.7% FA fixation solution: See Subheading 2.4.2, item 1.

Table 1
Hapten detection details

Hapten	Antibody layers	Supplier	Order number	Amount	Dilution
DNP[a]	Rabbit anti-DNP	Sigma	D9656	1–1.7 mg/mL	1/100
	Goat anti-rabbit Alexa488	Invitrogen	A-11034	2 mg/mL	1/200
	Rat anti-DNP	Monosan	MON 5070	1 mg/mL	1/500
	Donkey anti-rat Cy3	Jackson Immunoresearch	712-165-150	1.5 mg	1/200
	Rabbit anti-horse Cy3	Jackson Immunoresearch	308-166-003	1.5 mg	1/200
Biotin	Streptavidin[b] Cy3	Jackson Immunoresearch	016-160-084	1 mg	1/200
	Biotinylated anti-streptavidin	Vector	BA-0500	0.5 mg/mL	1/100
	Streptavidin Cy3	Jackson Immunoresearch	016-160-084	1 mg	1/200
DIG[c]	Sheep anti-DIG FITC	Roche	11-207-741-910	200 μg/mL	1/50
	Rabbit anti-sheep FITC	Vector	FI-6000	1.5 mg/mL	1/100
	Goat anti-rabbit FITC	Jackson Immunoresearch	111-095-144	1.5 mg	1/200

[a]Dinitrophenyl: a hapten commonly used in in situ experiments
[b]Streptavidin: a bacterial protein with a very high affinity for the hapten biotin
[c]Digoxigenin: a hapten commonly used in in situ experiments

6. PBS: See Subheading 2.3, item 1.

7. Ethanol series: See Subheading 2.6.1, item 5.

8. Nuclear counterstain in antifade medium: 4′,6-diamidino-2-phenylindole (DAPI) (Roche) is diluted to 1 µg/mL in Vectashield antifade mounting medium (Vector). Store at 4°C in 1-mL aliquots.

9. Parafilm: Cut to 22 × 50 mm strips (Appleton).

2.7. Gentle RNA-FISH

1. Solutions from Subheading 2.6.2 are all required.

2. 0.5% Triton-X (Calbiochem; RNase free): Make 0.5 % Triton-X solution (v/v) in PBS. Mix well and store at RT.

3. PBS: See Subheading 2.3, item 1.

4. 2× SSC: See Subheading 2.6.2, item 1.

5. RNA HM: See Subheading 2.6.1, item 4.

6. Probes: See Subheadings 2.5 and 3.5.

7. Hybridization chamber: See Subheading 2.6.1, item 14.

8. RNase-free coverslips: See Subheading 2.6.1, item 6.

2.8. RNA-DNA-FISH

1. Solutions in Subheadings 2.6.1 and 2.6.2 are all required here.

2. RNase (Sigma): 100 µg/mL in 2× SSC. Store in aliquots at −20°C.

3. 3.5 N HCl: 10-mL concentrated HCl (VWR; sp. gr. 1.18) and 23-mL MQW. Make fresh (see Note 13).

4. Ice-cold PBS: See Subheading 2.3, item 1.

5. DNA hybridization mix (DNA HM): 50% (v/v) formamide (Ambion; deionized), 10% (w/v) dextran sulphate (Sigma), 1% (v/v) Tween-20 (BDH) in 2× SSC, pH 7.0. Store in aliquots at −20°C.

6. 4× and 2× SSC: See Subheading 2.6.2, item 1.

7. 3% (w/v) BSA: Add 3-g BSA (Sigma; Fraction V) to 100 mL 4× SSC. Unused solution can be filtered through 0.45-µm filter and stored at 4°C for 2–3 months.

8. Washing solution (SSCT): 4× SSC with 0.05% (v/v) Tween-20 (BDH). To 1 L of 4× SSC add 500 µL Tween-20, mix well. Store at RT.

9. Probes: See Subheadings 2.5 and 3.5.

2.9. RNA-ImmunoFISH

1. Solutions in Subheadings 2.6.2 and 2.7 are required here.

2. RNAsin: Recombinant RNAsin (Invitrogen) (see Note 14). Use at 25 U/mL.

3. 1% (w/v) BSA blocking solution: 1-g BSA in 100-mL PBS. Filter sterilize, store at 4°C and use within 2–3 months.

4. Antibodies against proteins to be detected: Dilute in blocking solution.

5. RNase-free 22×50 mm coverslips: See Subheading 2.6.1, item 6.

2.10. Controls

1. RNase: See Subheading 2.8, item 2.

3. Methods

A general explanation of RNA-FISH methodologies can be found in most standard cell biology laboratory manuals (11, 28). The following fixation and pre-hybridization steps described here are based on techniques originally described by the Raap group (10). Many online resources give detailed RNA-FISH methodologies; one particularly useful resource is the Epigenome, Network of Excellence website, which is found at the following URL: http://www.epigenome-noe.net/index.php and the links therein.

3.1. Glassware, General Equipment, and Working Area Preparation

1. Spray all worksurfaces, pipettes, forceps, glassware, etc. needed with RNaseZap and leave for 5 min.

2. Rinse twice with DEPC MQW and allow to dry; worksurfaces do not require rinsing (see Notes 3 and 15).

3.2. Slide Cleaning and Preparation

1. Rack slides, place in large glass trough and immerse in RNaseZap for 10 min (see Note 16).

2. Rinse twice in DEPC MQW.

3. Dehydrate in 100% ethanol and air-dry in a tissue culture hood.

4. Working in a tissue culture hood, lay slides flat on a clean sheet of paper towel and flood with 500 μL 0.01% poly-L-lysine solution for 5 min.

5. Re-rack slides, place in glass trough and wash three times in DEPC MQW.

6. Dehydrate in 100% ethanol and air-dry (see Note 17).

3.3. Preparation of Cells for Fixation

The state of the cells is paramount to the success of the RNA-FISH. The cells must be treated optimally while culturing and should be handling as quickly as possible up to the point of fixation otherwise the experimental outcome will be compromised.

3.3.1. Primary Embryonic Blood Cells

1. Wash embryo in PBS and transfer to sterile, plastic Petri dish so that PBS forms a puddle around the embryo. Squeeze off placenta with forceps and let embryo bleed into PBS for about 30 s.

2. For foetal liver, disrupt 12 day livers in 100-μL PBS. For 13 day, use 200-μL PBS, 14 day use 300-μL PBS, etc.

3.3.2. Cultured Cells in Suspension

1. Spin cells at $200 \times g$ at RT for 5 min in 15-mL centrifuge tube.

2. Wash twice in PBS.

3. Resuspend in PBS (~100 μL for each 10 mL of original culture – vary this for density of culture).

3.3.3. Sorted Cell Populations

1. Sort cells using appropriate antibodies to delineate the cell population required (see Note 18).

2. Place sorted cell populations in appropriate culture medium and incubate at suitable conditions to allow recovery to the pre-sorting transcriptional state (e.g. ~ 6-h post-sorting incubation for human intermediate erythroblasts; this should be determined for each cell type used) (see Note 19).

3. Treat as suspension cultures (see Subheading 3.3.2) prior to fixation.

3.3.4. Solid Tissues

1. Solid, soft tissues (such as brain) can be sliced with a clean scalpel and dabbed directly onto poly-L-lysine coated slides (see Note 20).

3.4. Cell Fixation

3.4.1. For Standard RNA-FISH and DNA-RNA-FISH

1. Spot 20 μL of cell suspension per poly-L-lysine coated slide (see Subheading 3.2) in an area of around 1 cm².

2. Let cells settle on slide for 1 min 30 s (see Note 20).

3. Place slides flat in bioassay dish of fixative for 20 min at RT.

4. Place slides in Coplin or Hellendahl jar and wash in PBS at RT for three times for 5 min.

5. Store slides in 70% ethanol at −20°C in Stretton Young jar. Slides can remain suitable for use for up to 5 years although they are best used as soon as possible.

3.4.2. For Gentle RNA-FISH and RNA-ImmunoFISH

1. Spot 20 μL of cell suspension per poly-L-lysine coated slide (see Subheading 3.2) in an area of around 1 cm².

2. Let cells settle on slide for 1 min 30 s (see Note 20).

3. Place slides flat in bioassay dish of "gentle" fixative for 10 min at RT.

4. Place slides in Coplin or Hellendahl jar and wash in PBS at RT for three times for 5 min, then proceed to Subheading 3.7 or 3.9.

3.5. Probe Labelling

We describe the production of oligos and nick-translated genomic probes in this subheading. There are other types of probe that can be used in RNA-FISH such as riboprobes and LNA probes. Riboprobes are generated by cloning the target DNA to be detected into a transcription vector followed by transcription with the appropriate RNA polymerase to yield an RNA probe (10).

LNA probes are oligos that contain locked nucleic acid residues and have a higher hybridization efficiency than standard oligos (29). The website of Robert Singer's group gives an overview of oligos and riboprobes, probe labelling protocols, and RNA-FISH protocols (http://www.singerlab.org/protocols).

3.5.1. Oligonucleotide Probes

1. In a 1.5-mL Eppendorf mix the following: 4-μL TdT buffer, 4-μL CoCl$_2$, 4 μL 10 mM dNTP mix, 1 nmol hapten, 1-μL TdT (400 U), 100 pmol oligo; make up to 20 μL with ddH$_2$O (this is when using a TdT labelling kit from Roche).

2. Incubate at 37°C for 6 h.

3. Apply labelled oligos to Qiaquick nucleotide removal columns following the manufacturer's instructions; recover oligos at a final concentration of 25–50 ng/μL. Store at 4°C for 6 months – 1 year or store long term at –70°C.

3.5.2. Nick-Translated Genomic DNA

1. In a 1.5-mL Eppendorf, mix 5 μL each of 10× nick-translation buffer, 0.1 M β-mercaptoethanol, and 0.5 M dATP/dGTP/dCTP mix. Also add 1 nmol of hapten, 3 μL of 1:1000 diluted DNAse (see Note 21), and 10-U DNA polymerase I; make up to 50 μL with ddH$_2$O. Vortex and centrifuge briefly.

2. Incubate reaction at 16°C for 2 h, then put on ice to halt any further enzymatic action.

3. Run a 5-μL aliquot of the reaction mix on 2% TBE agarose gel at 100 V for about 1 h with an appropriate size marker. The DNA should be 200–300 bp in size. If the DNA requires further cutting, then add additional DNAse and reincubate.

4. When the size is correct clean up using an illustra™ G-50 MicroSpin™ column following the manufacturer's instructions.

5. Store at –20°C until required.

3.6. Standard RNA-FISH

3.6.1. Pre-treatment and Hybridization

1. Take slides from –20°C into fresh 70% ethanol at RT for 5 min.

2. Rinse in TS.

3. Incubate in 0.02% pepsin solution at 37°C for 5 min.

4. Prepare RNase-free 22 × 22 mm coverslips at this point; see Subheading 2.6.1.

5. Rinse slides in DEPC MQW for 1 min.

6. Post-fix in 3.7% FA solution at RT for 5 min.

7. Wash in PBS at RT for 10 min.

8. Dehydrate for 3 min in each 70, 90, and 100% ethanol in DEPC MQW and air-dry.

9. Add 12-μL RNA HM mix with probe added and cover with an RNase-free 22 × 22 mm coverslip and seal with vulcanizing rubber solution. The amount of probe required per hybridization area is as follows: for labelled oligos, take 25–50 ng of each oligo and add to RNA HM. Add a maximum volume of 1.2-μL oligos to 10.8-μL RNA HM. For nick-translated genomic probes, take ~150 ng labelled DNA and ethanol precipitate with 3 μg C_0t-1 DNA (Roche), 20-μg salmon testes DNA (Sigma), one-tenth volume of 3 M sodium acetate solution, and 2.25 volumes of ice-cold 100% ethanol. Microfuge at 15,700 × g for 30 min at 4°C. Wash precipitated pellet in ice-cold 70% ethanol and then air-dry. Resuspend precipitated DNA in RNA HM. For genomic probes, the DNA must be made single stranded; denature at 90°C for 8 min and pre-anneal at 37°C for 10 min. Oligo probes may require straightening if they form secondary structures or dimerize; heat oligo in HM at 90°C for 8 min, then place on ice until applied to slide.

10. Hybridize overnight in hybridization chamber at 37°C.

3.6.2. Detection

Do not allow the slides to dry out at any stage in the detection procedure as this can result in artefactual background.

1. Wash in 2× SSC at 37°C for 5 min to remove coverslips.

2. Wash three times in 2× SSC at 37°C for 10 min.

3. Wash in TST at RT for 5 min.

4. Apply 100-μL blocking solution under 24 × 50 mm Parafilm strip at RT for 30 min in a humid chamber.

5. Make up antibodies in blocking solution (see Table 1). Microfuge at 4°C for 20 min.

6. Rinse in TST.

7. Apply 100-μL antibodies at RT for 30 min similar to the blocking step.

8. Wash twice in TST at RT for 5 min between each antibody layer.

9. Following the final antibody wash, rinse in PBS.

10. Post-fix in 3.7% FA solution at RT for 5 min, then rinse in PBS.

11. Dehydrate through ethanol series and air-dry (optional).

12. Mount in 10-μL mounting medium with DAPI.

13. Store at 4°C to prolong fluorescence staining.

3.7. Gentle RNA-FISH

Preservation of cytoplasmic RNA ideally requires the absolute minimum level of permeabilization. We use the following protocol to preserve mature, cytoplasmic RNA or when combining RNA-FISH with other techniques requiring fresh cells. Continue from Subheading 3.4.2, step 5.

1. Permeabilize cells in 0.5% Triton-X in PBS on ice for 6 min (use a Petri or bioassay dish for this step) (see Note 22).

2. Wash slides three times for 5 min in PBS at RT. Rinse in 2× SSC.

3. Add RNA probe in 12-μL RNA HM to slide and cover with RNase-free 22×22 mm coverslip and seal with vulcanizing rubber solution. See Subheading 3.6.1, step 9 for probe details.

4. Hybridize overnight in a humid chamber at 37°C.

5. Detect as for standard RNA-FISH (see Subheading 3.6.2).

3.8. RNA-DNA-FISH

3.8.1. RNA-FISH

Perform as described in Subheading 3.6.1.

3.8.2. RNA-FISH Detection and DNA-FISH

Perform RNA-FISH detection as described in Subheading 3.6.2 up to step 10 then proceed as follows:

1. Rinse slides in 2× SSC at RT.

2. Apply 100 μL 100 μg/mL RNase solution to each slide, cover with Parafilm strip, and incubate at 37°C for 1 h.

3. Rinse in PBS, then denature DNA with 3.5 N HCl at RT for 20 min (see Note 23). Remove HCl, place immediately on ice, and rinse three times with ice-cold PBS.

4. Remove as much PBS as possible without allowing the cells to dry out then add denatured and pre-annealed DNA probe in 12-μL DNA HM, cover with 22×22 mm coverslip, seal with vulcanizing rubber solution, and hybridize overnight at 37°C. To prepare DNA probes, ethanol precipitate ~100-ng nick-translation labelled genomic probe with 3 μg C_0t-1 DNA (Roche) and 20-μg salmon testes DNA (Sigma) with one-tenth volume 3 M sodium acetate solution and 2.25 volumes of ice-cold 100% ethanol. Microfuge at $15,700 \times g$ for 30 min at 4°C. Wash precipitated pellet in ice-cold 70% ethanol and then air-dry. Resuspend precipitated DNA in DNA HM. Denature the DNA in DNA HM at 90°C for 8 min and pre-anneal at 37°C for 10 min.

3.8.3. DNA-FISH Detection

Do not allow the slides to dry out at any stage in the detection procedure as this can result in artefactual background.

1. Float off coverslips in 4× SSC, wash slides in 2× SSC at 37°C for twice for 30 min then 1× SSC at RT for 30 min.

2. Block non-specific binding sites on the slides in 3% BSA at RT for 30 min.

3. Prepare appropriate antibodies (see Table 1) in blocking solution, leave at 4°C for 10 min, and then spin at $15,700 \times g$ at 4°C for 20 min to remove any aggregates.

4. Incubate slides with 100 μL of antibody, cover with Parafilm strip, and leave at RT for 30 min.

5. Wash between each antibody layer in SSCT at RT for three times for 3 min.

6. Following the detection layers and final washes, rinse in PBS and then post-fix the DNA detection with 3.7% PFA in PBS at RT for 10 min. Wash twice for 5 min in PBS.

7. Mount slides in 10-μL mounting medium with DAPI.

8. Store at 4°C to prolong fluorescence staining.

3.9. RNA-ImmunoFISH

This method is adapted from the technique described by Chaumeil and colleagues (30), a detailed protocol is also given online at the following URL: http://www.epigenome-noe.net/researchtools/protocol.php?protid=3. In this method, the proteins are detected before the RNA-FISH. It is absolutely imperative therefore that all reagents, glassware, and equipment are RNase free to preserve the RNA throughout the protein detection stage. Continue from Subheading 3.4.2, step 5.

3.9.1. Permeabilization

1. Permeabilize cells by laying slides flat in 0.5% Triton-X in PBS in a Petri or bioassay dish on ice for 6 min (see Notes 22 and 24).

2. Wash slides in PBS at RT three times for 5 min.

3.9.2. Protein Detection

1. Block non-specific binding sites on the slide by adding 100 μL 1% BSA (with 25 U/mL RNasin added) and cover with RNase-free 22 × 50 mm coverslip; incubate at RT for 15 min (see Note 25).

2. Prepare appropriate antibodies in blocking solution.

3. Incubate slides with 100 μL of antibody, cover with 22 × 50 mm RNase-free coverslips, and leave at RT for 45 min.

4. Wash in between each antibody layer with PBS at RT four times for 5 min.

5. Post-fix protein detection using 3.7% PFA in PBS at RT for 10 min. Wash twice 5 min in PBS, then rinse in 2× SSC.

3.9.3. RNA-FISH and RNA-FISH Detection

1. Add prepared RNA probe in 12-μL RNA HM to slide (see Subheading 3.6.1, step 9 for probe details) and cover with 22 × 22 mm RNase-free coverslip.

2. Hybridize overnight at 37°C (see Note 26).

3. The following day proceed with the standard RNA-FISH detection as detailed in Subheading 3.6.2.

3.10. Suggestions for Controls

Negative control experiments can provide information on non-specific signal resulting from either the detection system or probe binding to cellular components (31).

1. No probe added. This negative control will give an idea of background fluorescence resulting from the detection system.

2. mRNA can be removed by RNase treatment. This will indicate the contribution to signal made by non-specific probe binding and also ensure that DNA has not denatured.

3. Non-expressing cell type. Hybridizing a gene-specific probe to cells where that gene is inactive will identify hybridization noise resulting directly from the probe.

4. Notes

1. Milli-Q™ water is defined as having a minimum resistivity of 18.0 MΩ/cm with maximum ten parts per billion total organic carbon. Use of Milli-Q water is not essential but is preferred as it provides a consistent basis for solutions.

2. Solutions should not be used for other laboratory techniques; dedicate for RNA work only.

3. RNaseZap is expensive. A less costly alternative to prepare RNase-free glassware is to bake cleaned glassware at 180°C or higher for several hours.

4. DEPC is a hazardous chemical, open only in a chemical fume hood.

5. Using PBS solution of the correct osmolarity when handling human vs. mouse cells is important in maintaining cells in an optimal state of viability.

6. FA is a hazardous chemical; read safety notes carefully. The ordering details are: formaldehyde solution 37% minimum, free from acid, for histology (Merck catalogue number 1.03999; this is not a standard catalogue item).

7. Purchase of ready-labelled oligos is very expensive and may only represent good value if RNA-FISH is a routine lab technique.

8. Wherever possible several oligos should be spaced across the gene in different introns.

9. Formula for conversion of amounts in μg to pmol: (μg of oligo $\times 10^6$)/(number of bases $\times 330$) = pmol of oligo.

10. DNA for use in nick-translation reactions must be prepared without RNase treatment.

11. Add pepsin to the pre-warmed HCl solution just before use as the pepsin autodigests at 37°C; pepsin pre-treatment should be optimized for each cell type used.

12. Store formamide at 4°C to preserve stability as it may decompose to release chemicals, which can degrade nucleic acids.

13. Mixing of HCl and H_2O is an exothermic process; therefore, allow time for the solution to cool to RT.

14. RNasin is an RNase inhibitor available from Promega that helps to prevent RNA degradation by non-covalently binding to RNases. Many other versions of RNase inhibitors are available from numerous commercial sources.

15. Prepared RNase-free glassware can be stored in a clean, closed cupboard until required.

16. A less costly alternative for slide preparation is to immerse slides in 10% HCl/70% ethanol for 2 min. Wash slides thoroughly in ddH_2O followed by 95% ethanol. Wrap the racked slides in aluminium foil and bake in oven at 150°C for 5 min. Slides are then RNase free.

17. Poly-L-lysine coated slides can be prepared in advance and stored racked in lidded glass troughs at RT until needed.

18. When sorting cells for use in RNA-FISH, the cells must be maintained in a viable state throughout the sorting process; for example, if certain factors are required to maintain cell viability then these should be added to solutions required both for preparation of the cells for sorting and to the medium into which the cells are sorted. No azide should be used in the sorting machine fluids.

19. We routinely save the culture medium in which the cells have grown prior to sorting (where required factors for differentiation will have accumulated) and use this (filtered twice through 0.2-μm filter) to allow optimal recovery of the sorted cells.

20. When fixing it is critical that the cells are not allowed to dry out before placing in the fixative solution.

21. DNAse dilution should be prepared just before use and then immediately discarded. The dilution of DNase should be titrated for each batch of DNAse as activity can vary.

22. Optimal permeabilization treatments will differ for each cell type and each probe used and in some cases they may not be required at all.

23. The normality of HCl required is dependent on cell type. Titrate for each cell type used; the concentration given here is that which is optimal for human intermediate erythroblasts.

24. When using an IgM primary antibody, fixation and permeabilization steps should be carried out simultaneously due

to the very large size of IgM antibodies. If fixation and permeabilization are carried out sequentially, it impedes the IgM antibody access to the nucleus.

25. Treat the incubation chambers with RNaseZap as there is a strong likelihood these are contaminated with RNases.

26. Ensure the hybridization temperature is 37°C and not higher as this can destroy delicate protein staining.

Acknowledgments

We thank the Grosveld Lab in Rotterdam, particularly Peter Fraser, Joost Gribnau, Tolleiv Trimborn, and Mark Wijgerde for initial support with setting up RNA-FISH as a technique in our laboratory. We thank Jackie Sloane-Stanley and Sue Butler for provision of cells. Our work is funded by the Medical Research Council.

References

1. Hargrave, M., Bowles, J., and Koopman, P. (2006) In situ hybridization of whole-mount embryos. *Methods Mol Biol* **326**, 103–113.

2. Asp, J., Abramsson, A., and Betsholtz, C. (2006) Nonradioactive in situ hybridization on frozen sections and whole mounts. *Methods Mol Biol* **326**, 89–102.

3. Iborra, F. J., Pombo, A., Jackson, D. A., and Cook, P. R. (1996) Active RNA polymerases are localized within discrete transcription 'factories' in human nuclei. *J Cell Sci* **109**, 1427–1436.

4. LeMaire, M. F., and Thummel, C. S. (1990) Splicing precedes polyadenylation during Drosophila E74A transcription. *Mol Cell Biol* **10**, 6059–6063.

5. Beyer, A. L., and Osheim, Y. N. (1988) Splice site selection, rate of splicing, and alternative splicing on nascent transcripts. *Genes Dev* **2**, 754–765.

6. Dirks, R. W., Daniel, K. C., and Raap, A. K. (1995) RNAs radiate from gene to cytoplasm as revealed by fluorescence in situ hybridization. *J Cell Sci* **108**, 2565–2572.

7. Xing, Y., Johnson, C. V., Dobner, P. R., and Lawrence, J. B. (1993) Higher level organization of individual gene transcription and RNA splicing. *Science* **259**, 1326–1330.

8. Xing, Y., and Lawrence, J. B. (1993) Nuclear RNA tracks: structural basis for transcription and splicing? *Trends Cell Biol* **3**, 346–353.

9. Johnson, C., Primorac, D., McKinstry, M., McNeil, J., Rowe, D., and Lawrence, J. B. (2000) Tracking COL1A1 RNA in osteogenesis imperfecta: splice-defective transcripts initiate transport from the gene but are retained within the SC35 domain. *J Cell Biol* **150**, 417–432.

10. Dirks, R. W., van de Rijke, F. M., Fujishita, S., van der Ploeg, M., and Raap, A. K. (1993) Methodologies for specific intron and exon RNA localization in cultured cells by haptenized and fluorochromized probes. *J Cell Sci* **104**, 1187–1197.

11. Dirks, R. W. (1996) RNA molecules lighting up under the microscope. *Histochem Cell Biol* **106**, 151–166.

12. van Raamsdonk, C. D., and Tilghman, S. M. (2001) Optimizing the detection of nascent transcripts by RNA fluorescence in situ hybridization. *Nucleic Acids Res* **29**, E42.

13. Takizawa, T., Gudla, P. R., Guo, L., Lockett, S., and Misteli, T. (2008) Allele-specific nuclear positioning of the monoallelically expressed astrocyte marker GFAP. *Genes Dev* **22**, 489–498.

14. Chess, A., Simon, I., Cedar, H., and Axel, R. (1994) Allelic inactivation regulates olfactory receptor gene expression. *Cell* **78**, 823–834.

15. Brown, J. M., Leach, J., Reittie, J. E., Atzberger, A., Lee-Prudhoe, J., Wood, W. G., Higgs, D. R., Iborra, F. J., and Buckle, V. J. (2006) Coregulated human globin genes

are frequently in spatial proximity when active. *J Cell Biol* 172, 177–187.

16. Shopland, L. S., Johnson, C. V., Byron, M., McNeil, J., and Lawrence, J. B. (2003) Clustering of multiple specific genes and gene-rich R-bands around SC-35 domains: evidence for local euchromatic neighborhoods. *J Cell Biol* 162, 981–990.

17. Brown, J. M., Green, J., Pires das Neves, R., Wallace, H. A. C., Smith, A. J. H., Hughes, J., Gray, N., Taylor, S., Wood, W. G., Higgs, D. R., Iborra, F. J., and Buckle, V. J. (2008) Association between active genes occurs at nuclear speckles and is modulated by chromatin environment. *J Cell Biol* 182, 1083–1097

18. Levsky, J. M., Shenoy, S. M., Pezo, R. C., and Singer, R. H. (2002) Single-cell gene expression profiling. *Science* 297, 836–840.

19. Capodieci, P., Donovan, M., Buchinsky, H., Jeffers, Y., Cordon-Cardo, C., Gerald, W., Edelson, J., Shenoy, S. M., and Singer, R. H. (2005) Gene expression profiling in single cells within tissue. *Nat Methods* 2, 663–665.

20. Chubb, J. R., Trcek, T., Shenoy, S. M., and Singer, R. H. (2006) Transcriptional pulsing of a developmental gene. *Curr Biol* 16, 1018–1025.

21. Levsky, J. M., Shenoy, S. M., Chubb, J. R., Hall, C. B., Capodieci, P., and Singer, R. H. (2007) The spatial order of transcription in mammalian cells. *J Cell Biochem* 102, 609–617.

22. Tumbar, T., Sudlow, G., and Belmont, A. S. (1999) Large-scale chromatin unfolding and remodeling induced by VP16 acidic activation domain. *J Cell Biol* 145, 1341–1354.

23. Dietzel, S., Zolghadr, K., Hepperger, C., and Belmont, A. S. (2004) Differential large-scale chromatin compaction and intranuclear positioning of transcribed versus non-transcribed transgene arrays containing beta-globin regulatory sequences. *J Cell Sci* 117, 4603–4614.

24. Binnie, A., Castelo-Branco, P., Monks, J., and Proudfoot, N. J. (2006) Homologous gene sequences mediate transcription-domain formation. *J Cell Sci* 119, 3876–3887.

25. Dirks, R. W., Van Gijlswijk, R. P., Vooijs, M. A., Smit, A. B., Bogerd, J., van Minnen, J., Raap, A. K., and Van der Ploeg, M. (1991) 3′-end fluorochromized and haptenized oligonucleotides as in situ hybridization probes for multiple, simultaneous RNA detection. *Exp Cell Res* 194, 310–315.

26. Lawrence, J. B., Taneja, K., and Singer, R. H. (1989) Temporal resolution and sequential expression of muscle-specific genes revealed by in situ hybridization. *Dev Biol* 133, 235–246.

27. Trembleau, A., and Bloom, F. E. (1995) Enhanced sensitivity for light and electron microscopic in situ hybridization with multiple simultaneous non-radioactive oligodeoxynucleotide probes. *J Histochem Cytochem* 43, 829–841.

28. Spector, D. L., Goldman, R. D., and Leinwand, L. (Ed) (1998) *Cells: A Laboratory Manual.* Cold Spring Harbor Laboratory Press, Cold Spring Harbor, NY.

29. Thomsen, R., Nielsen, P. S., and Jensen, T. H. (2005) Dramatically improved RNA in situ hybridization signals using LNA-modified probes. *RNA* 11, 1745–1748.

30. Chaumeil, J., Okamoto, I., and Heard, E. (2004) X-chromosome inactivation in mouse embryonic stem cells: analysis of histone modifications and transcriptional activity using immunofluorescence and FISH. *Methods Enzymol* 376, 405–419.

31. van de Corput, M. P., and Grosveld, F. G. (2001) Fluorescence in situ hybridization analysis of transcript dynamics in cells. *Methods* 25, 111–118.

Chapter 4

Fluorescence In Situ Hybridization Analysis of Formalin Fixed Paraffin Embedded Tissues, Including Tissue Microarrays

Brenda M. Summersgill and Janet M. Shipley

Abstract

Formalin fixed paraffin embedded (FFPE) material is frequently the most convenient readily available source of diseased tissue, including tumors. Multiple cores of FFPE material are being used increasingly to construct tissue microarrays (TMAs) that enable simultaneous analyses of many archival samples. Fluorescence in situ hybridization (FISH) is an important approach to analyze FFPE material for specific genetic aberrations that may be associated with tumor types or subtypes, cellular morphology, and disease prognosis. Annealing, or hybridization of labeled nucleic acid sequences, or probes, to detect and locate one or more complementary nucleic acid sequences within fixed tissue sections allows the detection of structural (translocation/inversion) and numerical (deletion/gain) aberrations and their localization within tissues. The robust protocols described include probe preparation, hybridization, and detection and take 2–3 days to complete. A protocol is also described for the stripping of probes for repeat FISH in order to maximize the use of scarce tissue resources.

Key words: Fluorescence in situ hybridization, Formalin fixed paraffin embedded, Tissue microarrays, Genomic aberrations

1. Introduction

Cytogenetic analysis requiring fresh tumor material has historically been used to identify chromosome aberrations but is impractical for analysis of large sample numbers, especially from solid tumors, from which metaphases can be difficult to obtain. Fresh or frozen material is also impractical for retrospective studies. Formalin fixed paraffin embedded (FFPE) tissue is frequently the most convenient, readily available material allowing retrospective screening of markers from a large number of well-annotated tumor

Joanna M. Bridger and Emanuela V. Volpi (eds.), *Fluorescence in situ Hybridization (FISH): Protocols and Applications*, Methods in Molecular Biology, vol. 659,
DOI 10.1007/978-1-60761-789-1_4, © Springer Science+Business Media, LLC 2010

samples, including those patients undergoing clinical trials (1, 2). However, the nucleic acids extracted from formalin fixed tissues are frequently degraded (3, 4), unlike those obtained from fresh or snap-frozen tissue samples and alcohol-fixed material. Formalin fixation and paraffin embedding preserves cellular architecture and therefore in situ based detection of nucleic acids has the advantage of detecting and quantifying genomic changes in individual cells within the context of specific tissue.

In situ hybridization describes the annealing, or hybridization, of labeled nucleic acid sequences, or probes, to detect and locate complementary nucleic acid sequences within fixed target cells and tissue sections. In situ hybridization probes were originally labeled using radioactivity although now labeling or detection is mainly achieved by fluorescence methods (fluorescence in situ hybridization (FISH)). Fluorescence detection for in situ hybridization has the advantage of including the ability to use multiple fluorochromes to allow simultaneous specific detection and assessment of multiple probes by virtue of their differential labeling and emission spectra. Two or more probes can be used to determine the linear order of markers or small interstitial deletions. FISH can also be used to detect structural (translocation/inversion) and numerical (deletion/gain) aberrations. The disadvantages of FISH are that fluorescence microscopy is required and that signal fading, background autofluorescence, and lack of morphological features can be problematic.

Novel probes can be produced from various cloning vectors, which contain lengths of DNA sequence, for example, bacterial artificial chromosomes (BACs) 100 kbp to 1 Mbp, PACs (P1 artificial chromosomes) 130–150 kbp, fosmids up to 50 kbp or cosmids 30–40 kbp. Probes containing the larger target sequences usually give the strongest FISH signals. Probes are designed by selecting clones either spanning or flanking the region of interest and contain sequences homologous to either specific repetitive or unique regions of the genome. Probes that detect tandemly repeated sequences for centromere and telomere visualization (e.g., alpha satellites, beta satellites, and telomere probes) are useful to deduce aneuploidy and to act as controls for copy number analysis of other specific regions. Bacteria containing constructs, from which DNA for FISH probes can be isolated, are commercially available at (http://www.ncbi.nlm.nih.gov/genome/cyto and http://genome.ucsc.edu). The sequence can be positioned on the human genome using the UCSC (http://genome.ucsc.edu and http://www.ensembl.org/index.html) databases and potential strong sequence homology with other regions of the genome excluded. The identified constructs can be grown and DNA isolated by standard protocols (5) or using commercially available kits. A large quantity of good quality probe can be produced by amplifying the DNA obtained before labeling. Labeling of probes can be either directly by incorporation of nucleotides conjugated to fluorochromes, or indirectly with other nucleotide conjugates, such as biotin, which are detected indirectly through secondary binding to high affinity molecules linked to fluorochromes. Fluorochromes frequently used include fluorescein isothiocyanate (FITC) and tetramethyl rhodamine that fluoresce at different wavelengths when excited by UV light. Generally, deoxyuridine 5'-triphosphate (dUTP), or deoxythymidine 5'-triphosphate (dTTP), is conjugated directly to the fluorochrome or to biotin or digoxigenin and used in the place of dTTP in a labeling reaction. Some FISH probes directly labeled with fluorochromes are available commercially. After hybridization, excess unbound probe is washed off at specific stringencies, followed by detection steps for indirectly labeled probes. The probe is localized using an epifluorescence microscope and analysis of digital images. The procedures are summarized in Fig. 1.

Tissue microarrays (TMAs) consisting of multiple cores from different FFPE tumor samples (6) on one slide are increasingly used to facilitate high throughput analysis of hundreds of samples simultaneously. Control tissues can also be incorporated into the array. Consecutive sections from TMAs can be screened using in

Protocol flowchart for FISH analysis of FFPE sections

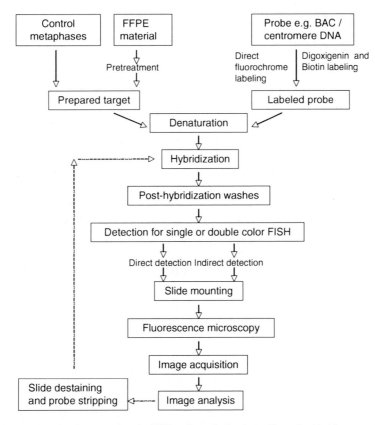

Fig. 1. A flow diagram showing the procedure for FISH on formalin fixed paraffin embedded tissue sections and TMA.

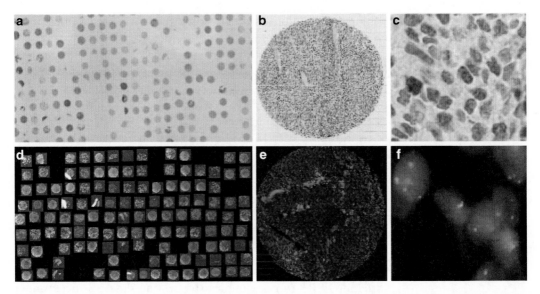

Fig. 2. Analyses of tissue microarrays of formalin fixed paraffin embedded tissues. (**a**, **b**) Tissue microarray used for immunohistochemical detection of the MYCN protein (visualized with 3,3′-diaminobenzidine (DAB) (*brown*) and hematoxylin as counterstain). (**a**) Shows rows of 0.6-mm diameter cores with (**b**) an enlargement of an individual core, and (**c**) an area of the core showing positively staining nuclei. (**d–f**) Tissue microarray analyzed by fluorescence in situ hybridization (FISH): (**d**) shows multiple rows of cores in the tissue microarray in a *black and white image* of the 4′-6-diamidino-2-phenylindole (DAPI) counterstained section, (**e**) is an image of a single core from this FISH experiment, and (**f**) shows individual *blue* DAPI stained nuclei with *green signal* (FITC) from a probe for *PAX3* at 2q35 and *red signal* from a Cy3-labeled probe encompassing the *FOXO1* gene at 13q14. Juxtaposition of the *red* and *green signals* indicates the presence of the *PAX3-FOXO1* fusion gene associated with alveolar rhabdomyosarcomas.

situ hybridization techniques for particular genomic changes previously identified using approaches such as array comparative genomic hybridization (aCGH) analyses. Parallel immunohistochemistry for expression of particular genes may also be relevant in studies. Analysis of multiple genetic and expression changes makes it possible to build up complex information of molecular markers in the same set of tumor specimens and rapidly identify and validate clinicopathological correlations. Examples of TMAs for immunohistochemical and FISH analyses are shown in Fig. 2.

This chapter describes robust FISH methodology for hybridization, and if necessary rehybridization, of DNA probes to DNA targets in FFPE material to evaluate specific genomic rearrangements and copy number imbalances.

2. Materials

2.1. DNA Probes and Amplification

1. DNA probes: Clones can be identified using UCSC Human Genome Browser (http://genome.ucsc.edu) or (http://www.ensembl.org/index.html) databases and obtained from

Sanger Institute Clone ordering system (http://www.sanger.ac.uk) or CHORI (Children's Hospital Oakland Research Institute) Clones Ordering page (http://bacpac.chori.org/order_clones.php).

2. GenomiPhi™ V2 DNA Amplification Kit (GE Healthcare). Aliquot and store at –70°C for up to 1 year.

3. 1:1 (v/v) phenol/chloroform: Phenol (Invitrogen Ltd.). Phenol is highly corrosive. Wear gloves when using. Dispose of using approved waste disposal protocol. After contact with skin, wash immediately with plenty of soapy water. Chloroform (VWR International) is toxic if swallowed and is a respiratory system irritant. Chloroform is a contact hazard; wear gloves when handling. Use in a chemical fume hood. Dispose of waste using approved disposal protocol. Make phenol/chloroform immediately before use.

4. Chloroform–isoamyl alcohol mix: 24 volumes of chloroform to 1 volume of isoamyl alcohol (Fisher Scientific). Make immediately before use.

2.2. Probe Labeling

1. BioPrime DNA Labeling System (Invitrogen Ltd.) Kit comprises 2.5× Random Primer solution, 10× dNTP biotinylate mixture, Control DNA, Klenow Fragment DNA polymerase, Stop Buffer, Distilled water. Store at –20°C for up to a year.

2. 10× digoxigenin dNTP random priming labeling buffer: 1 mM dGTP, 1 mM dATP, 1 mM dCTP, 0.65 mM dTTP in distilled water. Aliquot in 50 μL amounts. Can be stored at –20°C for up to a year.

3. Digoxigenin-11-dUTP (25 nmol; Roche Applied Science). Store at –20°C.

4. Human C_0t-1 DNA (Invitrogen Ltd.) (1 μg/μL) plus 10 μL Salmon Sperm DNA (10 μg/μL) buffer: Mix 500 μL C_0t-1 DNA (1 μg/μL) plus 10 μL Salmon Sperm (10 μg/μL). Reduce volume from 500 to 50 μL in a Microcon-YM30 (Millipore), or by precipitation. Aliquot and store at –20°C for up to 1 year.

2.3. Labeled Probe Purification

1. Microspin G50 columns (GE Healthcare). Store at room temperature.

2. Sephadex column washing buffer: Tris–EDTA (10 mM Tris–HCl/1 mM EDTA pH 7.0).

2.4. Target Slide Preparation and Pretreatment

1. Normal target metaphase slides (Abbott). Store at –20°C.

2. TMA blocks can be constructed and sectioned in-house. They can also be commissioned as slides ordered from commercial suppliers. TMA blocks consist of multiple 0.6-mm cylindrical

cores of FFPE tissue. Cores are punched from the donor block and inserted into precisely spaced holes in the recipient block to form the array. The region of interest in the tissue for coring is usually identified by a pathologist in a stained section from the donor block. Freshly cut FFPE tissue sections and/or TMA sections are cut at a thickness of 4–5 μm and mounted on positively charged slides – which help to prevent tissue detaching.

3. Polypropylene Coplin jar with screw cap (Sigma-Aldrich).

4. Xylene (VWR International Ltd.): CAUTION: Harmful if swallowed and a respiratory system irritant. Use in a chemical fume hood. Dispose of waste using approved disposal protocol. Xylene is flammable and is a contact hazard; wear gloves when handling. Less toxic alternatives to xylene are D-limonene-based products, e.g., Histo-Clear (RA Lamb) or CitriSolv (Fisher). However, we felt deparaffinization was achieved most efficiently when using xylene.

5. SPoT-Light Tissue Pretreatment Kit (Invitrogen Ltd.). Store at 4°C.

6. Digest-All-3 Pepsin Solution (Invitrogen Ltd.). Store at 4°C.

2.5. Hybridization

1. Human C_0t-1 DNA – 1 mg/mL.

2. Salmon sperm DNA – 10 mg/mL.

3. Prehybridization buffer: Mix 500 μL C_0t-1 DNA (1 mg/mL) plus 10 μL Salmon Sperm (10 mg/mL). Reduce volume from 500 to 50 μL in a Microcon-YM30 (Millipore) or by precipitation. Aliquot. Store at –20°C for up to 1 year.

4. 20× SSC: 3 M NaCl, 0.3 M sodium citrate. For 500 mL, 87.6-g NaCl, 44.1-g Na citrate. Add some distilled water. Adjust to pH 7.4 with concentrated HCl before adding distilled water to 500 mL. Autoclave. Can be stored for several months at 4°C.

5. Deionized formamide (NBS Biologicals Ltd.). Formamide is a carcinogen. Use in a fume hood. Wear gloves when handling. Dispose of using approved waste disposal protocol.

6. Hybridization buffer solution: Deionized formamide 60% (v/v), dextran sulfate 12% (w/v), 2.4× SSC, EDTA pH 8.0 (500 mM) 0.7 mM, salmon sperm DNA (10 mg/mL) 400 μg/mL. It is important that the pH of hybridization mix is adjusted to between pH 7.0 and 7.5. At less than pH 7.0, sensitivity is significantly reduced, whereas at higher than pH 7.5 background staining is significantly increased. Mix at 4°C overnight, heat to 42°C, while stirring, for 20 min. Aliquot and store at –20°C for up to 6 months. Deionized formamide should be used as it contains few impurities.

7. Denaturation solution and probe stripping buffer: 70% formamide/2× SSC pH 7.0 (v/v) in distilled water. Store at room temperature.

2.6. Posthybridization Washes

1. Wash solution: 50% formamide/2× SSC pH7.0. Store at room temperature.

2. Wash solution: 2× SSC pH 7.0. Autoclave. Store at room temperature.

3. SSCT solution: 4× SSC/0.05% Tween 20. The addition of detergent aids the removal of any unbound antibodies. Store at room temperature for up to 6 months.

4. PBS Type A solution: For 1 L, 10-g NaCl, 0.25-g KCl, 1.437-g KH_2PO_4, and 0.25-g Na_2HPO_4.

5. Dissolve all ingredients in distilled water. Adjust final volume to 1 L. Autoclave. This can be stored for 1 year at 4°C.

2.7. Preparation of Slides for Fluorescent Detection of Probes Labeled with Digoxigenin or Biotin

1. Blocking buffer: SSCTM. To 10-mL SSCT, add 0.5-g nonfat milk powder.

2. Dissolve at 42°C. Filter sterilize using 0.4-μm filter. Use fresh.

3. Antidigoxigenin-FITC 200 μg (Roche). Reconstitute with distilled water; aliquot in 0.5-mL microcentrifuge tubes. Can be stored with light excluded at −20°C for 1 year. Stable for a few months at 4°C. Before use, centrifuge at $10,000 \times g$ for 5 min to remove any protein complex precipitates, which may appear as speckled background staining and bright spots on the tissue when screening.

4. For each 22 mm×50 mm detection area, take 0.75 μL carefully from the surface of the antidigoxigenin-FITC and dilute in 150 μL SSCTM, giving a dilution of 1:200. Make this dilution on ice no more than 5 min before use.

5. Streptavidin-Cy3 conjugate 1 mg/mL (Sigma-Aldrich). For each 22 mm×50 mm detection area, mix 0.75 μL Streptavidin-Cy3 into 150 μL SSCTM. Make this dilution on ice no more than 5 min before use.

6. Antifade mounting medium and nuclear stain: Vectashield mounting medium with DAPI stain, 4′,6′-diamidino-2-phenylindole (Vector Laboratories Ltd., 1.5 μg/mL). Protect from light. Store at 4°C.

7. Microscope coverslips: #0 thickness; 22 mm×50 mm.

2.8. Image Acquisition and Analysis

Fluorescence microscope equipped with appropriate filters in combination with a cooled charge-coupled device (CCD) camera and suitable computer hardware and software.

3. Methods

3.1. Amplification and Purification of Probe DNA

1. If the DNA probe is good quality, amplification is not necessary and proceeding directly to labeling is possible. However, amplification can increase the amount of DNA available for labeling and also standardize quality – ensuring the DNA is not degraded or contaminated with RNA or protein. Amplification is carried out using GenomiPhi V2 amplification kit according to the manufacturer's instructions. A yield of 10–20 μg amplified DNA can be obtained from 1 μg of DNA amplified for 1.5 h in a 20-μL reaction.

2. The amplified DNA is then purified. Add 80 μL distilled water to each 20 μL of amplified sample in a microcentrifuge tube. Add 100 μL 1:1 phenol/chloroform at room temperature. Close the tube and mix well by inversion. Centrifuge at $10,000 \times g$ for 1–2 min at room temperature (see Note 1).

3. Transfer supernatant to a clean microcentrifuge tube containing 100 μL chloroform at room temperature. Close and mix well. Centrifuge at $10,000 \times g$ for 1–2 min at room temperature.

4. Transfer supernatant to a clean microcentrifuge tube containing 5 μL 4 M NaCl. Mix. Add 100 μL isopropanol to precipitate DNA. Mix well and leave at −20°C overnight.

5. Centrifuge at $10,000 \times g$ for 20 min at room temperature then wash the pellet with one volume of freshly made 80% ethanol and repeat the centrifugation.

6. The pellet can be difficult to see so carefully remove the ethanol and air-dry for approximately 30 min at room temperature. The pellet is dry when no liquid is visible in the tube. It is important not to over dry as this will make DNA resuspension more difficult.

7. Add 30–50 μL distilled water (see Note 2). It is important that the DNA is fully dissolved to allow accurate assessment of concentration. Add more water if necessary and leave on ice for 20 min. Flick tube and heat to 50°C for 5 min to aid resuspension. Samples can be rotated overnight at room temperature to ensure that DNA is fully resuspended.

8. Assess the concentration of amplified DNA by spectrophotometry, or gel electrophoresis using a 0.7% (w/v) agarose gel, comparing with known standards. The amplified DNA can be stored at −20°C for several months.

3.2. Probe Labeling

Probe DNA is labeled by random priming to incorporate biotin or digoxigenin (see Note 3).

1. Labeling with biotin is carried out with reagents included in the BioPrime DNA system. All DNA probes and labeling reagents should be kept on ice and centrifuged briefly before use.

2. Label each DNA probe using a 50-μL reaction. To a 0.2-mL reaction tube, add 300-ng amplified probe, adjusted to 24 μL with distilled water. On ice, add 20 μL 2.5× Random Primers solution. Mix well. Denature the DNA mixture by heating in hot block at 95°C for 5–10 min. Complete DNA denaturation is essential for efficient labeling. Cool immediately on ice to retain DNA as single strands.

3. Briefly centrifuge the tubes and then, on ice, add 5 μL 10× dNTP biotinylate mixture to label the denatured DNA with biotin, or add 1.75 μL digoxigenin-11-dUTP pH 7.5 and 5 μL 10× digoxigenin dNTP buffer for digoxigenin labeling. Adjust volume to 50 μL with distilled water.

4. Mix thoroughly but carefully to avoid shearing the DNA.

5. Add 5 U Klenow Fragment DNA polymerase. Mix gently but thoroughly then centrifuge at $10,000 \times g$ for 15–30 s.

6. Incubate at 37°C for 3 h.

7. Add 5 μL Stop Buffer and mix carefully.

8. The fragment size and label incorporation of the product can now be checked using a 0.7% (w/v) agarose gel (see Note 4). Product quantification should be carried out after purification.

3.3. Labeled Probe Purification

Purify the labeled probe using commercially available Microspin G50 columns following the manufacturer's instructions. This process takes approximately 1 h to complete (see Note 5).

1. Carefully pipette the labeled DNA into the center of the prepared column. It is important to ensure that all the liquid passes through the Sephadex so that all unincorporated nucleotides are removed.

2. Following centrifugation at $400 \times g$ for 5 min, collect the purified probe in a prelabeled tube.

3. The concentration can be checked using an agarose gel 0.7% (w/v). Probe concentration and quality can also be determined using, e.g., a nanodrop spectrophotometer. Probes labeled using random priming can be expected to yield 5–10 μg of DNA, which would represent a 10- to 20-fold increase. If there is no amplification, the labeling reaction has not worked.

4. The labeled DNA is now ready for hybridization or can be stored at –20°C for up to 4 years.

5. Probes should be checked for chromosomal location and quality of signal intensity (see Note 6) by hybridization to karyotypically normal metaphase target slides (7) to assess their suitability for hybridization to FFPE tissue. If available, probes can be tested using FFPE normal tissue with similar cellular density to the tumor tissue (see Note 7).

3.4. Preparation and Pretreatment of Target Formalin Fixed Paraffin Embedded Tissue Slides

1. The pretreatment of FFPE tissue, including TMA slides, can be carried out either using SPoT-Light tissue pretreatment buffer and Digest-All kits (as described below taking about 3 h) or by a standard pretreatment and protease digestion method (see Note 8).

2. Tissue pretreatment is important to maximize the accessibility of the target to the probe (8–13). Pretreatment will help break DNA cross-links formed during tissue fixation. However, variations in the processes of fixation and embedding mean that some tissue samples will not give good FISH results regardless of how pretreatments and proteolytic enzyme digestion are adjusted.

3. It is important that the tissue section does not dry, unless specified, during the following procedures.

4. Preheat 50-mL xylene in a lidded Coplin jar and a further 200 mL of xylene in a bottle to 55°C in a water bath within a fume hood. It is important that the temperature reaches 55°C.

5. Slides can be placed on a hotplate set at 100°C for 30 min to melt paraffin wax, immediately prior to deparaffinization in xylene. The complete removal of paraffin wax is important as residual wax can lead to weak probe signal.

6. Assemble the equipment for the buffer pretreatment of the slides. Put approximately 1 L of water in a 2-L glass beaker. Add a stirrer bar and a raised plastic platform that supports one or two Coplin jars. Fill the Coplin jar with distilled water and place on the platform above the stirrer bar. Cover the beaker with foil and place on the hotplate/stirrer. Set hotplate to maximum heat. It is important that the temperature of the water in the Coplin jar reaches at least 95°C.

7. Meanwhile deparaffinize the slides by washing 2–3 times for 5 min each in xylene preheated to 55°C. With this, and all procedures, it is essential that slides placed against the edge of the Coplin jar have the tissue section facing inwards.

8. Wash twice for 3 min in 100% ethanol at room temperature.

9. Place each slide individually on a hot block at 98°C, pipette 250 μL ethanol onto the tissue area, covering immediately with a 22 mm × 50 mm #1·5 glass coverslip. Heat for 30 s, or until ethanol has evaporated, when the tissue will appear white.

This procedure helps to chemically age the slide but it is important not to over dry as the coverslip will stick to the tissue.

10. Remove the coverslip and cool the slide briefly before rehydrating the tissue by soaking in a Coplin jar of distilled water for 2 min at room temperature. Wash twice more in distilled water for 2 min at room temperature. The slides can be kept moist in the emptied Coplin jar for a short while. The tissue should not be allowed to dry out.

11. Microwave 50 mL of SPoT-Light pretreatment buffer to boiling in a 200-mL bottle with lid resting loosely on top. Heating will take approximately 1–1.5 min. Carefully empty the hot water from the Coplin jar in the beaker of boiling water and replace with the heated pretreatment buffer; cap and allow to heat for 5 min. It is important not to boil the pretreatment buffer for too long before adding the slides as the buffer deteriorates at this temperature.

12. Incubate slides in pretreatment buffer at ≥98°C for 15 min. Treat a maximum of three slides at one time to avoid lowering the buffer temperature when they are added. Pretreatment of slides needs to be optimized. Inadequate tissue permeabilization can give rise to paraffin autofluorescence and weak probe signals as the probes are unable to access the target DNA sequences.

13. Following incubation, wash twice for 5 min in a clean Coplin jar with distilled water. Gently dab the slide edges with a tissue to remove excess water and place the slide on a rack over a waterbath set at 28°C to warm for 5 min. At the same time warm 300 μL of Digest-All-3 Pepsin solution per 22 mm × 22 mm hybridization area, in capped microcentrifuge tubes floated in the 28°C waterbath for 5 min.

14. Then add 300-μL warmed Pepsin solution to each slide, incubating on a level rack over the waterbath at 28°C for 4–10 min depending on tissue type (see Note 9).

15. Drain pepsin solution from slide and soak in distilled water for two 3 min washes. Dab off excess water from the slide edge with a tissue. The slide is now ready for immediate hybridization (see Note 10).

3.5. Hybridization

The procedure takes about an hour to complete.

1. Prepare probe for two-color FISH by combining 300-ng digoxigenin-labeled and biotin-labeled probes plus 0.5 μL 10 mg/mL $C_0 t$-1 DNA/10 mg/mL salmon sperm DNA mix plus 9.6-μL hybridization buffer per 22 mm × 22 mm hybridization area. The size of the hybridization area will depend upon the size of the section or array, and the amount of probe

added should be adjusted accordingly. It is very important that all components are thoroughly resuspended and mixed before usage. Store on ice until the slide is ready for hybridization. If the probe is incompletely dissolved in the hybridization buffer, this can give rise to bright background spots visible when screening.

2. Centrifuge the probe for 5 s at $10,000 \times g$ and carefully apply to the center of the 22 mm $\times$ 22 mm coverslip and invert the slide gently on to it (see Note 11). Carefully remove any air bubbles by gently depressing the coverslip, e.g., with a micropipette barrel, but take care not to force the hybridization buffer out of the side of the coverslip. Air bubbles will result in weak or no signal. Repeat for each hybridization area. Seal with rubber cement or place in a metal hybridization chamber humidified with 300 μL 6× SSPE or 2× SSC (see Note 12).

3. Co-denature probe and target DNA on a hot block set at 98°C for 7–10 min. If the slides were sealed with rubber cement, reapply as this would have dried (see Note 13). Complete DNA denaturation is essential for annealing and efficient labeling to occur. If using commercially labeled probes, denature at the temperature/time recommended by the manufacturer (see Note 14).

4. Place the rubber cement sealed slides in a humidified container on gauze or paper towel moistened with 6× SSPE or 2× SSC, protect from light, and incubate at 37°C overnight to hybridize the probe to target DNA. Hybridization can be for longer, but this may increase background staining.

5. Programmable denaturation and hybridization instruments are available.

3.6. Posthybridization Washes

Posthybridization washes are performed at specific stringencies to wash off excess unbound probe. Insufficiently stringent washes will allow nonspecific probe binding and background fluorescence; however, with less homologous targets care should be taken not to wash off probe. The washes described below are solutions containing formamide. Alternative high temperature, stringent SSC washes can be used (see Note 15). Posthybridization washing procedures take up to an hour to complete.

1. Prepare equipment and washes in advance. Warm 50-mL Coplin jars, freshly prepared 50% formamide/2× SSC pH 7.0 (v/v) and 2× SSC pH 7.0 (v/v) washes to 42°C in a water-bath. Check solution temperatures with a thermometer before using. Warm a humidified box to 37°C in an incubator. Prepare SSCT at room temperature and store SSCTM and antibodies on ice. Washes and incubations should be

performed in subdued light to minimize background staining. Insufficiently stringent washes may give background staining or nonspecific binding of probe.

2. Remove slide from hybridization chamber after overnight incubation and carefully peel off any rubber cement. The coverslip will often lift off at the same time, but if not, gently dip the slide in 2× SSC at 42°C so allowing the coverslip to fall off without scratching the tissue. With all washes, it is important that there is sufficient liquid in the Coplin jar to completely cover the slide.

3. Wash in 50% formamide/2× SSC (v/v) at 42°C for 5 min, agitating the slides gently during the wash or using a shaking waterbath. Repeat this wash twice more.

4. Monitor the wash solution temperature as it should not fall below 42°C.

5. Wash twice in 2× SSC at 42°C for 5 min, agitating as above.

6. Wash once in SSCT for 3 min at room temperature. Drain and blot excess liquid from the slide edge with adsorbent towel and proceed to detection.

3.7. Preparation of Slides for Indirect Fluorescent Detection of Probes Labeled with Digoxigenin or Biotin

This section of the protocol takes up to 3 h to complete. Indirect fluorescent probe detection amplifies the size/strength of the probe signal. The antibodies used here are antidigoxigenin-FITC and Streptavidin-Cy3. Conjugated fluorescent dyes are light sensitive and should be used and stored in subdued light.

1. Pipette 150 μL SSCTM blocking buffer on a 22 mm × 50 mm #1·5 glass coverslip. Invert the slide on to the coverslip with blocking buffer and immediately, carefully, reinvert. Gently remove any air bubbles. Repeat for each slide. This blocking step reduces nonspecific protein binding, which causes high background fluorescence.

2. Place slide in a humidified box protected from light and incubate at 37°C for 20 min. Ensure that the box is kept level so that the coverslip remains in place and the solution evenly distributed across the tissue.

3. Dip the slide a few times in SSCT to remove the coverslip without damaging the tissue and then transfer the slide to a clean Coplin jar containing SSCT for 3 min at room temperature.

4. Drain slide but do not allow to dry. Pipette 150 μL antidigoxigenin-FITC in SSCTM onto a 22 mm × 50 mm #1·5 glass coverslip. Invert the tissue section on to it. Carefully reinvert the slide. Incubate at 37°C for 20 min in a humid box, again ensuring the box is level.

5. Remove the coverslip as above. Wash three times for 2 min in Coplin jars containing SSCT at room temperature.

6. Repeat steps 4–5 with the second antibody, e.g., Streptavidin-Cy3.

7. Drain slide. Wash twice in PBS solution for 5 min at room temperature.

8. Drain. Dehydrate through an ethanol series of 70, 90, and 100% (v/v) for 3 min each.

9. Air-dry in subdued light, e.g., in a drawer or under a foil cover. Mount in Vectashield containing antifade medium and DAPI stain (see Note 16). Carefully pipette one 10 µL drop per hybridization area onto each half of a 22 mm × 50 mm #0 glass coverslip. Invert and lower the tissue on to mounting medium and then carefully reinvert.

10. Allow at least 1 h for DAPI stain to penetrate tissue before screening. Slides can be stored at 4°C, protected from light. Screen usually within 7 days, but slides have been successfully screened after storage for several weeks.

3.8. Image Acquisition and Analysis

1. The slides are viewed using a fluorescence microscope with filters appropriate to the fluorochromes used – excitation at 543 nm induces Cy3 red emission, FITC at 530 nm green emission, with excitation at 364 nm inducing DAPI fluorescence (blue emission). Optimal filters are important to obtain maximum signal intensity. The filters can degrade with age and usage, and they should be checked to ensure continued maximum light transmission. The microscope should be linked to a computer system with an appropriate software package for capturing digital images. The procedure for image capture will vary accordingly. FISH images are usually captured using 63× or 100× oil immersion lenses (see Note 17).

2. TMAs can be screened more easily using an automated capture system (14, 15). Automated systems usually capture using a low power objective (40× as opposed to 100×) and increase the image size digitally which can give lower resolution. The length of TMA scanning time required and therefore prolonged exposure to UV light can cause fluorochrome bleaching and degrading of the FISH signals, even with the use of DAPI antifade mounting medium.

3. Screening and analysis time depends on the quality of signal obtained (see Note 18) and the size of the sample. It is realistic to expect reliable results in excess of 80% of samples. However, if hybridization is poor or if further probes are to be hybridized to the same slide to maximize the use of the tissue, treatments may be employed for destaining and reduction of autofluorescence (16) (see Notes 19 and 20).

4. A minimum of 100 distinct tumor nuclei should be scored and compared with analysis of a similar number from appropriate normal control tissue. This is important to establish cut off values for different probes and signal patterns so as to guard against scoring events occurring by chance, or ploidy variation. Analysis problems may also include cellular heterogeneity in tissues, truncated cells, and differences in hybridization efficiencies between samples as well as tumor and control tissue. Areas where borders of individual nuclei are not clearly defined or where high cellular density causes nuclei to overlap should not be analyzed. Care should be taken not to overlook signal that may be in a different focal plane and to exclude background fluorescence in more than one channel, which may appear as signal. The results of scoring and analysis should be checked blind by a second experienced operator.

4. Notes

1. Amplified DNA should be purified to remove any residual kit components that may inhibit subsequent labeling reactions. DNA probe amplification and purification takes 3–24 h.

2. The DNA is dissolved in water as TE can inhibit DNases and could therefore inhibit enzymatic activity in FISH labeling reactions.

3. Random priming is the preferred labeling method as it can label even very short DNA and generate sensitive, homogeneously labeled DNA probes that can detect single copy genes in tissue including FFPE material. The amount of newly synthesized DNA can exceed the amount of template so effectively amplifying the amount of probe. The process takes approximately 4 h. Probe DNA can also be labeled by nick translation using biotin or digoxigenin, or nick translation using fluorophores conjugated to nucleotides.

4. A smear of DNA with fragment sizes between 200 and 500 bp should be visible (as determined by commercially available molecular markers such as a 1-kb ladder). Incorrect probe size will give weak probe labeling. The smaller the insert size, the more difficult the detection of signals will be, but probe size can be adjusted by varying the amount of enzyme in the labeling reaction. The labeling buffers used interfere with spectrophotometer readings, but DNA can be measured after clean up. The amount of probe is important; accurate measurement of the DNA concentration is essential.

5. As an alternative to commercially prepared columns, Sephadex G50 columns can be made as follows: (1) Plug the base of a 1-mL syringe barrel with glass wool and fill with Sephadex G50 soaked in TE until fully hydrated. (2) Rinse twice with TE, centrifuging at $400 \times g$ for 5 min. After addition of the probe to the column, place the column in a fresh tube and centrifuge at $400 \times g$ for 5 min to collect labeled probe.

6. Control normal metaphase slides need no pretreatment or enzymatic digest, as per the manufacturer's instructions. Warm slides on hotplate set at 60°C for 2 h before probe hybridization.

7. If the probe has not labeled well, in addition to checking probe size, check that all reagents used are in date and that the detection reagents are the correct dilution. Stronger signals may be obtained through extra rounds of signal amplification following labeling with molecules such as biotin. Probes can be tested using FFPE normal tissue with a similar cellular density to tumor tissue. However, if this material is incorporated into a TMA, hybridization to the normal tissue is often different to the tumor cores for which pretreatment times will have been optimized.

8. Sodium thiocyanate and pepsin can be used as alternative tissue pretreatment and protease digestion solutions to SPoT-Light and Digest-All kits: (1) Warm 1 M sodium thiocyanate to 45°C. Warm 9% NaCl pH 1.5 (w/v) to 37°C. Immediately before use, prepare 0.5% pepsin (w/v) solution using warmed 9% NaCl pH 1.5 (w/v). (2) Following deparaffinization, incubate slides in a Coplin jar containing 1 M sodium thiocyanate solution, warmed to 45°C for 15–30 min according to tissue type and section thickness. Agitate slide gently during incubation. Discard solution after use. (3) Rinse in distilled water. (4) Incubate slide at 37°C in freshly diluted 0.5% pepsin solution (w/v) for 15 min (pepsin is added to warmed 9% NaCl pH 1.5 (w/v) solution immediately before use). (5) Wash once in Hanks Buffered Salt solution or PBS at room temperature. (6) Rinse in distilled water. (7) Dehydrate through an ethanol series (70, 90, and 100% (v/v)) and proceed to hybridization.

9. Required protease treatment times vary with tissue type, e.g., 6 min for breast, 5 min for prostate, and 4.5 min for testis. Pepsin digestion treatment will help remove cytoplasm and permeabilize tissue, allowing improved target signal detection. Tissue unmasking is one of the most critical steps for sensitive detection but in each step of tissue preparation there are many variables that could affect sensitivity and increase background staining. Pepsin is used for the protease treatment as it is easily inactivated giving greater control of the reaction.

It better preserves cellular morphology and nuclear structure but overdigestion may lead to poor morphology, increased background, loss of signal, or loosening of tissue from the slide.

10. If immediate hybridization is not possible, the slide can be stored in a humid Coplin jar for 1 h. For periods up to 3 days, it can be stored in 80% ethanol, then immediately prior to hybridization, soak for two 3 min washes in distilled water and drain.

11. Placing the probe onto the coverslip then inverting the slide on to the coverslip has been shown to reduce nonspecific probe binding to the slide and so reduce background.

12. The humidified container must be sealed to prevent evaporation, which would cause the reagents to dry on the tissue, increasing background staining and reducing probe signal strength.

13. Denaturation of FFPE tissue slides and probe can be performed by co-denaturation. However, slide and probe can be denatured separately: (1) In a waterbath, preheat a Coplin jar containing 50 mL 70% formamide/2× SSC pH 7.0 (v/v) DNA denaturation solution until temperature of the liquid measures 72°C. (2) Prepare fresh 70% ethanol (v/v) and cool to 4°C for use in slide dehydration. (3) Incubate slide in heated denaturation solution for approximately 2 min – older slides may need denaturation up to 4 min. (4) Immediately transfer to a Coplin jar containing 70% ethanol (v/v) at 4°C for 3 min to prevent reannealing of the DNA. Complete the dehydration through 90 and 100% (v/v) ethanol series for 3 min, each at room temperature. (5) Air-dry slide and keep warm, e.g., on hotplate, at 37°C until probe is added. (6) Denature probe by heating tubes in a preheated waterbath, or hot block, at 75–80°C for 5 min. (7) Chill on ice for 30 s and microcentrifuge briefly and preanneal repetitive sequences at 37°C for 30 min – 3 h before applying to denatured slide. (8) Centrifuge the probe for 5 s at $10,000 \times g$ and carefully apply the probe/hybridization mix to the center of a 22 mm × 22 mm coverslip and invert the slide gently on to the coverslip. Repeat for each hybridization area. Remove any air bubbles by gently depressing the coverslip, e.g., with blunt forceps or micropipette barrel, but take care not to force the hybridization buffer out of the side of the coverslip. (9) Seal coverslips with rubber cement incubate at 37°C overnight in a humid box.

14. Correct denaturation temperature is important as denaturation of the target DNA on the slide should be optimal. If the temperature is too high the target DNA will be damaged, if too low hybridization will be poor due to poor exposure of the DNA.

A temperature of 98°C is too high for nonparaffin tissue and can cause cells to lift from the slide and poor hybridization. This tissue should be denatured between 75 and 80°C for 5 min. When screening, overdenaturation will appear as nuclei with wrinkled appearance or very bright DAPI counterstain. Under denaturation will lead to bright DAPI staining with very good cellular morphology but poor probe signal.

15. Alternative high temperature washes can be used for directly labeled probes, including those commercially labeled. (1) Carefully remove coverslip, incubate slide for 1–2 min in a Coplin jar containing 0.3% NP-40/2× SSC at 72°C. (2) Wash for 1 min in a Coplin jar containing 0.3% NP-40/2× SSC at room temperature. (3) Drain. Agitate briefly in detergent wash (PBS + 1% Igepal CA-630/Tween 20 (v/v)) at ambient temperature. (4) Rinse twice in distilled water, agitating at ambient temperature. (5) Drain. Dehydrate through an ethanol series of 70, 90, and 100% (v/v) for 3 min each. (6) Air-dry. Mount with 10 μL Vectashield with DAPI per hybridization area on a 22 mm × 50 mm #0 glass coverslip. Invert tissue on to coverslip. Allow at least 1 h for DAPI stain to penetrate tissue before screening.

16. Mount in antifade medium Vectashield containing DAPI stain. It is important to use correct antifade reagent to avoid fast fading of the fluorochromes. The thickness of the coverslip (#0) and amount of Vectashield added are important as too great a thickness of either will make focusing difficult when screening the slide.

17. Capturing cells individually is usually better than capturing them as a group. Some signals can be brighter than others and when captured as a group, the brighter signals tend to shorten the exposure time making the weaker signals harder to see. For FISH on paraffin sections at least 100 nuclei should be captured for each experiment. Statistical analysis is needed for a final result.

18. No signal observed on the slide usually indicates that the probe was not labeled well. The purity and amount of DNA used in labeling is critical.

19. Rehybridization of slides with further FISH probes takes between 24 and 48 h to complete, depending on the ease with which the coverslip can be removed without damaging the tissue. The following procedures can be followed to allow the re-use of slides for further FISH hybridizations. (1) Remove any immersion oil used in microscope analysis by incubating in a Coplin jar with 50-mL xylene at 50°C for 5 min. Agitate gently. Continue to soak the slide until the coverslip loosens and will slide off easily so as not to damage

the tissue. This can take 1–3 days at 50°C depending upon the age of the preparation. (2) Wash slide in a Coplin jar with room temperature 2× SSC twice for 20 min each. Agitate during this time. This soak removes previous DAPI staining, which can inhibit rehybridization. (3) Soak slide in a Coplin jar containing 2× SSC at 68°C for 2 min. (4) Transfer to a Coplin jar containing 70% formamide/2× SSC pH 7.0 (v/v) at 68°C for 4 min to remove the previous probes. (5) Transfer to a Coplin jar containing 2× SSC at room temperature. (6) Drain slide. Dehydrate in separate Coplin jars containing graded ethanol 70, 80, and 100% (v/v) at room temperature for 2 min each. The slide is now ready for reprobing. Hybridize according to Subheading 3.5 onwards. (7) The denaturation time required will vary. Previous enzyme treatment will have removed proteins which would have protected the DNA and so it will be more sensitive to degradation. (8) Slides can be reprobed five or six times but this can give rise to high background fluorescence. An autofluorescence reduction protocol, as described below, can be followed to reduce accumulated background fluorescence.

20. The autofluorescence reduction protocol, excluding time taken to remove the coverslip, takes approximately 1.5 h to complete. This method also allows FISH analysis of previously hematoxylin and eosin (H&E) and 3,3′-diaminobenzidine (DAB) stained slides (if the DAB staining is not in the nucleus). (1) Remove the coverslip by soaking in xylene at 50°C for 1–3 days for hard set mounting media. Soaking in 2× SSC may be adequate for Vectashield mounted slides. (2) Wash in PBS A twice for 5 min. (3) Transfer the slides to Coplin jar containing 0.25% ammonia/70% ethanol (v/v) at room temperature for 20 min (dilute stock of 25% NH_3OH 1 in 100 in 70% ethanol (v/v)). Work quickly as the stock ammonia solution will drip out of a pipette. Ammonia is caustic, use in a fume cupboard with eye protection. (4) Wash in PBS A for 5 min. (5) Make a fresh solution of 1% sodium borohydride (w/v) in PBS A. Mix. Make the solution in a fume cupboard. Weigh out the 0.5-g sodium borohydride, add to the PBS A, and stir. The solution fizzes and gives off flammable fumes. Store the powder in a box with dry silica gel sachets in a fume hood. (6) Soak the slides in a Coplin jar containing sodium borohydride solution for 40 min at room temperature. (7) Wash twice with distilled water and then PBS for 5 min. (8) The slides are now ready for hybridization (Subheading 3.5). H&E or antibody stained slides will need pretreatment with SPoT-Light buffer and pepsin digestion (Subheading 3.4) prior to hybridization.

Acknowledgments

The authors would like to thank Drs. Jeremy Clark and Sian Rizzo for their helpful comments.

References

1. Shipley, J. and Fisher, C. (1998) Chromosome translocations in sarcomas and the analysis of paraffin-embedded material. *J. Pathol.* **184**, 1–3.

2. Summersgill, B., Clark, J. and Shipley, J. (2008) Fluorescence and chromogenic in situ hybridization to detect genetic aberrations in formalin-fixed paraffin embedded material, including tissue microarrays. *Nat. Protoc.* **3**, 220–234.

3. Srinivasan, M., Sedmak, D. and Jewell, S. (2002) Effect of fixatives and tissue processing on the content and integrity of nucleic acids. *Am. J. Pathol.* **161**, 1961–1971.

4. Bramwell, N.H. and Burns, B.F. (1988) The effects of fixative type and fixation time on the quantity and quality of extractable DNA for hybridization studies on lymphoid tissue. *Exp. Hematol.* **16**, 730–732.

5. Sambrook, J., Fritsch, E. and Maniatis, T. (1989) Molecular Cloning: A Laboratory Manual, Vol. 1–3. Cold Spring Harbor Laboratory Press, Cold Spring Harbor, New York.

6. Kononen, J. Bubendorf, L., Kallioniemi, A., Bärlund, M., Schraml, P., Leighton, S., Torhorst, J., Mihatsch, M.J., Sauter, G. and Kallioniemi, O.P. (1998). Tissue microarrays for high-throughput molecular profiling of tumor specimens. *Nat. Med.* **4**, 844–847.

7. Smedley, D., Sidhar, S., Birdsall, S., Bennett, D., Herlyn, M., Cooper, C. and Shipley, J. (2000) Characterization of chromosome 1 abnormalities in malignant melanomas. *Genes Chromosomes Cancer* **28**, 121–125.

8. Lu, Y.J., Birdsall, S., Summersgill, B., Smedley, D., Osin, P., Fisher, C. and Shipley, J. (1999) Dual colour fluorescence in situ hybridization to paraffin-embedded samples to deduce the presence of the der(X)t(X;18)(p11.2;q11.2) and involvement of either the SSX1 or SSX2 gene: a diagnostic and prognostic aid for synovial sarcoma. *J. Pathol.* **187**, 490–496.

9. Birdsall, S., Osin, P., Lu, Y.J., Fisher, C. and Shipley, J. (1999) Synovial sarcoma specific translocation associated with both epithelial and spindle cell components. *Int. J. Cancer* **82**, 605–608.

10. Lambros, M.B., Simpson, P.T., Jones, C., Natrajan, R., Westbury, C., Steele, D., Savage, K., Mackay, A., Schmitt, F.C., Ashworth, A. and Reis-Filho, J.S. (2006) Unlocking pathology archives for molecular genetic studies: a reliable method to generate probes for chromogenic and fluorescent in situ hybridization. *Lab. Invest.* **86**, 398–408.

11. Schraml, P., Kononen, J., Bubendorf, L., Moch, H., Bissig, H., Nocito, A., Mihatsch, M.J., Kallioniemi, O.P. and Sauter, G. (1999) Tissue microarrays for gene amplification surveys in many different tumor types. *Clin. Cancer Res.* **5**, 1966–1975.

12. Brown, L.A. and Huntsman, D. (2007) Fluorescent in situ hybridization on tissue microarrays: challenges and solutions. *J. Mol. Histol.* **38**, 151–157.

13. Ventura, R.A., Martin-Subero, J.I., Jones, M., McParland, J., Gesk, S., Mason, D.Y. and Siebert, R. (2006) FISH analysis for the detection of lymphoma-associated chromosomal abnormalities in routine paraffin-embedded tissue. *J. Mol. Diagn.* **8**, 141–151.

14. Bayani, J. and Squire, J.A. (2007) Application and interpretation of FISH in biomarker studies. *Cancer Lett.* **249**, 97–109.

15. Shipley, J., Crew, J., Birdsall, S., Gill, S., Clark, J., Fisher, C., Kelsey, A., Nojima, T., Sonobe, H., Cooper, C. and Gusterson, B. (1996) Interphase fluorescence in situ hybridization and reverse transcription polymerase chain reaction as a diagnostic aid for synovial sarcoma. *Am. J. Pathol.* **148**, 559–567.

16. Baschong, W., Suetterlin, R. and Laeng, R.H. (2001) Control of autofluorescence of archival formaldehyde-fixed, paraffin-embedded tissue in confocal laser scanning microscopy (CLSM). *J. Histochem. Cytochem.* **49**, 1565–1572.

Chapter 5

Fluorescence *in situ* Hybridization (FISH) on Tissue Cryosections

Irina Solovei

Abstract

Recent progress in the understanding of the spatial organization of nuclear functions owes a lot to fluorescence in situ hybridization (FISH) methodology. The majority of studies using this technology have been carried out using cultured cells. However, nuclear processes in whole organisms, may be to a notable degree, different from those in cultured cells and actually not similar across different tissues. Therefore, for better understanding of nuclear processes in ex vivo organismal material, it is necessary to study nuclear organization in sections of tissue. FISH on sections is still not common in nuclear biology studies mostly due to methodological problems. The protocol suggested in this chapter is based on several years experience in hybridizing different probes on cryosections of various tissues.

Key words: Interphase nucleus, Tissue sections, Cryosections, Fluorescence in situ hybridization, Directly labeled probe, Immunostaining, Antigen retrieval

1. Introduction

Fluorescence in situ Hybridization (FISH) on histological sections allows visualization of whole chromosome territories or their subregions, as well as single copy genes, in cell nuclei within their native tissues (Fig. 1). Provided that an appropriate protocol is used, FISH on tissue sections is actually simpler than FISH on 3D-preserved cells (3D-FISH) (1–4). The first two issues important for FISH on sections are tissue fixation and the type of sections to be used. Most of the popular histological fixatives, in particular, Bouin fluid and glutaraldehyde, impede hybridization of nucleic acids. The fixative optimal for FISH is 4% formaldehyde. Paraffin and vibratome sections, when used for FISH, cause a number of serious problems, an impaired probe penetration and poor nuclear morphology, in the

Joanna M. Bridger and Emanuela V. Volpi (eds.), *Fluorescence in situ Hybridization (FISH):*
Protocols and Applications, Methods in Molecular Biology, vol. 659,
DOI 10.1007/978-1-60761-789-1_5, © Springer Science+Business Media, LLC 2010

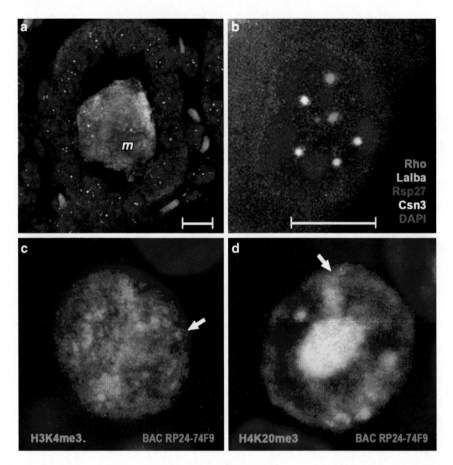

Fig. 1. Examples of FISH on cryosections. (**a**, **b**) Four-color FISH on alveolar cells from the mammary gland of lactating mouse: highly transcribed casein (*Csn3, white*) and α-lactalbumin (*Lalba, yellow*) genes, house-keeping ribosomal gene (*Rps27, red*), and transcriptionally silent rod opsin gene (*Rho, green*). BAC probes for the genes were labeled with FITC-dUTP (*Rsp27*), Cy3-dUTP (*Csn3*), TexRed-dUTP (*Lalba*), and Cy5-dUTP (*Rho*). *m*, milk droplet in the lumen of the alveolus, (**b**) shows a single alveolar cell nucleus with all eight signals at the larger magnification. (**c**, **d**) Simultaneous visualization of genomic region (*blue*) and histone modifications (green) characteristic for transcriptionally active euchromatin (**c**) and heterochromatin (**d**) in ganglion cells of mouse retina. The genomic region was visualized by FISH with BAC DNA (RP24-74F9) directly labeled with Cy3; histone modifications were visualized before FISH using rabbit-anti-H3K4me3 (**c**) and rabbit-anti-H4K20me3 (**d**) antibodies (Abcam) and secondary donkey-anti-rabbit antibodies conjugated to Alexa488 (Dianova). Note that hybridized genomic region (a gene-desert on MMU2; *arrows*) does not co-localize with euchromatin (**c**) but clearly co-localizes with heterochromatin (**d**). Nuclei were counterstained with DAPI [*blue* on (**a**, **b**) and *red* on (**c**, **d**)]. All images are maximum intensity projections of 1–4 μm long stacks acquired with Leica SP5 confocal microscope equipped with five lasers: Diode (laser line 405 nm), Ar (488 nm), DPSS (561 nm), HeNe (594 nm), and HeNe (633 nm). Scale bar on A is 10 μm; scale bar on B is 5 μm and applicable to (**c**, **d**).

first instance. In contrast, cryosections provide both a satisfactory nuclear morphology and a very good probe penetration (5). Furthermore, the use of thick (20–40 μm) cryosections assures that a reasonably high proportion of nuclei in the middle of the section remain intact, which is an important issue, for example, neurons with large nuclei. On the other hand, using thick sections can hinder permeation into the sample: accessibility for large molecules, as immunoglobulins, is reduced in this case. This can result in a gradient of signal intensity from top to bottom of a section due to poor penetration of antibodies used for detection. Experience with probes directly labeled with fluorochromes shows that accessibility for DNA/RNA probes generally is not a significant problem for cryosections and therefore, direct probe labeling is usually preferable to hapten labeling for FISH on tissue sections.

In contrast to FISH on chromosome spreads or 3D-preserved cells, FISH on sections includes an "unmasking" step, heating in sodium citrate buffer, in order to remove partially DNA-protein cross-linking. This treatment – known in immuno-cytochemistry as antigen retrieval – is often required for staining formalin fixed material (6, 7). Therefore, FISH on cryosections can be conveniently combined with immunostaining to map DNA and protein targets in the same sample (see Fig. 1c, d for examples). If immunostaining of the proteins of interest demands antigen retrieval, immunostaining can be performed either before or after FISH. Some proteins, however, need only very short antigen retrieval times or can even be damaged by heating – these then have to be stained before FISH.

Below we present the protocol for FISH on cryosections with directly labeled BAC probes and immunostaining before FISH. Modifications of this protocol for other probes and immunostaining after FISH are discussed in the Note section.

2. Materials

2.1. DNA Probe Preparation

1. High Pure Plasmid Isolation Kit (Roche Diagnostics).

2. GenomiPhi kit (GE Healthcare).

3. 0.1 M ß-Mercaptoethanol (Merck); aliquots are stored at −20°C.

4. dNTP mixture for labeling: Prepare mixture of 1 mM dATP, dGTP, dCTP from 100 mM dNTPs (Roche); pipette together 50 μL of 1 mM dA/G/CTP mixture, 25 μL of 1 mM dTTP, and 25 μL of 1 mM fluorochrome-conjugated dUTP, e.g., Cy3-dUTP (GE Healthcare) (see Note 1).

5. 10× reaction buffer supplied together with Polymerase I (Fermentas) could be replaced by 10× nick-translation buffer which contains 50 mM $MgCl_2$, 0.05% BSA, and 0.5 M Tris–HCl, pH 7.5.

6. DNase I (grade II, from bovine pancrease) (Roche) diluted to 1 mg/mL in 0.15 M NaCl in 50% and stored in aliquots at −20°C; working dilution is 1:200 in dH_2O (see Note 2).

7. Polymerase I (Kornberg Polymerase) (Fermentas).

8. Stop-mixture contains 0.1% bromphenol blue, 0.5% dextran blue, 0.1 M NaCl, 20 mM EDTA, 20 mM Tris–HCl, pH 7.5; aliquots are stored at –20°C.

9. C_0t 1 DNA, mouse or human (Invitrogen).

10. Salmon sperm DNA (Invitrogen).

11. Deionized formamide (Sigma–Aldrich), aliquots are stored at –20°C.

12. Master Mix contains 20% dextran sulphate (Amersham Biosciences) in 2× SSC; aliquots are stored at –20°C.

2.2. Tissue Fixation and Embedding

1. 1.4% formaldehyde in PBS, freshly prepared from paraformaldehyde powder by heating in PBS up to 60°C.

2. Sucrose solutions in PBS 10, 20, 30% (w/v).

3. Jung tissue freezing medium (Leica Microsystems).

4. Peel-A-Way® Disposable Embedding Molds (Polysciences Inc).

5. 100% ethanol chilled to –80°C.

2.3. Preparation and Storage of Cryosections

1. Cryostate (e.g., Leica Microsystems).

2. Plastic boxes for microscopic slides storage.

2.4. Pretreatment of Cryosections Before FISH

1. Coplin jar.

2. 10 mM sodium citrate buffer, pH 6.0.

3. 50% formamide (Merck) in 2× SSC.

4. Diamond cutter.

2.5. Hybridization Set Up

1. Glass coverslips.

2. Diamond cutter.

3. Nail polish.

4. Rubber cement.

5. Hot-block.

2.6. Posthybridization Steps

1. DAPI (4,6-diamidino-2-phenylindole) diluted to 500 µg/ mL in H_2O and stored as aliquots at –20°C; working solution of 0.05 µg/mL is prepared in 2× SSC.

2. Vectashield (Vectorlabs), stored at +4°C.

3. Colorless nail polish.

2.7. Combination of FISH and Immunostaining

1. Glass chambers from 18 × 18, 20 × 20, or 22 × 22 mm coverslips prepared as described in see step 1 of Subheading 3.5.

2. Blocking solution for antibody dilution is prepared on PBS and contains 0.1% Triton-X100, 1% BSA (bovine serum

albumin, ICN Biomedicals GmbH), and 0.1% Saponin (SERVA).

3. Humid dark box for slides incubation.

3. Methods

In the following protocol, we describe preparation of directly labeled BAC DNA probe (see Subheading 3.1), tissue fixation and preparation of cryosections (see Subheading 3.2–3.3), FISH setup (see Subheading 3.4–3.6), and possible combination with immunostaining before FISH (see Subheading 3.7).

3.1. DNA Probe Preparation

1. Prepare BAC clone DNA from bacteria using any commercially available kit.

2. Amplify prepared BAC clone DNA using GenomiPhi kit using manufacturer's instructions (see Note 3).

3. Label BAC DNA using nick-translation. Set up 50 μL of nick-translation reaction as follows:

   – 4 μL of amplified DNA (ca. 1 μg DNA)

   – 5 μL of 10× nick-translation buffer

   – 5 μL of 0.1 M mercaptoethanol

   – 5 μL of dNTP mixture which includes fluorochrome-dUTP

   – 29 μL of H_2O

4. Mix all components well using Vortex. Add 1 μL of diluted DNase (1:200) and 1 μL of DNA Polymerase I, mix gently, and incubate at +15°C for 90 min.

5. Place reaction mixture on ice, take a 3 μL aliquot, and run a 0.8% agarose gel to estimate length of the labeled probe.

6. If the length of the probe exceeds 2,000 bp, add 2 μL of freshly diluted DNase (1:200) and incubate other 20 min at RT.

7. Add an equal volume of stop-mixture (47 μL) and mix well (see Note 4). The labeled DNA can be stored at –20°C for years.

8. For precipitation of DNA and preparation of the probe for hybridization (20 μL), pipette together:

   – 100 μL of labeled BAC DNA

   – 20 μL of C_0t 1 DNA

  - 2 µL salmon sperm DNA
  - 300 µL of 100% ethanol

9. Mix well and chill at –80°C for 1 h.

10. Spin down in a top-bench centrifuge with maximum speed (~2,000 × *g*) for 30 min.

11. Remove supernatant as soon as the centrifuge stops; a blue-stained pellet should be visible at the bottom of the tube.

12. Dry the pellet in a SpeedVac for 5–7 min.

13. Load onto the top of the pellet 10 µL of 100% deionized formamide and incubate in a water bath at 50°C for 15 min.

14. Vortex vigorously to make sure that the pellet is dissolved: the blue stain helps to reveal the degree of solubilization. If the pellet is not dissolved, reheat the probe to 50°C and re-vortex.

15. Add 10 µL of prewarmed (50°C) well mixed Master Mix to the completely dissolved pellet and again vortex vigorously.

16. Incubate in a water bath at 50°C for a further 15 min and vortex well; probe can be stored at –20°C for years.

3.2. Tissue Fixation and Embedding

1. Fix tissue pieces of an appropriate size in freshly prepared 4% formaldehyde/PBS for 4–24 h (time of fixation depends on size of tissue samples, see Note 5).

2. Wash the tissue in PBS, 3 × 30 min. Tissue can be stored in PBS for few days at +4°C.

3. Incubate the tissue in a sucrose series with increasing concentrations: 10% for 1 h, 20% for 1 h, and 30% for 1–24 h. The incubation time also depends on the size of the tissue piece. When the tissue is equilibrated in the sucrose solution, it sinks to the bottom of the vessel – at this point, it can be moved to the next concentration. Tissue can be stored in 30% sucrose at +4°C up to a few days or can be frozen in this solution for longer storage at –20°C.

4. Fill embedding moulds with freezing medium and place the tissue at the mould bottom at the correct orientation: the tissue will be cut in the direction parallel to the mould bottom.

5. Prepare a bath with ethanol chilled to –80°C; add few pieces of dry ice to the ethanol to maintain the bath temperature.

6. Immerse the lower half of the embedding moulds into the bath using a pair of forceps; make sure that ethanol does not leak into the mould; wait until the whole freezing medium in the mould freezes (it becomes opaque and acquires a white color) and then place the block on dry ice.

7. Blocks with frozen tissue can be stored at –20°C or –80°C for years.

3.3. Preparation and Storage of Cryosections

1. Prepare cryosections using a cryostate with thickness of 15–30 µm (see Note 6).

2. Collect sections on SuperFrost®Plus slides and then immediately freeze them by placing slides in a storage box on dry ice.

3. Store slides in plastic microscopic slide storage boxes at –20°C or –80°C; in this way, cryosections can be stored for years without loosing antigen activity or degradation of DNA.

3.4. Pretreatment of Cryosections Before FISH

1. Remove slides with the cryosections from –80°C freezer, let them thaw and air-dry at RT for 30–60 min.

2. While the slides are drying, mark the future hybridization area as a rectangular on the back side of the slide using a diamond cutter. These marks will help to localize the hybridized area during the following procedures.

3. Re-hydrate sections in 10 mM sodium citrate buffer for 5 min.

4. Transfer slides into a Coplin jar with prewarmed (80°C) 10 mM sodium citrate buffer in a water bath and incubate there for 5–25 min (see Note 7).

5. Remove the Coplin jar from the water bath and let it cool down at RT.

6. Incubate the slides in 2× SSC for 5 min.

7. Transfer slides into 50% formamide in 2× SSC and incubate for at least 1 h; sections can be stored in this solution at 4°C for weeks.

3.5. Hybridization

All operations for this section of the protocol should be performed in gloves in order to protect skin from contact with formamide!

1. Prepare glass hybridization chambers as follows (Fig. 2a, b): cut glass strips from a coverslip (e.g., 8 × 8 mm) using a diamond cutter and a ruler; glue glass strips parallel on two opposite borders of an intact coverslip of the same size using nail polish; size of the chamber depends on a section size: chamber should fit to the section size as closely as possible in order to save the expensive DNA probe. Let the nail polish completely dry before use.

2. Remove slide from the formamide and wipe the liquid from the back of the slide and from the face side of the slide, carefully avoiding marked section area.

3. Place prepared hybridization chamber above the section and carefully press it against the microscopic slide.

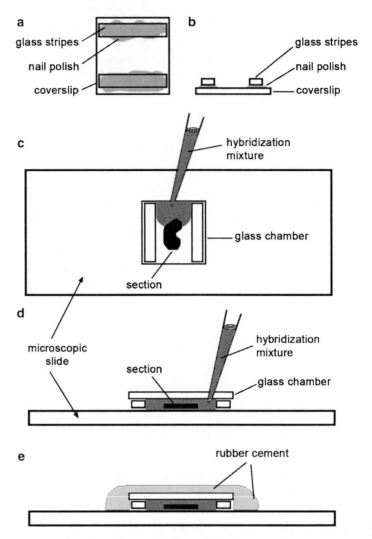

Fig. 2. Preparation and usage of hybridization glass chambers. (**a**, **b**) Preparation of the chamber, *top* (**a**) and *side* (**b**) *views*. Glass strips cut from a coverslip (e.g., 8 × 8 mm) are glued parallel on two borders of an intact coverslip with colorless nail polish. (**c–e**) Hybridization setup (see Subheading 3.5). Glass chamber is placed over a cryosection on the microscopic slide and filled with hybridization mixture from an open side (**c**, **d**) and then sealed with rubber cement (**e**) to prevent evaporation of hybridization mixture during DNA denaturation and hybridization. Similar chambers of a larger size (e.g., 20 × 20 mm) are used for incubation with antibody; they do not need sealing (see Subheading 3.7).

4. Load the probe under chamber from its open side so that probe fills in the whole space above the section (Fig. 2c, d) and seal the chamber with rubber cement. It is important that rubber cement covers the entire chamber periphery or even the whole chamber (see Fig. 2e).

5. Let the rubber cement dry at RT; from this step, protect slide from light!

6. Place slide on a hot-block (or water bath) with temperature 45°C for 1–3 h in order to allow infiltration of the section with the probe.

7. Place slide on a hot-block at 80°C for 5 min for the probe and section denaturation of DNA.

8. Hybridize slide for 2 or 3 days at 37°C (in a waterbath or temperature controlled oven).

3.6. Posthybridization Steps

Throughout all the following steps protect slides from light!

1. Remove rubber cement together with the chamber using a pair of fine forceps.

2. Wash slides in 2× SSC at 37°C, 3×15 min (preferably with shaking) to remove hybridization mixture.

3. Perform stringency washes in 0.1× SSC at 60°C, 2×5 min to remove nonspecific hybrids.

4. Equilibrate in 2× SSC for 2 min.

5. Counterstain section DNA with 0.05 µg/mL DAPI for 30 min (see Note 8).

6. Place a drop of Vectashield on the top of each section and cover with a coverslip; gently remove excess Vectashield with soft tissue and seal the coverslip with nail polish.

7. After the nail polish has dried, the preparation is ready for examination with the microscope. Sections become fragile after the FISH; therefore, strong pressure on the coverslip should be avoided (e.g., pressing the coverslip with the microscope objective when focusing often completely destroys the sample).

3.7. Combination of FISH and Immunostaining

1. Follow the above protocol see Subheading 3.4 up to the step 6 (see Note 9).

2. Transfer sections into PBS for 5 min.

3. Prepare glass chambers for incubation with antibody as following (see Note 10, Fig. 2): cut glass strips from a coverslip (e.g., 20×20 mm) using a diamond cutter. Glue glass strips parallel onto the two opposite borders of an intact coverslip of the same size using nail polish. Size of the chamber can exceed a section size. Allow wet nail polish to completely dry before use; the chambers can be re-used after washing with water.

4. Pull slide out of PBS; using soft tissue, wipe PBS from the back side and from the face side of the slide, carefully avoiding marked section area; place slide in a dark humid box.

5. Place prepared glass chamber over the section and gently press it against the microscopic slide.

6. Load primary antibody diluted in blocking solution under the chamber from its open side and incubate for 12–24 h at RT.

7. Remove the glass chamber and wash the slide with PBS/0.05%Triton X-100, 3 × 15 min at 37°C.

8. Repeat steps 4–6 for the incubation with secondary antibody. After this step, protect slides from light.

9. Postfix section with 2% formaldehyde for 10 min.

10. Wash in PBS, 3 × 5 min.

11. From this step, the slide could be treated as for FISH procedure – starting from see step 6 of Subheading 3.4.

4. Notes

1. The following fluorochrome-dUTPs can be successfully used for probe labeling: Cy3-dUTP, TAMRA-dUTP, TexasRed-dUTP, FITC-dUTP, and Cy5-dUTP. Labeled nucleotides are expensive and preparing them in the lab reduces costs of probes 5–10 times often with a gain in quality (8–10).

2. The working dilution of DNase should be estimated experimentally for every DNase batch; it usually varies between 1:100 and 1:1,000.

3. If large amounts of pure DNA is available (e.g., from a maxiprep), DNA amplification is not needed.

4. Bromphenol blue and dextran blue in stop mixture facilitate the monitoring of the solubilization of the DNA pellet during the later steps. If dyes are for any reason undesirable, prepare 0.5 M EDTA and use 1 μL to stop a 50 μL nick-translation reaction.

5. In order to achieve good fixation in deeper layers of a tissue, the size of the tissue sample should be as small as possible, e.g., 1 × 1 mm or 2 × 2 mm. When samples of larger size are necessary, fixation time should be longer.

6. Thickness of cryosections depends mostly on the size of the nuclei: the smaller their size, the thinner sections can be in order to permit imaging of several (or many) whole nuclei in a single image stack. For example, for most of the mouse cell types with nuclei of ~10 μm, sections of 15–20 μm is sufficient. For large mouse cells, such as neurons with the cell body of ~15 μm in diameter or strongly elongated cells, such as smooth muscle cells which are 30 μm long, cryosections of ~50–60 μm are required.

7. The length of time of heating material strongly depends on tissue and probe; therefore, it needs adjustment for every sample type. Repetitive sequences and BAC clone DNA probes typically require 5–10 min of heating. For chromosome painting, a longer heating time (up to 25 min) is usually needed. Heating at 80°C for more than 30 min can damage the morphology of nuclei.

8. DAPI is not the only dye that can be recommended for staining nuclear DNA in sections. Depending on the probe labeling scheme and the wavelengths available for microscopy, one can also use, e.g., (1) propidium iodide (Sigma), red fluorescence, working concentration 5 µg/mL; (2) TO-PRO-3 (Invitrogen), far red fluorescence, working concentration 1 µM; (3) SYTO 16 (Invitrogen), green fluorescence, working concentration 10 µM. Since all these dyes stain both DNA and RNA, preliminary digestion with RNase before DNA counterstaining is recommended. RNase treatment can be performed simultaneously with incubation with some antibody.

9. All antibodies for nuclear proteins tested in our laboratory when applied to cryosections required antigen retrieval, although to a different degree for each. For example, anti-H3K4me3 (Abcam) staining required 5 min of antigen retrieval while staining with anti-H4K20me3 (Abcam) is better after 20 min, etc. To define an optimal time, set up a time series, e.g., 5, 15, 30 min, and compare the results.

10. Although incubation with antibodies can be performed simply under a coverslip, using chambers is sensible: they prevent drying of the section and notably save antibodies.

Acknowledgments

The author thanks Yana Feodorova and Süleyman Kösem who, as graduate students, carried out some of the immuno-FISH experiments described in this protocol and Mike Fessing (University of Bradford) for providing BAC DNA for Rps27. This work was supported by DFG grant SO1054/1.

References

1. Cremer, M., Grasser, F., Lanctot, C., Müller, S., Neußer, M., Zinner, R., Solovei, I., and Cremer, T. (2008) Multicolor 3D Fluorescence In Situ Hybridization for Imaging Interphase Chromosomes, in Methods in Molecular Biology: *The Nucleus* (Hancock, R., Ed.) pp 205–239, Humana Press.

2. Cremer, M., Müller, S., Köhler, D., Brero, A., and Solovei, I. (2007) in *CSH Protocols* pp doi:10.1101/pdb.prot4723.

3. Solovei, I., and Cremer, M. (2010) Combination of Immunostaining and 3D-FISH on Cultured Cells, in Methods in Molecular Biology: *FISH* (Bridger J. M., V. E., Ed.) Humana Press.

4. Solovei, I., Walter, J., Cremer, M., Habermann, H., Schermelleh, L., and Cremer, T. (2002) FISH on Three-Dimensionally Preserved Nuclei, in *FISH* (Beatty B., Larson S. M., J. Squire, Ed.) pp 119–157, Oxford University Press, Oxford.

5. Solovei, I., Grasser, F., and Lanctôt, C. (2007) in *CSH Protocols* pp doi:10.1101/pdb.prot 4729.

6. Shi, S. R., Cote, R. J., and Taylor, C. R. (1997) Antigen retrieval immunohistochemistry: past, present, and future. *J Histochem Cytochem* **45**, 327–43.

7. Shi, S. R., Cote, R. J., and Taylor, C. R. (2001) Antigen retrieval techniques: current perspectives. *J Histochem Cytochem* **49**, 931–7.

8. Henegariu, O., Bray-Ward, P., and Ward, D. C. (2000) Custom fluorescent-nucleotide synthesis as an alternative method for nucleic acid labeling. *Nat Biotechnol* **18**, 345–8.

9. Müller, S., Neusser, M., Köhler, D., and Cremer, M. (2007) in *CSH Protocols* pp doi:10.1101/pdb.prot4730.

10. Nimmakayalu, M., Henegariu, O., Ward, D. C., and Bray-Ward, P. (2000) Simple method for preparation of fluor/hapten-labeled dUTP. *Biotechniques* **28**, 518–22.

Chapter 6

Multiplex Fluorescence *in situ* Hybridization (M-FISH)

Rhona Anderson

Abstract

Multiplex *in situ* hybridization (M-FISH) is a 24-color karyotyping technique and is the method of choice for studying complex interchromosomal rearrangements. The process involves three major steps. Firstly, the multiplex labeling of all chromosomes in the genome with finite numbers of spectrally distinct fluorophores in a combinatorial fashion, such that each homologous pair of chromosomes is uniquely labeled. Secondly, the microscopic visualization and digital acquisition of each fluorophore using specific single band-pass filter sets and dedicated M-FISH software. These acquired images are then superimposed enabling individual chromosomes to be classified based on the fluor composition in accordance with the combinatorial labeling scheme of the M-FISH probe cocktail used. The third step involves the detailed analysis of these digitally acquired and processed images to resolve structural and numerical abnormalities.

Key words: 24-color karyotyping, Complex aberrations, Marker chromosomes, M-FISH

1. Introduction

Multiplex in situ hybridization (M-FISH), first described by Speicher (1) enables the simultaneous visualization of all chromosomes in a single hybridization. The technique, which essentially paints the entire genome in multiple colors, was developed in the same year as the related SKY technique (2) and represented a major achievement in molecular cytogenetics. Technologically, both M-FISH and SKY combined the significant advances that had been made in fluorescent probe labeling strategies (3), digital imaging fluorescence microscopy, and image processing capability, to create revolutionary karyotypic tools for the analysis of structural and numerical abnormalities (4). The power of M-FISH (and SKY) lies in its ability to resolve complex karyotypes and

Joanna M. Bridger and Emanuela V. Volpi (eds.), *Fluorescence in situ Hybridization (FISH): Protocols and Applications*, Methods in Molecular Biology, vol. 659,
DOI 10.1007/978-1-60761-789-1_6, © Springer Science+Business Media, LLC 2010

identify the origin of marker chromosomes as evidenced by its many applications in tumor diagnostics and research, for example (5, 6), evolutionary cytogenetics (7) and in the study of chemical and radiation-induced aberrations (8). Impressively, even quite complex karyotypes of individual cells within a nonclonal population can be resolved with a high degree of confidence (see Note 1) (9, 10).

A multiplex probe set for M-FISH consists of whole chromosome probes (WCP) each labeled with one of a finite number of spectrally distinct fluorophores. For example, the majority of commercially available human probe sets (e.g., human M-FISH (Genetix) or 24 X-Cyte (Metasystems)) consist of 52 WCP directly and indirectly labeled with one of five fluorophores such that each chromosome is uniquely labeled entirely down its whole length with one, two, or three different fluors (Table 1). It is this combinatorial labeling strategy of fluorophores that forms the basis of M-FISH chromosome classification.

M-FISH relies on the use of single band-pass filters to achieve color separation and optimal signal:noise ratio for all fluorophores employed. Accordingly, the direct visualization of metaphase cells through any individual filter will only reveal that fraction of the genome painted with WCP labeled with its corresponding fluorophore (Fig. 1a–e). This fraction is determined by the combinatorial labeling scheme of the probe kit used (e.g., Table 1). What this means is that the analysis of M-FISH painted metaphase spreads cannot be performed by fluorescence microscopy alone, rather dedicated software is required for the digital acquisition of each fluorescence channel and for their subsequent processing into a merged or composite M-FISH pseudocolored image. A fluorescence microscope with the capacity for holding six different fluorescence filters (five fluors and DAPI) linked to a digital camera and driven by specialized software is therefore a fundamental requirement for this technique. Further, to ensure the acquisition of high-resolution images and accurate registration of all six acquired images, it is crucial that fluorescence exposure times are short and filter changes during the sequence acquisition are quick and smooth with minimal pixel shift. This is best achieved using a microscope fitted with a high-resolution camera (e.g., charged-coupled device (CCD) camera) and a motorized filter wheel (see Note 2). Overall therefore, a significant investment in hardware and software is necessary to build a good quality M-FISH system.

The process of M-FISH image acquisition for each metaphase spread involves the digital capture of five fluorescence channels (Fig. 1a–e) and the DAPI counterstain in sequence (Fig. 1f). The order of this sequence should be set up to minimize any photobleaching of the fluorophores in your probe set and should ideally end with DAPI (see Note 3). Modifications to the focal plane and

Table 1
A combinatorial labeling scheme

Chromosome	Fluor 1	Fluor 2	Fluor 3	Fluor 4	Fluor 5
1					X
2	X				
3				X	
4		X			
5			X		
6		X			X
7	X				X
8				X	X
9			X		X
10	X	X			
11		X		X	
12		X	X		
13	X			X	
14	X		X		
15			X	X	
16	X	X			X
17		X		X	X
18		X	X		X
19	X			X	X
20	X		X		X
21			X	X	X
22	X	X		X	
X	X	X	X		
Y	X		X	X	

exposure times will be necessary between each fluorophore meaning that, when added to the time taken to initially locate each metaphase spread, image acquisition can take between 3 and 5 min of an operator's time. This can be quite restrictive, especially when many hundreds of metaphase spreads need to be acquired per sample (11). Importantly however, M-FISH image acquisition does lend itself to full automation enabling operators to commit the majority of their time to M-FISH analysis (see Note 4).

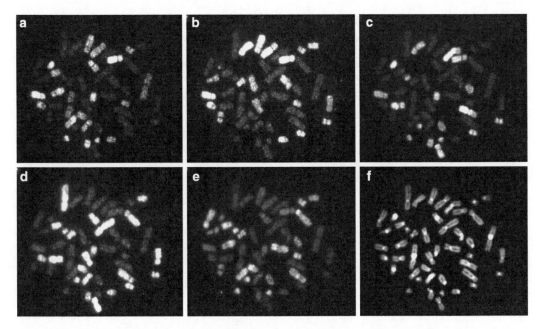

Fig. 1. Raw image files of positive and negative painted chromosomes for each fluorescence channel (**a–e**) and DAPI counter-stain (**f**).

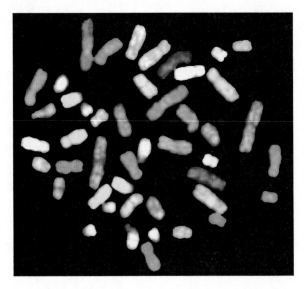

Fig. 2. Metaphase cell showing pseudocolor composite.

Once this image acquisition sequence has been completed, M-FISH dedicated software processes all six raw images by registering each image on top of each other to produce a merged or composite pseudocolor M-FISH image (Fig. 2). The identifying color classification of each homologous chromosome pair is determined by the presence or absence of each fluor detected on a pixel-pixel basis, in accordance with the combinatorial labeling scheme (e.g., Table 1).

M-FISH is a very powerful technique for detecting interchromosomal alterations (~3–10 Mb in size) (12) (see Note 5) of varying complexity, throughout the genome. At its most simplistic, structural rearrangements are identified as discontinuities and changes in M-FISH pseudocolor down the length of DAPI counterstained chromosomes. Theoretically therefore, it is possible to analyze each metaphase using just the M-FISH color karyotype alone (Fig. 3f). However, this is not recommended for a number of reasons. Firstly, chromosome classification is determined on a pixel by pixel basis and is therefore very sensitive to subtle changes in the signal:noise ratio of component fluors across and down the

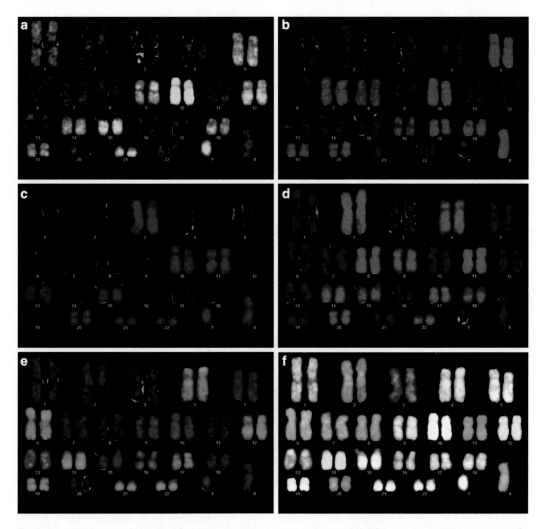

Fig. 3. Karyotype images of individual fluorescence channels (**a–e**) and merged M-FISH pseudocolored chromosomes (**f**).

length of individual chromosomes. Secondly, flaring can occur at the color junctions of chromosome exchanges depending upon the combination of fluors of participating chromosomes (12). This flaring may result in the formation of a false color that could be misinterpreted as an additional insertion leading to misclassification of the actual aberration present. Thirdly, a limitation of M-FISH is that it cannot detect inversions, amplifications, deletions, or homologous pair rearrangements, unless there are obvious changes in centromere position or size of chromosome. Instead, what is recommended is the detailed analysis of each processed single-color karyotype (Fig. 3a–e) in a stepwise fashion before ultimately confirming this assessment with the M-FISH pseudocolored karyotype (Fig. 3f). Thus by virtue of the mode of image acquisition, each fluor can be reviewed separately allowing paint coverage of all fluors on each individual chromosome to be ascertained and true color junctions representing structural alterations to be discriminated (Fig. 4). Further, analysis of the DAPI image in an inverted grey-scale mode will result in the visualization of bands and thereby provide a level of detail on the intrachromosomal structure (see Note 6). This ability to gain information by the review of all acquired images (Figs. 1 and 3) is particularly important for the analysis of nonclonal aberrations as it reduces the risk of bias and potential over-interpretation of color changes visualized from the M-FISH pseudoimage. As an example, we routinely analyze M-FISH painted metaphase spreads by initially examining the DAPI raw image allowing centromere and/or chromosome number to be counted and chromatid-type aberrations (relevant for genomic instability) to be scored. We then karyotype using the enhanced DAPI mode. Taking each fluorescence channel in turn (Fig. 3a–e), we assess the paint coverage down the length of each chromosome in detail, looking for discontinuities, before finally confirming these observations using the composite M-FISH image (Fig. 3f). A chromosome is classified as apparently normal if no breaks or color junctions are observed down its entire length.

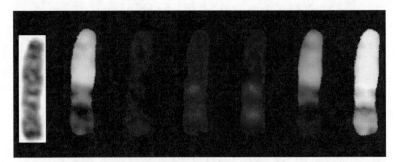

Fig. 4. Derived chromosome 12 (*green+gold*): color junction between C12 and C11 (*red+blue*) with insertions of C11 and C12 material down length, terminating in colur junction between chromosome 7 (*far-red*) and 11 at terminal.

Ultimately though, since M-FISH relies on the analysis of digital images, the standard of analysis achievable is only as good as the quality of each of the six raw images (Fig. 1a–f) initially acquired. In turn, the quality of images acquired is determined by the quality of metaphase spread, optimal FISH biology, microscope/camera hardware employed, and optimal fluorescence exposure times, image focus, and image registration.

2. Materials

2.1. Harvesting of Cell Culture and Slide Preparation

1. KaryoMax colcemid solution (0.5 µg/mL).
2. Hypotonic solution (0.075 M potassium chloride).
3. Fixative solution (3:1 methanol (Analar grade): glacial acetic acid).
4. Superfrost microscope slides.
5. Silica crystals.
6. Slide storage boxes.

2.2. Hardening of Chromosomes

1. Freshly prepared fixative solution.
2. Ethanol series (70, 90 and 100% ethanol).
3. Acetone.

2.3. Pretreatment of Chromosomes

1. 2× SSC (17.6 g sodium chloride and 88 g sodium citrate in 1 L of distilled water).
2. RNase (100 µg/mL in 2× SSC).
3. Phosphate buffered saline (PBS), can be purchased as tablets.
4. Pepsin (1:20,000 in 10 mM HCl).
5. 50 mM $MgCl_2$ in PBS.
6. 1% formaldehyde in 50 mM $MgCl_2$/PBS.

2.4. Hybridization and Posthybridization Washes

2.4.1. Formamide Denaturation of Chromosomal DNA

1. 2× SSC.
2. 70% formamide (Analar Grade, BDH) in 2× SSC.
3. Ice-cold 70% ethanol.
4. 90 and 100% ethanol at room temperature.

2.4.2. Nonformamide Denaturation of Chromosomal DNA

1. Ethanol series (30, 50, 70, 100%).
2. 0.1× SSC.
3. 0.07 N NaOH.

Materials for posthybridization washes will vary depending on the supplier of the multiplex probe kit, e.g., human M-FISH (Genetix) or 24 X-Cyte (Metasystems).

2.5. Digital Image Acquisition and Processing

To acquire and process M-FISH painted metaphase spreads, you will require at a minimum:

1. A fluorescence microscope illuminated by a mercury arc burner (see Note 7) capable of holding six different epifluorescence filter cubes (e.g., 6-position Olympus BX51 or 8-position Zeiss Axioplan II).
2. Single band-pass excitation/emission filters specific for fluorophores in M-FISH probe kit of choice.
3. High-resolution camera (e.g., CCD camera (Photometrics Sensys CCD)).
4. Dedicated M-FISH software, e.g., CytoVision (Genetix) or ISIS (Metasystems).

3. Methods

3.1. Harvesting of Cell Culture and Slide Preparation

M-FISH can be performed on metaphase spreads produced from any exponentially growing cell type. Cell culture methodology and length of time in culture before harvest will vary depending on the cell type used.

1. Add colcemid (0.1–0.5 µg/mL) to the culture medium and incubate at 37°C for 20 min–1 h (see Note 8).
2. Harvest the cells from the flask and centrifuge at $200 \times g$ for 10 min.
3. Discard the supernatant and resuspend the cell pellet.
4. Add 5–10 mL of hypotonic solution and incubate for ~15 min at 37°C (see Note 9).
5. Add ~3 drops of ice-cold 3:1 fresh fixative solution and centrifuge at $200 \times g$ for 10 min.
6. Discard the supernatant and resuspend the pellet.
7. Add ~1 mL of ice-cold 3:1 fresh fixative solution slowly in a drop-wise fashion with continual mixing.
8. Top up to 8–10 mL total with ice-cold 3:1 fresh fixative solution and place at –20°C for at least 20 min.
9. Centrifuge at $200 \times g$ for 10 min.
10. Discard the supernatant and resuspend pellet.
11. Repeat steps 8–10 until the cell suspension is clear.

12. Drop 8–12 µl of cell suspension onto a clean glass slide and air-dry.

13. Check the number and quality of metaphase spreads by phase contrast microscopy (see Note 10).

14. Slides can be left at room temperature and used for M-FISH within a 12 h period. Alternatively, dropped slides can be stored "fresh" at –20°C for months if sealed in storage boxes in the presence of silica gel (see Note 11).

3.2. Hardening of Chromosomes

1. Incubate slide in freshly prepared 3:1 fixative for 1 h at room temperature.

2. Dehydrate through an ethanol series for 2 min each.

3. Bake slides at 65°C for 20 min.

4. Incubate in acetone for 10 min and air-dry.

3.3. Slide Pretreatment

Given the cost associated with purchasing commercial probes (~£130 per 22 × 22 mm slide area) and the importance for all 52 WCP to be efficiently hybridized, only high quality preparations should be used. Pretreatments to maximize probe penetration must be balanced with the need to maintain chromosome morphology and achieving this can often be difficult due to the cost-restricted number of experimental variables you have. Pretreatment conditions can therefore be optimized for your cell type by initially hybridizing with another WCP before additional tests with the M-FISH probe are performed using small test areas.

3.3.1. RNAse Pretreatment (See Note 12)

1. Flood slide with 100 µg/mL RNase, cover with parafilm and incubate at 37°C for 1 h in an humidified chamber.

2. Wash slides in 2× SSC and PBS for 5 min each.

3. Proceed directly to Subheading 3.3.2 or dehydrate slide through an ethanol series for 2 min and proceed to Subheading 3.4.

3.3.2. Protease Pretreatment (See Note 13)

1. Incubate slide in pepsin (1:20,000 in 10 mM HCl) at 37°C for 5–10 min.

2. Wash slides for 5 min in PBS at room temperature.

3. Wash slides for 5 min in 50 mM $MgCl_2$/PBS at room temperature.

4. Fix chromosomes by incubating in 50 mM $MgCl_2$/1% formaldehyde/PBS for 5 min at room temperature.

5. Wash slides in PBS for 5 min.

6. Dehydrate through an ethanol series (70, 70, 90, 90, 100%) for 2 min each.

3.4. Hybridization and Posthybridization Washes

To ensure efficient hybridization of all 52 WCP in an M-FISH probe set, it is critical that temperature, pH, and salt concentration are optimized (see Note 14).

Denature chromosomes using either formamide or nonformamide methodology (see Note 6).

3.4.1. For Formamide Denaturation of Chromosomal DNA

1. Denature chromosomes by incubating slide in 70% formamide in 2× SSC at 72°C for 3 min.
2. Quench slide in ice-cold 70% ethanol.
3. Dehydrate slide in 90 and 100% ethanol, sequentially, for 1 min each.
4. Dry slide by heating briefly on a hot-block at ~40°C.

3.4.2. Nonformamide Denaturation of Chromosomal DNA

1. Rehydrate slides 1 min each in 100, 70, 50, and 30% ethanol (1 min each).
2. Incubate slide in 0.1× SSC for 1 min at room temperature.
3. Incubate slide in 2× SSC for 30 min at 70°C.
4. Remove Coplin jar containing slide from water bath and allow to cool on the bench ~15–20 min until ~37°C.
5. Incubate slide in 0.1× SSC for 1 min at room temperature.
6. Incubate slide in 0.07 N NaOH for 1 min at room temperature.
7. Incubate slide in 0.1× SSC for 1 min at 4°C.
8. Incubate slide in 2× SSC for 1 min at 4°C.
9. Dehydrate slide through ethanol series (30, 50, 70, 100%) for 1 min each and air-dry slide.

3.4.3. Denaturation of M-FISH Probe Set

1. In parallel to denaturation of material (see Note 15), denature the M-FISH probe set as described by supplier (see Note 16).
2. Spot the light-sensitive M-FISH probe set onto 22 × 22 mm coverslip and immediately overlay with the marked area of the slide.
3. Seal the coverslip with rubber cement and incubate in the dark in a humidifed chamber for 2–4 days at 37°C (see Note 17).
4. Wash and perform any detection and amplification steps in the dark as described by supplier.
5. Counterstain cells with DAPI (see Note 18), overlay with 22 × 40 mm coverslip, and seal with nail varnish.
6. Store slides in the dark at −20°C (see Note 19).

3.5. Digital Image Capture

When setting up an M-FISH system, it is important to understand how each chromosome is identified and displayed as an unique color. For instance, M-FISH software will assign a pseudocolored label to each fluorophore in your probe set of choice and will color match this to the appropriate filter block position in the microscope. It will also assign a pseudocolored label to each homologous pair of chromosomes and match this to the combination of fluors for that chromosome, as defined by the classifying labeling scheme of your probe set (e.g., Table 1). It is therefore crucial to check that the probe set, filter position, and pseudocolored assignment are all aligned in advance of acquiring any M-FISH images.

1. Scan slides using an 63× oil immersion objective lens, using a filter that will minimize fluorescence "bleed-through". For example, for SpectraVision™ probe set, the gold filter is recommended.

2. Find a suitable metaphase and then center and focus under the 100× oil immersion objective lens (see Note 20).

3. Check the efficiency of hybridization and signal:noise ratio of each fluorophore by directly looking down the microscope and moving the microscope filter wheel. Fluorescence painting of all fluorophores must be of sufficient quality to ensure the generation of quality M-FISH images for analysis (see Note 21).

4. Direct the fluorescence light to the camera, focus the image, and set the exposure time for the first fluorophore. Digitally capture the image according to the software instructions.

5. Move the filter wheel on the microscope (see Note 22) to visualize the next fluorophore and repeat step 4.

6. Continue capturing images until all five fluorophores and DAPI have been captured for that metaphase (Fig. 1) (see Notes 23 and 24).

7. Return objectives to the 63× position and scan slide for the next suitable cell.

3.6. M-FISH Analysis

1. Select a metaphase spread for analysis and process the raw images (each cell has six separate raw image files) according to the software instructions.

2. Assess the automatic threshold quality by comparing the processed pseudocolored image (Fig. 2) with the original DAPI image (Fig. 1f) to ensure that all chromosomal materials like small fragments are included.

3. Analyze the DAPI image (Fig. 1f) and score any visible chromatid-type breaks and gaps (This is optional).

4. Karyotype the cell in the inverted grey-scale enhanced DAPI banding mode according to the software instructions.

5. Analyze the karyotype of each fluorescence channel separately (Fig. 3a–e) to determine the coverage of paint down the length of each individual chromosome. Use both the raw and the processed images of each fluorophore.

6. Confirm your findings with that performed automatically by the M-FISH software by assessing the m-FISH pseudocolored karyotype (Fig. 3f).

7. A metaphase spread is classified as being apparently normal if all 46 chromosomes are observed to have their appropriate combinatorial paint composition down their entire length.

8. Structural chromosome abnormalities are identified as color junctions down the length of individual chromosomes and/or by the presence of chromosomal fragments. Use the labelling scheme (Table 1) of the M-FISH probe kit to classify the chromosomes involved.

9. Minimize the false classification of insertions at color junctions by assigning the breakpoint position (color junction) for each fluorophore involved separately (Fig. 4).

4. Notes

1. It is not necessary to verify M-FISH results using other FISH-based techniques, unless the aberration is at the limit of M-FISH resolution (see Note 5). The filter-based design of M-FISH enables direct observation of the original hybridization signals using the "raw" images.

2. Fluorescence microscopes with fully automated capability and motorized filter wheel/stage include Olympus BX61 and Zeiss Axioplan II.

3. Automated image acquisition systems use DAPI to scan for metaphase spreads and are therefore an exception to this. Reduce photobleaching by inserting neutral density filters and ensuring scanning is performed at low (×10 or ×20) objective power.

4. Both Metasystems and Genetix distribute software packages (Metafer and Cytovision, respectively) that enable automated metaphase scanning and M-FISH image sequence acquisition. A fully motorized microscope is required to support these packages.

5. The ability to detect the presence of rearranged material is dependent on chromosome condensation and fluorophores involved. For example, if the alteration involves a positive paint signal, the limit of detection is ~3–4 Mb; however, if the

alteration involves a negative paint signal, then the limit of detection can be as high as 10 Mb.

6. Metasystems recommend denaturing chromosomal DNA by a nonformamide methodology. In our opinion, M-FISH hybridization results are consistently excellent, however, the DAPI-stain banding pattern is not produced. Inverted grey-scale DAPI bands are extremely useful during the analysis of M-FISH painted metaphase spreads.

7. A mercury arc burner is required to produce an evenly illuminated field. Follow the microscope operating instructions to ensure the lamp is correctly centered and focused.

8. Optimal colcemid concentration and exposure time will vary depending on cycling time and synchronicity of cell population used.

9. The volume of hypotonic solution used usually increases with pellet size.

10. Aim to achieve a maximum number of quality metaphase cells within a marked 22×22 mm area.

11. When defrosting the slides, leave the box unopened on the bench until room temperature is reached. This is to avoid condensation that would impact on the quality of target material.

12. We pretreat all slides with RNase A as a matter of course, however, this stage is optional.

13. This step is optional and removes excess cytoplasmic and nuclear proteins. Exposure time needs to be optimized for different cell types to achieve the balance between optimal probe penetration and loss of chromosomal structure.

14. Only one to two slides are recommended to be processed at any one time. In this way, information on the quality of previous hybridizations can be considered and subtle adjustments are made as appropriate.

15. Plan the chromosomal and probe denaturation steps to ensure that they are completed at the same time.

16. The conditions during chromosome denaturation and posthybridization washing can be altered depending on the cell type you are using and previous results. However, it is recommended that you follow the methodology as described by the supplier of your M-FISH probe set of choice in the first instance.

17. We regularly set up a hybridization on a Friday afternoon and carry out the posthybridization washes on a Monday morning.

18. We recommend using the DAPI/antifade produced by the suppliers of your probe kit of choice. This is to ensure optimal concentration of DAPI for the fluorophores in that kit.

19. Leave slides for 1–2 days prior to image capture to achieve optimal DAPI staining. It is advisable to capture all required metaphase cells from each slide within 1 week of hybridization to ensure all cells are imaged at their highest fluorescence signal intensity.

20. Incomplete cells or cells with excessive overlapping of chromosomes should be avoided.

21. Poor hybridization and/or signal:noise ratio due to photobleaching of individual, multiple, or all fluors will limit the accuracy of M-FISH analysis. Standard troubleshooting for optimal FISH, e.g., alterations in posthybridization temperatures may need to be carried out. To reduce exposure to UV, use the neutral density filter and minimize exposure to DAPI filter.

22. Do not disturb the microscope during the acquisition sequence to ensure accurate registration of all six images. Microscopes with incorporated automatic filter wheel are recommended.

23. Select the order of image acquisition that has minimal impact for all fluorophores in your probe set. For SpectraVision™ probe set, this is gold, far-red, aqua, red, green, DAPI.

24. Batches of captured images should be archived on removal discs for backup.

References

1. Speicher MR, Gwyn Ballard S, Ward DC. (1996) Karyotyping human chromosomes by combinatorial multi-fluor FISH. *Nat Genet* **12**; 368–375.

2. Schrock E, du Manoir S, Veldman T, Schoell B, Wienberg J, Ferguson-Smith MA, Ning Y, Ledbetter DH, Bar-Am I, Soenksen D, Garini Y, Ried T. (1996) Multicolor spectral karyotyping of human chromosomes. *Science* **273**; 494–497.

3. Nederlof PM, Robinson D, Abuknesha R, Wiegant J, Hopman AH, Tanke HJ, Raap AK. (1989) Three-color fluorescence in situ hybridization for the simultaneous detection of multiple nucleic acid sequences. *Cytometry* **10**; 20–27.

4. Kearney L. (2006) Multiplex-FISH (M-FISH): technique, developments and applications. *Cytogenet Genome Res* **114**; 189–198.

5. Berrieman HK, Ashman JN, Cowen ME, Greenman J, Lind MJ, Cawkwell L. (2004) Chromosomal analysis of non-small-cell lung cancer by multicolour fluorescent in situ hybridisation. *Br J Cancer* **90**; 900–905.

6. Williams SV, Adams J, Coulter J, Summersgill BM, Shipley J, Knowles MA. (2005) Assessment by M-FISH of karyotypic complexity and cytogenetic evolution in bladder cancer in vitro. *Genes Chromosomes Cancer* **43**; 315–328.

7. Bahia H, Ashman JN, Cawkwell L, Lind M, Monson JR, Drew PJ, Greenman J. (2002) Karyotypic variation between independently cultured strains of the cell line MCF-7 identified by multicolour fluorescence in situ hybridization. *Int J Oncol* **20**; 489–494.

8. Anderson RM, Stevens DL, Goodhead DT. (2002) M-FISH analysis shows that complex

chromosome aberrations induced by alpha - particle tracks are cumulative products of localized rearrangements. *Proc Natl Acad Sci U S A* **99**; 12167–12172.

9. Loucas BD, Cornforth MN. (2001) Complex chromosome exchanges induced by gamma rays in human lymphocytes: an mFISH study. *Radiat Res* **155**; 660–671.

10. Anderson RM, Tsepenko VV, Gasteva GN, Molokanov AA, Sevan'kaev AV, Goodhead DT. (2005) mFISH Analysis reveals complexity of chromosome aberrations in individuals occupationally exposed to internal plutonium: a pilot study to assess the relevance of complex aberrations as biomarkers of exposure to high-LET alpha particles. *Radiat Res* **163**; 26–35.

11. Anderson RM, Stevens DL, Sumption ND, Townsend KM, Goodhead DT, Hill MA. (2007) Effect of linear energy transfer (LET) on the complexity of alpha-particle-induced chromosome aberrations in human CD34+ cells. *Radiat Res* **167**; 541–550.

12. Azofeifa J, Fauth C, Kraus J, Maierhofer C, Langer S, Bolzer A, Reichman J, Schuffenhauer S, Speicher MR. (2000) An optimized probe set for the detection of small interchromosomal aberrations by use of 24-color FISH. *Am J Hum Genet* **66**; 1684–1688.

Chapter 7

Optical Mapping of Protein–DNA Complexes on Chromatin Fibers

Beth A. Sullivan

Abstract

Immunofluorescence (IF) and Fluorescence in situ Hybridization (FISH) are conventional methods used to study the structure and organization of metaphase chromosomes and interphase nuclei. Using these techniques, the locations of whole chromosome territories, chromatin subdomains, and specific DNA sequences can be evaluated at kilobase or megabase resolution. Even higher resolution of the spatial relationships of proteins and DNA can be achieved using combined IF-FISH on extended chromatin fibers. This method of optical mapping is a powerful system for localizing molecular probes along released chromatin fibers and visualizing small (<20 kb) or large (20–5,000 kb) chromosomal domains. Chromatin fiber analysis can fill the gaps in resolution between classical chromosome studies and molecular analyses, such as chromatin immunoprecipitation (ChIP) that evaluates chromatin organization at the level of single or multiple nucleosomes. In this chapter, the conceptual and technical aspects of chromatin fiber IF-FISH are presented, along with examples of successful applications.

Key words: Chromatin, Immunofluorescence, In situ hybridization, Histone, Centromere, Heterochromatin, Euchromatin

1. Introduction

The interphase nucleus is nonrandomly organized into functional compartments that contain chromosomes and machinery required for processing various nuclear events such as replication, transcription, and translation (1, 2). Colocalizing proteins and DNA sequences in less compacted, three-dimensionally preserved interphase nuclei can overcome the limited resolution achieved by performing immunofluorescence-FISH on highly compacted metaphase chromosomes (3, 4). Even still, higher order packaging within the nucleus cannot afford a view of chromatin organization

Joanna M. Bridger and Emanuela V. Volpi (eds.), *Fluorescence in situ Hybridization (FISH):*
Protocols and Applications, Methods in Molecular Biology, vol. 659,
DOI 10.1007/978-1-60761-789-1_7, © Springer Science+Business Media, LLC 2010

on a linear scale. Extraction of nuclei with salt and detergents instead can be used to extend chromatin 50–100× its interphase length (5, 6). In this method, unfixed cells or isolated nuclei are immobilized on a glass slide and lysed or extracted by immersion in a solution containing salt and Triton X-100. The slide is positioned vertically during lysis so that the recession of the meniscus and the force of gravity cause the released chromatin to attach along the glass as it is slowly removed from the solution. Low concentrations of urea, a powerful protein denaturant, can also be used to increase the solubility of proteins and to further open the chromatin fiber. When combined with IF-FISH, a long-range view of DNA–protein complexes is achieved on a large linear scale (kilobases to megabases). This technique can be used to order DNA sequences in genomic efforts, to evaluate the relative positions of proteins within large chromosomal domains (5, 7–10), map replication origins along single chromosomes (11, 12), and compare higher order chromatin organization in various cell types (13, 14). The investigator should consider the biochemical properties of the protein(s) to be studied and the effects of extraction conditions on the strength/integrity of the protein interactions with the chromatin and/or DNA. The protocols outlined here are most suitable for core, variant, and modified histones, centromere proteins, nuclear and chromosomal scaffold proteins, and for detecting repetitive and/or unique DNA sequences after immunofluorescence (5, 13, 15). However, lysis conditions can be adapted to visualize many other chromatin or chromosome-associated proteins.

2. Materials

2.1. Cell Types and Cell Culture

1. The following cell types have been successfully used for chromatin fiber preparations: HeLa, HT1080, RPE1, human lymphoblastoid, primary human dermal fibroblasts, hTERT immortalized human fibroblasts, HCT116, H1299, A549, mouse L929 cells, mouse-human somatic cell hybrids, hamster-human somatic cell hybrids, S2 *Drosophila melanogaster* cultured cells, DM2 *D. melanogaster* cells, *D. melanogaster* neuroblast tissues, and *Caenorhabditis elegans* embryos (Fig. 1).

2. Dulbecco's Modified Eagle's Medium (DMEM) supplemented with 10% fetal bovine serum (1× antibiotic–antimycotic or penicillin–streptomycin).

3. Modified Eagle's Medium Alpha (MEM-α) supplemented with 10% fetal bovine serum and antibiotics (1× antibiotic–antimycotic or penicillin–streptomycin).

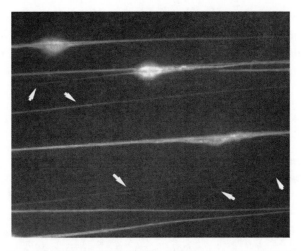

Fig. 1. Chromatin fibers produced by high salt and detergent lysis and stained with DAPI. Field of view (taken at 60×) of fibers lysed for 12 min. Fibers stretch horizontally across the field. Partially lysed nuclei appear as brightly staining with DAPI and are often scattered across the slide. Very thin, presumably single fibers are denoted by *arrowheads*. These are the fibers most ideal for mapping protein and DNA colocalization.

4. Roswell Park Memorial Institute 1640 (RPMI-1640) supplemented with 15–20% fetal bovine serum and antibiotics (1× antibiotic–antimycotic or penicillin–streptomycin).

2.2. Cytocentrifugation of Cells onto Glass Slides

1. Shandon Cytospin 4 (Thermo Fisher Scientific).

2. Plastic cytofunnels (single or double chambered) and metal cytoslips OR all-in-one disposable plastic cytofunnel-clips (Shandon EZ Cytofunnels).

3. Phosphate buffered saline.

4. Hypotonic solutions:
   Solution I: 75 mM KCl (for lymphoblast cells).

   Solution II: 0.8% Sodium Citrate (for mouse fibroblast and *Drosophila* tissue culture and neuroblast cells).

   Solution III: 1:1:1 (v/v/v) 75 mM KCl: 0.8% Na Citrate: dH_2O (for human primary and immortalized fibroblasts and rodent-human somatic cell hybrid cells).

5. Dissociation buffer for *Drosophila* neuroblasts: 1 vol 0.8% Sodium Citrate: 0.5 vol dispase: 0.1 vol collagenase.

6. Dispase stock solution: Dispase is resuspended in water at 2 U/mL and stored in small aliquots at –20°C.

7. Collagenase stock solution: Stock is 0.8% in water (2,160 U/mL) and stored in small aliquots at –20°C.

2.3. Lysis of Nuclei and Fixation of Chromatin Fibers

1. Lysis buffer: Prepare fresh each time 2.5 mM Tris–HCl pH 7.5, 0.5 M NaCl, 1% Triton X-100, 0.2–0.5 M Urea. Dilute Urea from a fresh stock of 1 M Urea.

2. Paraformaldehyde: Prepare a 2–4% (w/v) solution in PBS. The solution may need to be heated carefully to dissolve. Solution should be brought to room temperature before use.

3. Permeabilization solution: 0.1% (v/v) Triton X-100 in PBS.

2.4. Immuno-fluorescence

1. Blocking solution and antibody dilution buffer: Prepare a 250–500 mL stock of PBS containing 0.5% BSA and 0.01% Triton X-100. Supplement with 0.02% sodium azide and keep at 4°C for long-term storage.

2. Wash buffer: 0.05% (v/v) Tween-20 in PBS (PBST).

3. Secondary antibodies: I recommend antirabbit, antimouse, and antigoat IgG conjugated to Alexa-Fluor 488 and Alexa-Fluor 594 (Molecular Probes, Invitrogen) or antirabbit, antimouse, and antigoat IgG conjugated to Cy3 or Cy5 (Jackson Laboratories).

4. Cross-linking solution: 8–10% formalin (3.7% formaldehyde) (10% solution available from Sigma-Aldrich).

5. Parafilm coverslips cut into ~15 mm^2 squares.

2.5. FISH

1. Salmon sperm DNA: 10 mg/mL in water; autoclave to reduce the size of DNA; cool and store in small aliquots at –20°C.

2. C_0t–1 DNA (Invitrogen or Abbott Laboratories).

3. 3 M sodium acetate, pH 5.5: Dissolve in water and pH using acetic acid.

4. 100% ethanol.

5. Denaturing solution: 70% formamide/2× SSC, pH 7.0.

6. Probe Hybridization solutions:
   Unique sequence probes: 50% formamide/2× SSC, pH 7, 20% dextran sulfate, 0.1% Tween-20.

   Repetitive probes: 65% formamide/2× SSC, pH 7.0, 20% dextran sulfate, 0.1% Tween-20.

7. 12 mm circular glass coverslips (Fisher Scientific).

8. Rubber cement.

9. Slide warmer or hot plate set to 37–45°C.

10. Posthybridization washing solutions:
    Unique sequence probes: 4 × 70 mL of 50% formamide/2× SSC, pH 7.0 and 4 × 70 mL 2× SSCT (SSC + 0.05% Tween-20).

Repetitive probes: 4 × 70 mL of 65% formamide/2× SSC, pH 7.0 and 4 × 70 mL 2× SSCT.

11. Wash buffer: 4× SSCT.

12. Blocking buffer: 4× SSC + 5% nonfat milk (purchased at any supermarket or grocery store); spin for 5 min at $13,000 \times g$ before use.

13. Nuclear stain and mounting media: Vectashield (Vector Laboratories) supplemented with 5–10 μg/mL DAPI (4,6-daimidino-2-phenylindole in water, Sigma).

3. Methods

Extended chromatin fibers can be prepared from many different tissue culture cells and from dissociated cells of animal tissues. For immunofluorescence following fiber production, it is important to consider the protein that is being detected and to use highly specific antibodies. Proteins that are tightly associated with DNA will be most efficiently preserved after extraction of nuclei, although less tightly associated proteins can be detected by modifying the protocol, although at the expense of superior fiber morphology.

3.1. Preparation of Cells for Cytocentrifugation

1. Determine how many antibodies and/or FISH probes will be used in the experiment and plan to prepare at least two slides per antibody (see Note 1).

2. All the materials for cytocentrifugation, cell lysis, and fixation are made ready. Label each slide with the date and a distinguishing number. Fit each slide into a metal or plastic cytoclip, placing a single- or double-chambered cytofunnel on top of slide before closing the clip. It is recommended that the lysis buffer and the solution of paraformaldehyde that has resuspended in PBS are prepared fresh each time. Solutions are poured into glass or plastic Coplin jars (~70 mL).

3. Cells should be removed from culture vessels by trypsinization (see Note 2). Retain media from dish or flask in 50 mL Falcon centrifuge tube. Wash cells once with 2 mL of PBS and add PBS wash into tube with retained media. Add 0.5–1 mL of trypsin to detach cells. Use retained media to inactive trypsin; add 15–20 mL of PBS to the cell/trypsin mixture to increase the volume (see Note 3).

4. Count cells using a haemocytometer. Centrifuge cells at $200 \times g$ in an Eppendorf 5810R for 5 min at room temperature (RT).

5. Resuspend cells in hypotonic solution and incubate briefly at RT for 2–5 min (see Note 4).

6. Add 500 µL of cells to hypotonic (250 µL per side of double cytofunnel; 500 µL in single cytofunnel) solution. We use caps that are provided with cytofunnels to prevent fibers from filter paper and other debris from becoming affixed to the area containing the cells.

7. Spin for 4 min at $200 \times g$ with high acceleration (see Note 5).

3.2. Preparation and Fixation of Chromatin Fibers

1. Immediately remove slides from cytoclips and immerse in fiber lysis buffer for 12–15 min (see Note 6). Do not agitate or move the slides in the Coplin jar during this incubation.

2. Slowly remove slides from lysis buffer. Remove slides at a steady rate (~15–20 s/slide), keeping the slide in a vertical position. As the meniscus recedes, the chromatin will stream down the slide and stretch into long fibers that are contained within the region of the 9 mm circle. Steady, constant motion will ensure that the fibers will be unidirectional and largely unbroken.

3. Place slides in Coplin jar containing 2–4% formaldehyde in PBS. Fix for 10 min at room temperature. Do not move or agitate slides during this step.

4. Extract fibers in permeabilization solution for 5–10 min at room temperature (see Note 7).

5. Proceed to immunofluorescence or store slides in Coplin jar containing PBST or PBS at 4°C (see Note 8).

3.3. Immunofluorescence

1. The slides are placed in blocking solution in a Coplin jar for 15–30 min at room temperature. Blocking at 4°C is also acceptable.

2. Add 15–25 µL of primary antibody diluted in blocking solution onto the circular area containing the chromatin fibers. Cover the area with a ~15 mm² parafilm coverslip (see Note 9).

3. Incubate at 4°C for 24–48 h in a humidified chamber (see Note 10).

4. Place slides vertically in wash solution in a glass or plastic Coplin jar. Do not remove the parafilm coverslips, but allow them to float off in the wash solution. Do not agitate slides or put Coplin jar on orbital shaker. This preserves the fiber preparations that can be easily scratched or dislodged by mechanical forces exerted by forceps, shaking, or even manual parafilm removal. Wash slides three times for 5 min each, carefully moving the slides to a new Coplin jar containing fresh wash solution after each wash period. The third (last) jar of wash solution may be

saved and used as the first wash after secondary antibody incubation (step 15).

5. Add 15–25 µL of secondary antibody freshly diluted in blocking solution onto the circular area containing the chromatin fibers. Cover the area carefully with a fresh 15 mm² parafilm coverslip. Incubate at RT for 1–2 h (see Note 11).

6. Wash as in step 13, taking care to float off coverslips and to avoid agitating slides.

7. Crosslink antibody–protein complexes by immersing slides in a Coplin jar containing 8–10% formalin for 10 min at RT.

8. Place slides in Coplin jar containing PBST and immediately proceed to FISH or store covered Coplin jar at 4°C for up to 1 week until FISH is performed (see Note 12).

3.4. Fluorescence In Situ Hybridization

Both repetitive and single copy DNA probes can be successfully hybridized to chromatin fibers. It is necessary to increase the amount of each probe used in an individual experiment by at least 50% to ensure visualization on fibers. In general, we use ~100–150 ng of repetitive probes per slide (50–75 ng per 9 mm area) and 300–500 ng of unique sequence probes (150–250 ng per 9 mm area). Probe labeled by incorporation of nonfluorescent haptens (biotin, digoxygenin) are acceptable to use in FISH experiments and can be detected with fluorophore-conjugated streptavidin/avidin or antidigoxygenin. However, further amplification of signals using antiavidin can lead to significant generalized fluorescent background. DNA probes directly labeled with fluorophore-conjugated dUTPs are advantageous in that they are associated with low background, but we have found that more probe, particularly for unique sequences (BACs, PACs, P1s), is required to easily visualize fluorescent probe signals.

3.4.1. Probe Preparation and Denaturation

1. Probes are labeled by incorporating modified dUTP via nick translation using DNA polymerase I (see Section A, Core protocols 1 and 2). We recommend removing unincorporated nucleotides after the nick translation reaction using G-50 sephadex columns.

2. Precipitate the appropriate amount of probe with 1 µL salmon sperm DNA (20 mg/mL), 0.1 volume sodium acetate, and 2.5 volumes of 100% ethanol at –20°C overnight or –80°C for at least 30 min. When precipitating unique sequence probes, include 10–20 µg of C_0t-1 DNA to suppress repetitive sequences.

3. Centrifuge at top speed $(14,000 \times g)$ in a refrigerated microcentrifuge set to 4°C for 20 min. Carefully aspirate ethanol, taking care not to dislodge the pellet.

4. Wash pellet with 200 mL of 70% ethanol. Centrifuge at top speed for 5 min at 4°C. Aspirate ethanol and allow pellet to air-dry. Alternatively, the pellet can be dried in a speed vac for ~5 min, but ensure that the pellet does not over-dry.

5. Resuspend pellet in the 15 mL of the appropriate hybridization buffer (see Note 13). Vortex/mix briefly. Collect probe at the bottom of tube by brief centrifugation using a benchtop mini centrifuge.

6. For unique sequence probes, denature at 78°C for 10 min and place tube in 37°C for 15–60 min to preanneal unlabeled C_0t-1 DNA and repetitive sequences within the denatured probe. Time the preannealing period to coincide with denaturation of chromatin fibers (see below, Subheading 3.4.2, step 3).

3.4.2. Denaturation of Chromatin Fibers/Target

1. Prepare reagents needed: circular coverslips, small Kimwipes folded into small squares, rubber cement, 3 or 5 mL syringe, 18 or 21 gauge needle.

2. If fibers have been stored in PBST/PBS solution at 4°C, allow Coplin jar with slides to equilibrate to RT.

3. Prepare 140 mL of denaturing solution and place glass Coplin jars in 37°C waterbath for 10–15 min. Step the jars into a bath set to 42°C before finally placing them into a waterbath set to 80°C (see Note 14). Check the temperature of denaturing solution in each Coplin jar. It should be least 78°C before proceeding with denaturing of chromatin fibers.

4. Denature one slide per Coplin jar. Remove slide from jar of PBST, quickly wipe the back of the slide to remove excess solution, and place slide in denaturing solution for 8–10 min.

5. Simultaneously, place a tube containing labeled probe in hybridization mixture in 78°C waterbath to denature DNA (i.e., for repetitive probes that do not require preannealing) for 9 min.

6. During this time, prepare 3 or 5 mL syringe containing rubber cement and attach 18 or 23 gauge needle. Replace protective plastic needle cap until ready to use and set aside.

7. After 9 min, place tube of denature probe on ice and get pipette with clean pipette tip ready. Set aside.

8. Remove slide from denaturing solution and using folded kimwipe, wipe back of slide, then quickly wipe excess formamide from the front off slide, taking care to wipe around the circular areas containing the fibers.

9. Set slide fiber side up on slide warmer and quickly add 5–7 μL of denatured probe to each circular area of fibers. Quickly,

but carefully cover each probe-fiber area with 12×12 round or 18×18 mm square coverglass.

10. Quickly seal area with rubber cement by pushing on syringe plunger ensuring that the coverslip does not move from area containing fibers. Put slides in humidified chamber covered with aluminum foil and place chamber in 37°C incubator for 16–48 h.

11. For preparations that were cross-linked and fixed by the alternative method (Note 10), see Note 15 for denaturation conditions.

3.4.3. Posthybridization Washes and Counterstaining

1. Prepare 280 mL of posthybridization washing buffer and 280 mL of 2× SSC/0.05% Tween-20 (2× SSCT) in Wheaton or Pyrex glass bottles and place in waterbath to equilibrate to 42–45°C. For unique sequence probes, postwash solution should be 50% formamide/2× SSC/0.05% Tween-20 pH 7, and for repetitive probes, use 65% formamide/2× SSC/0.05% Tween-20 pH 7. Place two glass Coplin jars with lids in the waterbath so that they heat up to temperature of the bath. We typically use the Coplin jars that can hold eight slides facing the same way. However, 12 slides can be fitted into these jars with four slides placed in a back-to-back position. This configuration will not damage the side of the slides that contains the fibers/specimen.

2. Pour 70 mL of washing solution into each Coplin jar.

3. Remove the humidified tray from the 37°C incubator. Using fine tip forceps, gently peel rubber cement from the edges of the coverglass. In many cases, the coverglass will stick to the rubber cement and be removed. However, sometimes the coverglass will remain attached to the slide when all of the rubber cement is removed. Do not try to pry the coverglass off as this will damage the fibers underneath and potentially scrape them from the slide.

4. Place the slide into a Coplin jar containing the wash solution. Repeat for each slide, placing each slide into the groove designated for an individual slide. Incubate for 5 min without agitating the slides.

5. Carefully transfer slides to the second Coplin jar containing wash solution. At this point, a coverglass that was attached when the rubber cement was removed should have detached in the wash solution. If it has not, gently swirl the slide in the first Coplin jar of wash solution; this will facilitate removal of the coverglass.

6. Incubate 5 min. Meanwhile, pour out the wash solution in the first Coplin jar and replace with 70 mL of fresh wash

solution (see Note 16). Repeat washes two more times for 5 min each.

7. After third formamide wash, replace formamide solution in one Coplin jar with 2× SSCT.

8. Wash slides four times for 2 min each in 2× SSCT.

9. Transfer slides to a plastic staining jar of 4× SSCT (4× SSC/0.05% Tween-20) at RT for at least 5 min (see Note 17).

10. For probes that have been directly labeled, counterstain with DAPI and mount with antifadent. We use Vectashield mounting medium containing 5–10 µg/mL DAPI (see Note 18).

11. Place 24×50 mm coverglass on slide, taking care to remove air bubbles without pressing directly on the areas containing fibers. Seal around edges of coverslip with plastic sealing solution or rubber vulcanizing solution (see Note 19).

12. For probes that have been indirectly labeled with biotin or digoxygenin, add 100 µL of blocking buffer to slides. Cover with a parafilm coverslip cut to fit the 24×50 mm slide area and incubate at RT for 10–30 min.

13. While slides are in blocking solution, dilute appropriate antibodies in blocking buffer and centrifuge to clear unconjugated fluorophores from solution (see Note 20).

14. Carefully lift off parafilm coverslip using fine tip forceps. Lift from the frosted end of the slide.

15. Add 25–50 µL of antibody to the circular area containing the fibers. Cover each circular area with parafilm coverslips cut to ~15×15 mm. Incubate 8–24 h at 4°C in a humidified chamber (see Note 21).

16. Place slides in plastic staining jar containing 40–50 mL of wash buffer (4× SSCT). Take care not to place the face of slide containing fibers against the side of the staining jar. Parafilm coverslips will float off in solution as before during the first wash (Subheading 3.4). Wash for 5 min at RT without agitating slides.

17. Transfer slides to second plastic staining jar of wash buffer. Wash twice more for 5 min each.

18. Counterstain/mount slides in Vectashield containing DAPI as in step 10.

3.5. Microscopy

Slides that have been mounted and counterstained with Vectashield-DAPI can be viewed using standard wide-field epifluorescence microscopes, as well as high-resolution confocal and deconvolution microscopy systems that are connected to cooled charged-coupled devices and digital cameras. Usually, exposure times for fiber images are a few seconds more than those used for standard metaphase or interphase fluorescence. Also, because fibers are often variably stretched, fluorescent antibody or probe signals from a more densely compacted region of the fiber may saturate the image and the less bright signals. It is recommended that the investigator use manually selected exposure times when capturing digital images to ensure that all antibody or probe signals are captured effectively.

Antibody and FISH signals should overlap with DAPI-stained fibers. The signals will appear as punctate dots of fluorescence (Fig. 2). Gaps between dots are usually small breaks in the fibers that have occurred during the stretching process. To gain confidence in the location or colocalization of fluorescent signals (either two proteins or a protein and FISH probe), many fibers should be analyzed. We routinely analyze more than 50 fibers for each experiment to evaluate if fluorescent signals are consistent in their patterns of localization. Fluorescent background usually appears to be higher on chromatin fiber slides than is noticeable

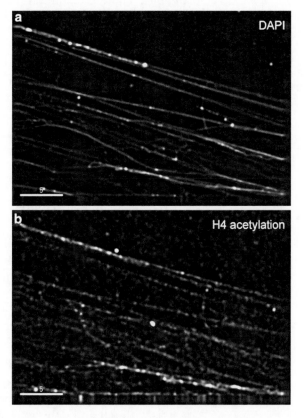

Fig. 2. Immunofluorescence on extended chromatin fibers. (a) DAPI staining of chromatin fibers from human cells lysed with salt, detergent, and urea. (b) Antibodies to acetylated histone H4, a histone modification associated with euchromatin, detected on the chromatin fibers. Note the dotted appearance on the thinner fibers.

on standard metaphase or interphase IF-FISH; so it is important to analyze many fibers and to ensure that fluorescent signals overlap with DAPI-stained fibers. We do not score or analyze any fluorescent signal that is not coincident with a DAPI-stained fiber. It is possible that fluorescent signals that do not correspond with the location of DAPI staining are present on very thin fibers, but by using higher concentrations of DAPI in the antifadent/mounting solutions, it is ensured that these thin fibers are detected.

Ideal chromatin fibers are often quite long, and on standard microscopes, they extend over multiple fields of view, especially if viewed with 100× objectives. Many software packages that are used to capture digital images, such as the SoftWoRx Resolve3D capture system by Applied Precision, Inc., have an option to capture multiple fields and "stitch" them together into a single image for analysis and presentation purposes. In the absence of this option, it is possible to capture images of the whole fiber at lower magnification (i.e., 60×) or to take multiple images at higher magnification and manually fit the separate images together in Photoshop or other graphics programs. There are computer algorithms that are able to create composites from multiple fields of view of standard light or fluorescence microscopy images (10).

4. Notes

1. All solutions should be prepared in water that is molecular grade (18.2 MΩ–cm) water. This standard is referred to as "water" or dH_2O in this text.

2. One 10 cm dish or one 25 cm² flask containing cells that are over 75% confluent is sufficient to immobilize cells onto at least 6–12 glass slides.

3. For *Drosophila* neuroblasts, dissect 20 brains per 2 mL of dissociation buffer. Incubate brains in dissociation buffer for 30–60 min at room temperature. We have also dissociated the brains for 30 min at 37°C. Pass dissociated cell mixture through a 27-gauge needle 5–10 times. Load 500 μL of solution into single-chambered cytofunnels or 250 μL per side of double cytofunnels. Spin for 4 min at 200 g.

4. Cell density is the single most important factor in cytospinning (i.e., centrifuging cells or nuclei onto glass slides). Too few cells will produce extended chromatin fibers prior to lysis with salt and detergent, but many fibers will lack directionality and will stretch heterogeneously. Too many cells will prevent complete lysis and may promote the displacement of cells during washing steps and other manipulations

(such as adding and removing coverslips). Based on over 10 years of experimentation, the following cell concentrations have been determined to be optimal for the following cells types: $1.5–2 \times 10^5$ cells/mL for *Drosophila* S2 or Kc cells – single chamber; 3.5×10^5 cells/mL for *Drosophila* S2 or Kc cells – double chamber; 1.8×10^5 cells/mL for human lymphoblasts – single chambers; 2.2×10^5 cells/mL for human lymphoblasts – double chambers; 4×10^4 cells/mL for mammalian primary fibroblasts – single chambers; $6.0–8.2 \times 10^4$ cells/mL for mammalian primary fibroblasts – double chambers; 6.1×10^4 cells/mL for mammalian immortalized fibroblasts – single chambers; $7.4–8.2 \times 10^4$ cells/mL for mammalian immortalized fibroblasts – double chambers.

5. It is imperative to use the high acceleration setting on the Shandon cytospin. This ensures that cells are deposited onto the glass slides at high impact while maintaining morphology.

6. Cells will be visible as small, opaque 8–9 mm circles on the glass slides that may be allowed to air-dry briefly. However, once cells have been placed in lysis buffer, they should remain hydrated throughout the rest of the immunofluorescence procedure. Determination of correct lysis time is absolutely key to producing superb chromatin fibers. This step is the one that gives investigators the most trouble because lysis time often varies among cell lines. Start with a lysis time of 12 min, but if >60% of the nuclei remain unlysed or appear bundled (Fig. 3), increase time in lysis buffer starting with 1 min increments. Alternatively, one can cytospin multiple slides and lyse each slide for a different amount of time. Once cells are fixed, the fibers can be stained with DAPI and evaluated to determine which time produced optimal lysis and the greatest number of long, straight fibers. This is a worthwhile step to include each time one makes chromatin fibers to ensure that fiber quality is satisfactory. In general, lymphoblast cells require less time in lysis buffer (12–13 min) than fibroblast cells (15–16 min). An example of appropriately lysed fibers staining with DAPI is shown in Fig. 1.

7. In lieu of staining with DAPI (see Note 6) and to determine the extent of lysis, slides can be examined using a phase microscope after they have been placed in permeabilization buffer. This step must be done quickly to prevent slides from drying. If only nuclei, bundled fibers or a few short fibers are present, longer lysis is required (Fig. 3) (see Note 6). This is an important step to include each time one makes chromatin fibers to ensure that fiber quality is satisfactory. However, I recommend incorporating this practice of not using DAPI to evaluate fibers only after the investigator has mastered the protocol

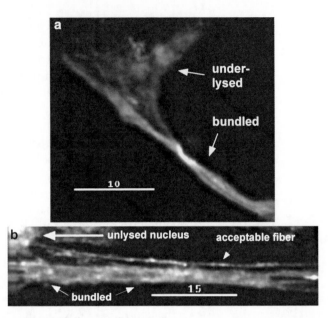

Fig. 3. Inappropriately lysed chromatin fibers. (**a**) DAPI staining chromatin fibers from cells that have not been lysed for long enough will appear bundled rather than as independent fibers. (**a**) Bundled fibers appear as thick or twisted masses of chromatin. Underlysed nuclei appear as diffuse DAPI staining. (**b**) Partial field showing heterogenous lysis, representative of general under-lysis. An acceptable fiber, bundled fibers, and an unlysed nucleus are denoted by *arrows*.

and feels comfortable recognizing fibers very quickly under the phase microscope.

8. Fiber preparations can be stored in PBS or PBST at 4°C for up to 2 weeks before immunofluorescence is performed. It should be noted that fiber quality decreases with increased time in buffer, and some fibers will detach from slides after prolonged periods in solution.

9. The appropriate dilution of primary antibody must be determined empirically. However, as a general guideline, we find that we dilute our antibodies by 1/5–1/10th of that used for metaphase or interphase immunofluorescence. For instance, we use H3K4me2 antibodies (Abcam ab7766) at 1:500–1:1,000 for metaphase chromosomes and at 1:100–1:200 for chromatin fibers.

10. For our humidified chambers, we use 245 cm² plastic dishes (Corning), covering the outside of the lid and base with aluminum foil. Inside, we place a sheet of water-saturated blotting paper (Whatman). Slides are placed on top of 5 mL falcon plastic pipettes manufacturer is Whatman cut to fit the tray. Break the pipettes around the 3 mL mark and discard the tapered ends. Two truncated pipettes are bound together

with laboratory tape on each end, leaving ~2.5–3 inches of space between them so that they create a platform on which to rest the slides. In between uses, leave the lid off of the tray to prevent microbial growth and replace the filter paper and tape around the pipettes as needed.

11. Our choice of secondary antibodies are Molecular Probes' (Invitrogen) Alexa Fluor conjugated secondaries, including Alexa Fluor 488 (FITC equivalent), Alexa Fluor 555 (Cy3 equivalent), Alexa Fluor 594 (Texas Red equivalent), and Alexa Fluor 647 (Cy5 equivalent). We also recommend Jackson Immunoresearch secondary antibodies (IgG, streptavidin, and digoxin) conjugated to AMCA, FITC, Cy3, and Cy5.

12. An alternative cross-linking and fixation step may be used. After cross-linking with formalin, slides can be dipped briefly (<1 min) in a Coplin jar containing dH_2O to wash off the formalin, then fixed with 3:1 (v/v) methanol:acetic acid for 15 min at RT. Slides are removed, air dried, and stored in slide boxes with dessicant packs or under vacuum until FISH is performed. The acid fixation step sometimes reduces or destabilizes protein–antibody complexes and should be tested for specific antibodies. It does not affect the retention of centromere proteins and antibodies or scaffold protein–antibody interactions.

13. Precipitated repetitive probes are usually resuspended in hybridization mix containing 65% formamide. However, we have also successfully used lower concentrations of formamide (60%) without obtaining cross-hybridizing signals. PNA probes are resuspended in hybridization mixture containing 20% formamide.

14. As a safety measure, we introduce the glass Coplin jars incrementally to higher temperature baths to prevent cracking and shattering of the glass jars, which often occurs if RT jars are placed directly into 80°C bath.

15. Denature slides for 8–10 min in 70% formamide/2× SSC pH 7 at 78°C. Unlike the slides that did not undergo acid fixation, it is acceptable to denature two of these slides at a time in the Coplin jar of denaturing solution. Remove slides from denaturing solution and place through a cold ethanol series (70%/85%/100% or 70%/95%/100%) for 1–2 min in each ethanol. Air-dry all slides on a slide warmer. Apply 5–7 µL of denatured probe (or denatured/preannealed probe, if hybridizing unique sequence probes) to circular area containing fibers, carefully apply 12 × 12 mm circular coverglass, seal with rubber cement, and hybridize for 16–48 h at 37°C in humidified chamber. Postwashing and secondary detection is performed as described in Subheading 3.4.3.

16. Formamide is a hazardous chemical that should not be disposed of by pouring down the drain. Wear gloves and a laboratory coat when working with formamide. Dispose of solutions containing formamide in secondary containers that are clearly marked with the date, time, amount, and concentration of formamide.

17. We use plastic staining jars at this and subsequent steps to protect fluorescence on antibodies from immunofluorescence and direct-labeled probes from light.

18. Vectashield is an aqueous mountant that does not solidify but remains viscous on the slide. This mountant inhibits photobleaching of many fluorochromes but should be tested for your specific fluorescent protein and/or probe if using anything other than common fluorochromes. Vectashield Mounting Medium can be purchased containing DAPI, but we prefer to buy it without DAPI and add an increased concentration of DAPI (5–10 μg/mL). This concentration allows the detection of very thin chromatin fibers, ensuring that the investigator is viewing single fibers rather than bundles of multiple chromatin fibers that will fluoresce brightly.

19. We use a rubber/silicone solution used to repair leaks in tires, called Truflex/Pang. This solution is slightly viscous and dries into a rubbery opaque film that creates a seal between the coverglass and slide. This seal is strong enough to keep the coverglass from moving, but, unlike nail varnish, is not permanent and allows us to remove the coverglass in case we need to amplify the fluorescent probe signal or wish to do further hybridizations or other experiments on the fibers.

20. Incubation at 4°C at this step is important. Since fluorescent signals on chromatin fibers must be discriminated from potential background fluorescence, it is important to take all precautions to minimize baseline fluorescence that might be generated by secondary antibodies.

21. Biotin-labeled probes are detected with streptavidin conjugated to fluorophores, such as fluorescein (FITC), Alexa Fluors 488 (green), 555 (Cy3, red), 594 (Texas Red), 350 (AMCA, blue), 647 (Cy5, far red), texas red, or rhodamine, to name a few. They can also be detected with antibiotin antibodies conjugated to the same fluorophores, although the avidin-biotin interaction represents the highest known affinity between a protein and a ligand. Dioxygenin-labeled probes are detected using antidigoxgenin or digoxin antibodies conjugated to FITC or fluorescein (green), Cy3, texas red, or rhodamine (red), AMCA (blue), or Cy5 (far red). Dilute antibodies in blocking buffer and spin at $13,000 \times g$ for 5 min at 4°C to remove unconjugated fluorescent molecules.

References

1. Foster, H. A., and Bridger, J. M. (2005) The genome and the nucleus: a marriage made by evolution. Genome organisation and nuclear architecture. *Chromosoma* **114**, 212–229.

2. Parada, L., and Misteli, T. (2002) Chromosome positioning in the interphase nucleus. *Trends Cell Biol* **12**, 425–432.

3. Trask, B. J., Allen, S., Massa, H., Fertitta, A., Sachs, R., van den Engh, G., and Wu, M. (1993) Studies of metaphase and interphase chromosomes using fluorescence in situ hybridization. *Cold Spring Harb Symp Quant Biol* **58**, 767–775.

4. Brandriff, B., Gordon, L., and Trask, B. (1991) A new system for high-resolution DNA sequence mapping interphase pronuclei. *Genomics* **10**, 75–82.

5. Haaf, T., and Ward, D. C. (1994) Structural analysis of alpha-satellite DNA and centromere proteins using extended chromatin and chromosomes. *Hum Mol Genet* **3**, 697–709.

6. Michalet, X., Ekong, R., Fougerousse, F., Rousseaux, S., Schurra, C., Hornigold, N., van Slegtenhorst, M., Wolfe, J., Povey, S., Beckmann, J. S., and Bensimon, A. (1997) Dynamic molecular combing: stretching the whole human genome for high-resolution studies. *Science* **277**, 1518–1523.

7. Haaf, T., and Ward, D. C. (1994) High resolution ordering of YAC contigs using extended chromatin and chromosomes. *Hum Mol Genet* **3**, 629–633.

8. Sullivan, B. A., and Karpen, G. H. (2004) Centromeric chromatin exhibits a histone modification pattern that is distinct from both euchromatin and heterochromatin. *Nat Struct Mol Biol* **11**, 1076–1083.

9. Blower, M. D., Sullivan, B. A., and Karpen, G. H. (2002) Conserved organization of centromeric chromatin in flies and humans. *Dev Cell* **2**, 319–330.

10. Appleton, B., Bradley, A. P., and Wildermoth, M. (2005) *in* "DICTA 2005", Vol. 1, IEEES Computer Society Press, Cairns, Australia.

11. Pasero, P., Bensimon, A., and Schwob, E. (2002) Single-molecule analysis reveals clustering and epigenetic regulation of replication origins at the yeast rDNA locus. *Genes Dev* **16**, 2479–2484.

12. Haaf, T. (1996) High-resolution analysis of DNA replication in released chromatin fibers containing 5-bromodeoxyuridine. *Biotechniques* **21**, 1050–1054.

13. Haaf, T., and Ward, D. C. (1995) Higher order nuclear structure in mammalian sperm revealed by in situ hybridization and extended chromatin fibers. *Exp Cell Res* **219**, 604–611.

14. Heng, H. H. (2000) Released chromatin or DNA fiber preparations for high-resolution fiber FISH. *Methods Mol Biol* **123**, 69–81.

15. Haaf, T., and Ward, D. C. (1996) Inhibition of RNA polymerase II transcription causes chromatin decondensation, loss of nucleolar structure, and dispersion of chromosomal domains. *Exp Cell Res* **224**, 163–173.

Chapter 8

3D-FISH on Cultured Cells Combined with Immunostaining

Irina Solovei and Marion Cremer

Abstract

Fluorescence in situ hybridization on three-dimensionally preserved nuclei (3D-FISH), in combination with immunocytochemistry and 3D fluorescence microscopy, is a key tool to analyze the functional organization of the interphase nucleus. In the last decade, 3D-FISH on cultured cells has become a routine technique and is now widely used in nuclear biology. This method allows visualization of chromosome territories, chromosome subregions, single genes, and RNA transcripts preserving their spatial positions in the cell nucleus. In many cases, it is desirable to combine 3D-FISH and immunostaining to map DNA/RNA and protein targets in the same cells. Some steps of the FISH procedure, however, may interfere with immunostaining and special efforts have to be done to combine FISH and antibody staining successfully. The protocol suggested in this chapter describes three variants of combined 3D-FISH and immunostaining which have been successfully used in our laboratory for many years.

Key words: Interphase nucleus, Cultured cells, 3D-preservation of nuclear morphology, Fluorescence in situ hybridization (FISH), 3D-FISH, Immunostaining, Immuno-FISH

1. Introduction

This chapter describes the fluorescence in situ hybridization (FISH) of DNA probes on three-dimensionally preserved cells, so called 3D-FISH, provided an appropriate microscopy method (preferably, confocal microscopy); this technique allows three-dimensional visualization of specific DNA targets within nuclei and provides information about the intranuclear arrangement of chromosome territories (CT) as well as the positions of chromatin regions and genes within the CTs, for example, (1–10). With specific modifications, the same methods also allow mapping of RNA transcripts, though more specific protocols for this purpose also exist (Chapter 3). The main difficulty of 3D-FISH is the need

Joanna M. Bridger and Emanuela V. Volpi (eds.), *Fluorescence in situ Hybridization (FISH): Protocols and Applications*, Methods in Molecular Biology, vol. 659,
DOI 10.1007/978-1-60761-789-1_8, © Springer Science+Business Media, LLC 2010

to satisfy two rather contradictory conditions, spatial preservation of native chromatin arrangement and making nuclear DNA accessible for probes. Probe labeling, hybridization, and posthybridization steps, as well as equipment necessary for 3D-FISH are basically the same as for FISH on metaphase chromosomes (2D-FISH). Since detailed protocols for 2D-FISH are well-known and, in particular, provided in this book (Chapter 1), this protocol is focused on the fixation methods of both adherently growing cells and cells growing in suspension and on the pretreatment of fixed cells.

FISH can be combined with the detection of cellular, and in particular, nuclear proteins (11–13). The difficulty of such combined staining depends strongly on the protein to be detected. Some antigens are stable enough to withstand permeabilization pretreatments and DNA denaturation, whereas instability of some other antigens makes combination of FISH with immunostaining a tedious procedure. Therefore, quite different strategies should be employed for different proteins. In this chapter, we describe a "modular" protocol (Fig. 1) that can be adapted for successful immuno-FISH on a wide range of targets (Fig. 2).

2. Materials

2.1. Fixation of Adherently Growing Cells

1. Phosphate Buffered Saline (PBS), pH 7.4 prepared from 20× PBS stock (1× PBS: 140 mM NaCl, 2.7 mM KCl, 6.5 mM Na_2HPO_4, 1.5 mM KH_2PO_4).
2. 4% Paraformaldehyde in PBS, pH 7.3, freshly prepared from paraformaldehyde powder (Merck) by heating in PBS up to 60°C.
3. 0.5% and 0.05% Triton X-100 (Merck) in PBS.

2.2. Fixation of Cells Growing in Suspension

1. Cell counting chamber.
2. Polylysine-coated coverslips prepared as follows: Incubate coverslips on drops of 1 mg/mL poly L-lysine (Sigma–Aldrich) in H_2O loaded on Parafilm for 30 min, rinse well in dH_2O, and air-dry.
3. 50% Fetal Calf Serum (FCS) in RPMI (Biochrom KG).
4. 0.4× PBS.
5. 4% Paraformaldehyde in 0.4× PBS, pH 7.3, freshly prepared from paraformaldehyde powder by heating in PBS up to 60°C.

2.3. Permeabilization of Nuclei

1. 20% Glycerol (Merck) in PBS.
2. Styrofoam container filled with liquid nitrogen.

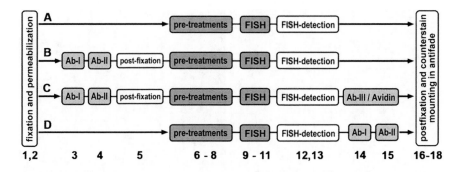

step		brief specification	protocol
1	fixation	4% formaldehyde, 10 min	*3.1. – 3.2.*
2	permiabilization-1: membranes extraction	0.5% Triton X100, 10 min	*3.3.1.*
3	primary antibody	in blocking solution / PBST, 45 min	
4	secondary antibody-fluorochrome or secondary antibody-hapten	in blocking solution / PBST, 45 min	*3.6.2 – 5.*
5	postfixation	2% formaldehyde, 10 min	*3.6.6 – 7.*
6	permeabilization-2: freezing / thawing	incubation in 20% glycerol, freezing in liquid nitrogen / thawing, 5x	
7	permeabilization-3: protein removing	0.1M HCl, 5-20 min	*3.3.2 - 7.*
8	further pretreatments	2xSSC, 5 min 50% formaldehyde, 30 min - months	
9	probe loading and denaturation	denaturation at 80°C for 1 – 3 min	*3.4.1 – 4.*
10	hybridization	at 37°C for 1-3 days	
11	post-hybridization washings	2xSSC at 37°C, 3 x 10 min 0.1xSSC at 60°C, 2 x 5 min	*3.5.1 – 3.*
12	hapten(s)-detection layer-1 *	in blocking solution / 4xSSCT, 45 min	
13	hapten(s)-detection layer-2 *	in blocking solution / 4xSSCT, 45 min	*3.5.5 – 8.*
14	primary antibody	in blocking solution / PBST, 45 min	
15	secondary antibody-fluorochrome	in blocking solution / PBST, 45 min	*3.7.2 – 6.*
16	postfixation	2% formaldehyde, 10 min	*3.5.9.*
17	counterstain	DAPI in PBS	
18	mounting in antifade	Vectashield and coverslip sealing	*3.5.10 - 11.*

* in case of probes labeled with haptens

Fig. 1. Basic workflow (*top*) and step description (*bottom*) for four typical variants of 3D-FISH (A) and 3D-immuno-FISH (B–D) protocol. Variants B and C are recommended for unstable antigens which do not withstand heating or HCl-treatment (e.g., histone modifications). Variant D is recommended for stable antigens (e.g., lamins B1/B2, pB23, SC-35). Images shown in Fig. 2a–d show samples prepared using the respective protocol variants shown in this figure. *Ab* antibody.

3. 0.1 N HCl (Merck) in H_2O.

4. 2× SSC prepared by dilution of 20× SSC stock (1× SSC: 0.15 M NaCl and 0.015 M sodium citrate).

5. 50% Formamide (Merck) in 2× SSC.

2.4. FISH Protocol

1. Rubber cement (e.g., Fixogum).

2. Hot block with exact temperature control (e.g., Techne dry hot block DB 2D).

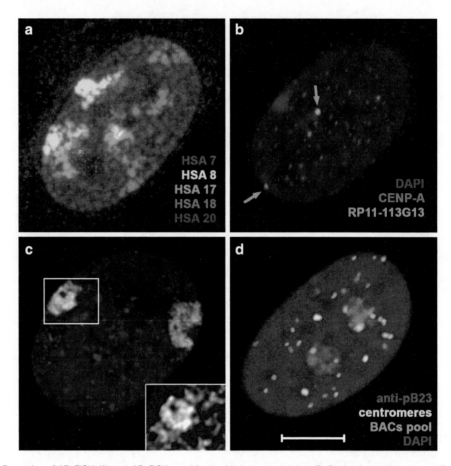

Fig. 2. Examples of 3D-FISH (A) and 3D-FISH combined with immunostaining (B–D) corresponding to workflows A–D shown on Fig. 1. (**a**) Five-color chromosome painting. Paint probes for human chromosomes were labeled and detected as follows: HSA 7 directly labeled with TAMRA; HSA 8 labeled with DNP and detected with Alexa514-conjugated antibodies; HSA 17 labeled with biotin and detected with Alexa488-conjugated avidin; HSA 18 directly labeled with *Texas Red*; HSA 20 labeled with digoxigenin and detected with Cy5-conjugated antibodies. (**b**) Immunostaining before FISH (workflow B). Kinetochores (*red*) were stained using primary anti-CENPA and Cy3-conjugated secondary antibody before FISH with BAC RP11-113G13 (mapping to HSA 4q22.1); the BAC DNA was labeled with biotin and hybrids were visualized by Alexa488-conjugated avidin (*green arrows*). (**c**) Immunostaining before and after FISH (workflow C). Sites of facultative heterochromatin (*red*) were labeled with primary antibody against histone modification H3K27me3 and secondary antibody conjugated to biotin prior to FISH with chromosome paint for HSA X (*green*) labeled with digoxigenin. During FISH detection, biotin in H3K27me3 sites was detected using avidin–Alexa488, while digoxigenin labeled hybrids were detected using anti-digoxigenin-Cy3. Insertion shows intense staining with antibodies in the region of X inactive chromosome. (**d**) Immunostaining after FISH (workflow D). Nucleolar protein pB23 (*red*) was stained with primary and secondary antibodies after FISH with pancentromeric probe (*yellow*) directly labeled with Cy3-dUTP and pooled BACs probe (*green*) for late-replicating genomic regions directly labeled with FITC-dUTP. All shown cells are diploid human fibroblasts. Nuclear DNA was counterstained with DAPI (not shown on A). Images are maximum intensity projections of confocal stacks. Bar is 5 μm and applicable to all panels.

2.5. Posthybridization Washing, Hybrids Detection, Counterstain, and Mounting in Antifade

1. 0.1× SSC, 2× SSC, and 4× SSC prepared by dilution of 20× SSC stock.

2. 4× SSCT is 4× SSC with 0.2% Tween 20 (Merck).

3. Blocking solution is 4× SSCT with 4% Bovine Serum Albumin (BSA).

4. 2% Paraformaldehyde in PBS, pH 7.3, freshly prepared from paraformaldehyde powder.

5. DAPI (4,6-diamidino-2-phenylindole) (Sigma–Aldrich) diluted to 500 µg/mL in H_2O and stored as aliquots at –20°C; working solution of 0.05 µg/mL is prepared in 2× SSC.

6. Vectashield (Vector), stored at 4°C.

7. Colorless nail polish.

2.6. Combination of FISH and Immunostaining

1. PBST is PBS with 0.2% Tween 20 (Merck).

2. Blocking solution is PBST with 4% BSA.

3. Methods

For 3D-FISH, adherently growing cells have to be subcultured on coverslips (0.17 mm thick) with a size not less than 15 × 15 mm. Cells that grow in suspension – e.g., peripheral lymphocytes, EBV-transformed lymphoblastoid, or any other hematopoetic cell lines – must be attached to a coverslip so that their 3D morphology is preserved. Accessibility of the DNA probe to the intranuclear target DNA usually should be increased by further pretreatment of fixed cells such as incubation in nonionic detergent Triton X-100, repeated freezing/thawing after incubation in glycerol, and deproteinization using 0.1 N HCl.

3.1. Fixation of Adherently Growing Cells

1. Grow cells on coverslips (e.g., 18 × 18 mm) under the standard conditions.

2. Rinse coverslips with PBS prewarmed to 37°C.

3. Transfer coverslips in 4% formaldehyde for 10 min at RT. During the last minute of fixation, add one to two drops of 0.5%Triton X-100/PBS.

4. Wash coverslips in PBS/0.05% Triton X-100, 3 × 5 min.

3.2. Fixation of Cells Growing in Suspension

1. Calculate the concentration of cells in cell culture using a cell counting chamber. Allow ca. 1×10^6 cells per coverslip, transfer the required volume of cell suspension to a plastic tube, and centrifuge at $170 \times g$ for 10 min.

2. Resuspend cells in 50% FCS/RPMI at a concentration of ca. 3×10^6 cells/mL.

3. Transfer 0.3 mL of this cell suspension to the polylysine-coated coverslips (e.g., 18 × 18 mm) and allow cells to attach in a CO_2 incubator at 37°C for 15–45 min. Monitor attachment of cells under the microscope: cells are attached when

lymphocytes (or lymphoblastoid cells) have formed a flat thin cytoplasmic rim lined to the surface of the coverslip.

4. Drain off the medium and transfer coverslips into 0.4× PBS for 1 min (avoid shorter or longer incubation).

5. Transfer coverslips into 4% formaldehyde/0.4× PBS for 10 min at RT. During the last minute of fixation, add one to two drops of 0.5% Triton X-100/PBS.

6. Wash coverslips in PBS/0.05% Triton X-100, 3×5 min.

3.3. Permeabilization of Nuclei

1. Incubate coverslips with fixed cells in PBS with 0.5% Triton X-100 for 10 min.

2. Transfer into 20% glycerol/PBS for 30–60 min. At this step, cells can be left at 4°C for up to a few days.

3. Using forceps, remove coverslip from the glycerol solution, dip into liquid nitrogen, and wait until the coverslip is completely frozen. Place the coverslip on a paper towel with cells facing up and wait until frozen glycerol is thawed. Repeat the freezing/thawing treatment four to six times. Before each freezing, briefly soak coverslips in 20% glycerol/PBS. If handled with care, coverslips withstand freezing–thawing and do not break.

4. Wash in PBS/0.05% Triton X-100, 3×5 min.

5. Briefly rinse coverslip in 0.1 N HCl and then incubate in a fresh portion of 0.1 N HCl for 10 min (see Note 1).

6. Wash in PBS with 0.05% Triton X-100, 3×5 min.

7. Equilibrate in 2× SSC for 5 min and then in 50% formamide/2× SSC for 30 min at RT before hybridization. Cells may be stored in 50% formamide/2× SSC at 4°C for up to 3 months.

3.4. FISH Protocol

1. Load hybridization solution (see Note 2) on a clean microscopic slide, allowing e.g., 5 μL of probe for a 18×18 mm coverslip. Pull coverslip out of 50% formamide/2× SSC, quickly drain the excess of fluid, and turn the coverslip with cells upside down on the drop of hybridization mixture. This operation should be done very quickly to avoid drying of cells.

2. Wipe the excess fluid around the coverslip with soft paper and seal the coverslip with rubber cement. Let the rubber cement dry up completely.

3. After rubber cement is dried up, place slides on a hot block and denature at 75°C for 3 min (hence cellular and probe DNA are denatured simultaneously). The exact temperature and exact time of denaturation are important! (see Note 3).

4. Hybridize at least overnight, better for 2–3 days, at 37°C. We recommend incubation in a metallic box floating in a 37°C water bath.

3.5. Posthybridization Washing, Hybrids Detection, Counterstaining, and Mounting in Antifade

1. Remove rubber cement from the preparation using fine forceps and carefully lift up the coverslip.

2. Wash in 2× SSC, 3 × 10 min at 37°C, shaking.

3. Wash in 0.1× SSC, 2 × 5 min at 60°C, shaking. Temperature and molarity of SSC buffer for stringency wash varies depending on the probe used (as for 2D-FISH).

4. If directly labeled probes are used, rinse coverslip in 2× SSC and proceed with postfixation, nuclear DNA counterstain and mounting in antifade (steps 9–11). When hapten detection is needed, place coverslips in 4× SSCT and proceed to step 5.

5. Dilute antibodies and/or avidin in blocking solution/4× SSCT, mix well, and centrifuge at $16,000 \times g$ for 3 min. Allow 100 µL of antibody solution per coverslip. If several haptens are used, apply all antibodies mixed in one solution.

6. Load a drop of antibody solution on a Parafilm in a dark humid chamber. Wipe the back side of the coverslip with soft tissue and place the coverslip on the drop with cells facing down. The coverslip will float on the drop. Incubate at RT for 45 min.

7. Wash in 4× SSCT at 37°C, 2 × 5 min.

8. Repeat the steps 5–7 for each successive layer of antibodies.

9. After last wash in 4× SSCT, equilibrate cells in PBS and postfix preparation in 2% formaldehyde/PBS for 10 min.

10. Stain with DAPI/PBS for 5 min and briefly wash in PBS prior to mounting in antifade (see Note 4).

11. Load a drop of Vectashield on a microscopic slide and place the coverslip with cells up side down on the drop. Be careful not to dry or squash cells. Remove excess of antifade medium around the coverslip with a soft paper, seal the coverslip with nail polish, and let it dry in darkness. After nail polish is completely dried up, the preparation is ready for microscopic observations (see Note 5).

3.6. Combination of FISH and Immunostaining: Methods B and C (For Fragile Antigens)

1. Fix cells as described in Subheading 3.1, step 1 to Subheading 3.3, step 1.

2. Apply primary antibodies diluted in PBST and incubate in dark humid chamber at RT for 45 min as described in Subheading 3.5, step 6.

3. Wash in PBST, 3 × 5 min.

4. Apply secondary antibodies diluted in PBST and incubate in dark humid chamber at RT for 45 min. The antibody can be conjugated either to a fluorochrome (e.g., Cy3) or to a hapten (e.g., biotin).

5. Wash in PBST, 3×5 min.

6. Postfix cells in 2% formaldehyde for 10 min at RT.

7. Wash in PBST, 2×5 min.

8. Proceed with permeabilization protocol (Subheading 3.3, step 2) and FISH setup (Subheading 3.4).

9. If the secondary antibodies were conjugated to a fluorochrome, proceed with FISH protocol as described in Subheading 3.5.

10. If the secondary antibody was conjugated to a hapten, it has to be detected after hybridization with avidin (for biotin) or an appropriate antibody conjugated to a fluorochrome; this can be done together with detection of hybrids (Subheading 3.5, steps 5–6).

3.7. Combination of FISH and Immunostaining: Method D (For Robust Antigens)

1. Fix, pretreat, and hybridize cells as described in Subheadings 3.1–3.4.

2. After posthybridization washings and hybrid-detection (Subheading 3.5, steps 1–7), equilibrate cells in PBST for 5 min.

3. Apply primary antibodies dissolved in blocking solution/PBST and incubate in dark humid chamber at RT for 45 min.

4. Wash in PBST, 3×5 min.

5. Apply fluorochrome-conjugated secondary antibodies in PBST blocking solution and incubate in dark humid chamber for 45 min at RT.

6. Wash in PBST, 3×5 min.

7. Proceed to the Subheading 3.5, steps 9–11.

4. Notes

1. This step is needed to partially remove proteins. Time of incubation in 0.1 N HCl varies from 5 to 20 min depending on the cell type. A too short incubation can cause weak hybridization signal; a too long one can result in damaged nuclear morphology after denaturation. If after deproteinization with HCl for 20 min only a weak signal is observed, then additional digestion with pepsin is recommended: incubate cells in 0.002% pepsin in 0.01 N HCl at 37°C for 3–10 min,

wash with PBS (2×5 min), postfix with 1% formaldehyde/ PBS for 10 min, and wash in PBS, 3×5 min. Note that deproteinization with pepsin might strongly interfere with immunostaining.

2. It is important always to check the quality of probes by FISH on metaphase spreads before 3D-FISH experiments.

3. Simultaneous denaturation on hot block is optimal for 3D-FISH, and we recommend to avoid separate denaturation of the probe and cell DNA when it is possible. Note that too short denaturation may result in a poor hybridization, while over-denaturation damages nuclear morphology. In our experience, a 3 min denaturation at 75°C is optimal for freshly fixed cells; for cells which were stored in 50% formamide/ $2 \times$ SSC longer than 1 month, denaturation time should usually be reduced to 2 or even 1 min.

4. An appropriate counterstaining is always necessary to test for preservation of normal nuclear morphology after FISH. As a certain proportion of nuclei always gets damaged due to freezing/thawing and denaturation, counterstaining is also necessary to select well-preserved nuclei for the further analysis. The choice of the counterstaining dye is determined by the fluorochromes used, as well as by the wave lengths available for microscopy. DAPI (producing a blue fluorescence) is highly specific to DNA and is therefore the most popular counterstain. For red fluorescence, one can use propidium iodide (Sigma, working concentration 5 µg/mL); for far red fluorescence TO-PRO-3 (Invitrogen, working concentration 1 µM); for green fluorescence SYTO 16 (Invitrogen, working concentration 10 µM). The three dyes mentioned above stain both DNA and RNA, and therefore, preliminary digestion with RNase before DNA staining is required. RNase (200 µg/ mL) can be added to antibody solutions.

5. Sealing with nail polish is strongly recommended: sealed coverslips do not move during microscopy thereby preventing damaging of the sample and mixing of antifade medium with immersion oil. Note that sealing does not prevent unmounting and for example, restaining of preparation: nail polish can be easily peeled off in PBS buffer.

Acknowledgments

The authors thank Jeff Craig and Andy Choo (Murdoch Childrens Research Institute, Australia) for providing BAC RP11-113G13 DNA and anti-CENPA serum. This work was supported by DFG grants SO1054/1 (to IS) and CR59/28 (to MC).

References

1. Bolzer, A., Kreth, G., Solovei, I., Koehler, D., Saracoglu, K., Fauth, C., Muller, S., Eils, R., Cremer, C., Speicher, M. R., and Cremer, T. (2005) Three-dimensional maps of all chromosomes in human male fibroblast nuclei and prometaphase rosettes. *PLoS Biol* **3**, e157.

2. Brown, J. M., Green, J., das Neves, R. P., Wallace, H. A., Smith, A. J., Hughes, J., Gray, N., Taylor, S., Wood, W. G., Higgs, D. R., Iborra, F. J., and Buckle, V. J. (2008) Association between active genes occurs at nuclear speckles and is modulated by chromatin environment. *J Cell Biol* **182**, 1083–1097.

3. Finlan, L. E., Sproul, D., Thomson, I., Boyle, S., Kerr, E., Perry, P., Ylstra, B., Chubb, J. R., and Bickmore, W. A. (2008) Recruitment to the nuclear periphery can alter expression of genes in human cells. *PLoS Genet* **4**, e1000039.

4. Goetze, S., Mateos-Langerak, J., Gierman, H. J., de Leeuw, W., Giromus, O., Indemans, M. H., Koster, J., Ondrej, V., Versteeg, R., and van Driel, R. (2007) The three-dimensional structure of human interphase chromosomes is related to the transcriptome map. *Mol Cell Biol* **27**, 4475–4487.

5. Kupper, K., Kolbl, A., Biener, D., Dittrich, S., von Hase, J., Thormeyer, T., Fiegler, H., Carter, N. P., Speicher, M. R., Cremer, T., and Cremer, M. (2007) Radial chromatin positioning is shaped by local gene density, not by gene expression. *Chromosoma* **116**, 285–306.

6. Meaburn, K. J., and Misteli, T. (2008) Locus-specific and activity-independent gene repositioning during early tumorigenesis. *J Cell Biol* **180**, 39–50.

7. Meaburn, K. J., Newbold, R. F., and Bridger, J. M. (2008) Positioning of human chromosomes in murine cell hybrids according to synteny. *Chromosoma* **117**, 579–591.

8. Osborne, C. S., Chakalova, L., Mitchell, J. A., Horton, A., Wood, A. L., Bolland, D. J., Corcoran, A. E., and Fraser, P. (2007) Myc dynamically and preferentially relocates to a transcription factory occupied by Igh. *PLoS Biol* **5**, e192.

9. Takizawa, T., Gudla, P. R., Guo, L., Lockett, S., and Misteli, T. (2008) Allele-specific nuclear positioning of the monoallelically expressed astrocyte marker GFAP. *Genes Dev* **22**, 489–498.

10. Teller, K., Solovei, I., Buiting, K., Horsthemke, B., and Cremer, T. (2007) Maintenance of imprinting and nuclear architecture in cycling cells. *Proc Natl Acad Sci U S A* **104**, 14970–14975.

11. Bridger, J. M., and Lichter, P. (1999) Analysis of mammalian interphase chromosomes by FISH and immunofluorescence, in *Chromosome structural analysis: A practical approach* (Bickmore, W. A., Ed.) pp 103–121, Oxford University Press, Oxford.

12. Ersfeld, K., and Stone, E. M. (2000) Simultaneous in situ detection of DNA and proteins, in *Protein localization by fluorescence microscopy* (Allan, V. J., Ed.), Oxford University Press, Oxford.

13. Solovei, I., Walter, J., Cremer, M., Habermann, H., Schermelleh, L., and Cremer, T. (2002) FISH on three-dimensionally preserved nuclei, in *FISH* (Beatty B., Mai, S., Squire, J., Eds.) pp 119–157, Oxford University Press, Oxford.

Part II

Technical Advancements and Novel Adaptations

Chapter 9

The Comet-FISH Assay for the Analysis of DNA Damage and Repair

Graciela Spivak

Abstract

In this chapter, I describe the alkaline single-cell gel electrophoresis (Comet assay) combined with fluorescence in situ hybridization (FISH) technology, used in our laboratory, to study the incidence and repair of lesions induced in human cells by ultraviolet light. The Comet-FISH method permits the simultaneous and comparative analysis of DNA damage and its repair throughout the genome and in defined chromosomal regions. This very sensitive approach can be applied to any lesion, such as those induced by chemical carcinogens and products of cellular metabolism that can be converted to DNA single- or double-strand breaks. The unique advantages and limitations of the method for particular applications are discussed.

Key words: DNA damage, DNA repair, Transcription-coupled repair, Comet, FISH, Fluorescence, Ultraviolet, Cyclobutane pyrimidine dimers, Single-cell gel electrophoresis, Human fibroblasts

1. Introduction

The genomic stability of living organisms is constantly threatened by endogenous and environmental agents. Several mechanisms have evolved to restore damaged DNA to its original status, such as nucleotide excision repair (NER), base excision repair (BER), mismatch repair (MMR), and pathways dedicated to the repair of single- and double-strand breaks. Complex lesions such as interstrand cross-links may require the coordination of two or more pathways. NER, an ubiquitous mechanism for repair of a large variety of lesions that cause distortions of the DNA structure, has two operational modes: the global genomic repair (GGR) mode, which can detect and repair lesions in the entire genome, and

Joanna M. Bridger and Emanuela V. Volpi (eds.), *Fluorescence in situ Hybridization (FISH):*
Protocols and Applications, Methods in Molecular Biology, vol. 659,
DOI 10.1007/978-1-60761-789-1_9, © Springer Science+Business Media, LLC 2010

transcription-coupled repair (TCR), a specialized pathway for repair in the transcribed strands of active genes. TCR has been demonstrated unequivocally for bulky DNA lesions such as cyclobutane pyrimidine dimers (CPD) induced by ultraviolet (UV) irradiation in bacteria, yeast, and mammalian cells (reviewed by Mellon (1)). In vitro studies have demonstrated that CPD induce complete arrest of transcription by RNA polymerases from phage, bacteria, and mammals (2, 3). TCR requires active transcription (4); thus, the working hypothesis is that the arrest of an RNA polymerase in the elongation mode serves as a signal to recruit repair complexes. Although TCR was discovered because CPD are removed at a faster rate in transcribed sequences than in the genome overall, it should be noted that the operation of TCR does not always result in more rapid repair (see Note 1).

Mutations in genes that encode for DNA repair enzymes can result in serious illness or premature aging. Several rare hereditary human diseases, such as xeroderma pigmentosum (XP), Cockayne syndrome (CS), trichothiodystrophy, and UV-sensitive syndrome (UV^{S}S) have defective NER; in CS and UV^{S}S, the respective mutations only affect the TCR subpathway (5). Individuals affected with any of the above diseases are extremely sensitive to sunlight; however, only XP patients are highly susceptible to skin cancer.

Methods for measuring the incidence and repair of CPD were greatly enhanced when the CPD-specific activity of T4 endonuclease V was characterized (6). The Southern blot method for analysis of TCR of UV-induced CPD, which was used to first define TCR, is suitable for investigating damage and repair in the individual DNA strands of specific restriction fragments up to 30 kb in length, and it requires induction of an average of one lesion in each strand of the fragment of interest for Poisson statistical analysis. Methods such as the ligation-mediated polymerase chain reaction have been used for analysis of damage and repair at the nucleotide level in much shorter sequences. Both approaches for measuring TCR require relatively high exposures to damaging agents that might be too toxic to the cells or the organism under study (7).

The Comet assay combined with fluorescence in situ hybridization (FISH), or Comet-FISH, is the method of choice for the simultaneous analysis of GGR and TCR in cells exposed to low doses of genotoxic agents; the FISH probes can be designed to allow a direct comparison of damage and repair in the genome and in two or more specific regions in the same cell. Moreover, the assay can be used to examine locus-specific genetic instability in, for example, cancer cells (8–10). To assess the effect of transcription on damage and repair, the investigator can select probes for transcriptionally active and silent genomic regions, or can treat cells with transcription inhibitors such as α-amanitin, dichloro-beta-D-ribofuranosylbenzimidazole (DRB), or 1-(5-isoquinolinylsulfonyl)-3-methylpiperazine (H7) that specifically inhibit mammalian RNA polymerase II (4, 11).

In the Comet assay, cells are mixed with agarose and layered on microscope slides, where they are gently lysed and subjected to electrophoresis; staining with fluorescent dyes permits microscopic visualization of individual cells. The molecular events that occur during processing of the cells and DNA to generate Comets have been discussed in several papers (reviewed in (12)). The model we favor is based on the assumption that DNA in chromatin is arranged in "loops" that are anchored via "matrix attachment sites" to the nuclear matrix. In undamaged, nondividing cells, the loops are tightly supercoiled. Upon mild detergent treatment to release histones and other DNA binding proteins, intact DNA remains supercoiled and associated with the matrix, a loosely defined structure composed of inextractable nuclear structure. A single-strand break releases the superhelix tension in a loop, which then unwinds and extends out from the nucleus. Upon application of an electrical field to the agarose-embedded cells, the DNA in loops migrates toward the anode forming the Comet "tail," while the large size DNA in the "heads" of the Comet is retained within the nuclear membrane skeletons. Quantification of the relative amount of DNA in tails and in heads of Comet provides an estimate of the frequency of strand breaks.

This extremely sensitive assay in principle can detect one single-strand cut in each DNA loop, which have been estimated to range between 25 and 200 kb in length (see Note 2). The limit of sensitivity has been estimated to be 50 strand breaks per diploid mammalian cell, and up to 10,000 breaks per cell can be resolved (13). Other advantages of the method include the low number of cells required and the ability to observe variations in responses between cells from the same population; for other methods such as those mentioned above, as the DNA from many cells is pooled, the information about differential responses is lost.

Our recent review (12) summarizes various applications of the Comet-FISH assay as carried out by different research groups, both in human and in plant cells. For example, electrophoresis can be performed in alkaline or in neutral solutions, which allow detection of single- and double-strand DNA breaks, respectively. For lesions other than frank single-strand breaks or alkali-sensitive lesions such as apurinic sites, the method requires enzymatic or physical means for nicking the DNA at or near the lesions; the nicking activity should be specific for the lesions to be analyzed and should have low background activity on undamaged DNA. The DNA incisions resulting from cellular repair processes per se can be used to quantify damage and repair, by inhibiting the repair synthesis step, thus DNA breaks accumulate over time (14). On the other hand, DNA crosslinks prevent unwinding of the complementary DNA strands; the presence of these toxic lesions can be assessed by their inhibitory effect on the electrophoretic migration of DNA strands damaged with a standard dose of an agent that induces single-strand breaks, such as X-rays or hydrogen peroxide (15). Comets prepared at different times after infliction of DNA damage usually contain less DNA in the tails as the lesions are repaired; thus, the time course of repair can be determined (see Note 3).

There are disadvantages inherent to the Comet assay, such as the inability to measure the sizes of fragments between lesions or the number of lesions. Some treatments induce a variety of lesions, which might range widely in toxicity. In such cases, relying on the induction of single-strand breaks might be misleading, particularly when assessing the genotoxic potential of complex mixes. And finally, the procedure often suffers from lack of reproducibility. This may result from the high sensitivity with which single-strand breaks can be detected by the assay and the constantly varying degrees of endogenous, metabolically induced breaks. Maintaining identical conditions, including culture media and environment, buffer composition, laboratory temperature and humidity, etc. is critical to ensure reproducible results.

The protocol described below has been optimized for human primary skin fibroblasts grown in adherent cultures; in principle, it could be applied to cells from different tissues and organisms, even bacteria, for which the Comet assay has been described in the literature (reviewed in Dhawan et al. (16)). Moreover, only the alkaline Comet assay conditions will be described; the neutral Comet-FISH protocol requires slight modifications, as described elsewhere (8, 17).

2. Materials

2.1. Slide Preparation

1. Slides: Erie Scientific or VWR SuperFrost Plus with frosting on one side (the frosting helps to anchor the agarose, see Note 4).

2. Agarose: 1% solution of regular melting point "ultra pure" agarose in double-distilled water, melted by autoclaving or in a microwave oven (use low power and swirl frequently; beware of the superboiling effect).

3. Water bath equilibrated at 50°C.

4. Coplin jar.

5. Marker pens with alcohol-insoluble ink (Securline®).

2.2. Cell Culture

1. Phosphate buffered saline (PBS): 137 mM NaCl, 2.75 mM KCl, 15.25 mM Na_2HPO_4, and 1.45 mM KH_2PO_4.

2. Cell culture medium, serum, and supplements appropriate for the cells to be analyzed.

3. Trypsin solution: 0.1% trypsin, 0.5 mM EDTA in PBS, store at −20°C.

4. Sterile cell culture plasticware.

5. Water-jacketed tissue culture incubator at 37°C and 5% CO_2 atmosphere.

6. Laminar flow cell culture hood with a vacuum system for aspiration of fluids; receptacle and rubber stopper fitted with two rubber or plastic tubes for connection to the vacuum system at one end, and to pipettes used for aspiration at the other end.

2.3. Cell Harvest and Embedding in Agarose

1. Clinical centrifuge for 15 or 50 mL tubes.

2. Microcentrifuge for 1.5 mL tubes.

3. 1.2% Low melting point agarose in PBS: Can be prepared in large batches and aliquoted to centrifuge tubes; store at room temperature.

4. Heat block equilibrated at 37°C.

5. 24 × 60 mm coverslips.

2.4. Lysis

1. Lysis solution: 2.5 M NaCl, 0.1 M EDTA, and 10 mM Tris–HCl. Adjust pH to 10.0 by adding concentrated NaOH dropwise. Store at 4°C.

2. Triton X-100.

3. Dimethyl sulfoxide (DMSO).

4. Nonreactive (plastic, enameled metal, or glass) container(s) with flat bottom for slide incubations.

2.5. Enzyme Treatment

1. Lesion-specific enzyme stock.

2. 1 L 1× enzyme buffer.

3. 22 × 22 mm coverslips.

4. Incubator at 37°C.

5. Humidifying box (a container with a layer of moist paper towels on the bottom).

2.6. Electrophoresis

1. Electrophoresis solution: 0.3 M NaOH, 1 mM Na_2EDTA; the pH should be >13.0; prepare fresh the day of use and store at 4°C. Caution: this high pH solution is corrosive; handle with gloves and wear protective gear.

2. Horizontal electrophoresis apparatus.

3. Level device.

4. Power supply.

5. Cold room, or a large pan, or a buffer recirculation system (see Note 5).

2.7. Hybridization

1. Neutralization buffer: 0.4 M Tris-HCl pH 8.0.

2. 20× SSPE: 22 mM EDTA, 0.2 M NaH_2PO_4, 3.6 M NaCl, and 0.22 N NaOH, pH 7.8–8.0.

3. Hybridization buffer: If not provided by the probe's manufacturer, a typical hybridization solution would consist of 30–50% formamide, 2× SSC (0.3 M NaCl, 30 mM Na citrate), and 10% dextran sulphate.

4. Posthybridization wash solutions: (a) 2× SSPE, 50% formamide in water; (b) 2× SSPE in water; and (c) 1× SSPE in water.

5. Probes: DNA and peptide nucleic acid (PNA) probes prelabeled with fluorescent dyes allow direct detection and can be purchased from suppliers such as Vysis, DakoCytomation, and others. Probes can also be synthesized in the lab by PCR in the presence of digoxigenin-11-dUTP (18) or biotinylated by nick translation (19); these approaches require incubations with fluorescence-labeled antibodies or avidin, respectively, for detection. Single-stranded DNA oligonucleotide probes were used by Horvâthovâ and colleagues (20). See Note 6.

6. Moist chamber.

7. Incubator at 37°C.

8. For amplification of the signal with fluorescence-tagged antibodies, follow the conditions specified by the antibody manufacturers. For washes between incubations, prepare 4× SSPE or SSC, 0.05% Tween 20.

2.8. Staining

1. Stain: There are several fluorescent dyes with affinity for DNA, such as 1 mg/mL 4′,6-diamidine-2-phenylindol dihydrochloride (DAPI) (blue, maximum excitation 358 nm, maximum emission 461 nm), 0.5 mg/mL Hoechst 33258 (blue, 352/461), 20 mg/mL ethidium bromide (red, 518/605), or 2.5 mg/mL propidium iodide (red, 535/617). These dyes can be prepared in distilled H_2O and stored at −20°C; however, they are mutagens, thus it is preferable to purchase them as solutions. A safer option is SYBR green, also available as a solution (green, 488/522). The color should be chosen to contrast with those of the probes. DAPI is reportedly best for long-lasting and low background fluorescence.

2. Antifade mounting solution: 20 mM Tris-HCl pH 8.0, 0.5% N-propyl gallat, and 90% glycerol; store at 4°C. Mixtures of

antifade and stain are available from commercial sources, such as Vectashield Hard Set with DAPI (Vector Laboratories).

3. 22 × 22 mm coverslips.

4. Storage box for slides.

2.9. Microscopy, Image Acquisition

1. Epifluorescence microscope equipped with 25× and 40× high aperture oil objectives.

2. Digital CCD camera: An 8-bit black-and-white camera is adequate, but higher sensitivity and pixel resolution is desirable.

3. Appropriate filters, such as DAPI (blue), CY3 (orange), Rhodamine (red), and FITC (green) filters.

4. Image acquisition software.

5. Computer equipped with sufficient memory and power (>200 GB for the hard drive, >2 GB RAM), appropriate operating system (Windows is preferable to Apple, see below), and a good quality video card; a large screen is not essential but it helps to prevent eye fatigue.

6. Image acquisition software: Most applications are only compatible with Windows-powered computers.

7. CD, DVD, or optical disc burner for storage of data.

8. Immersion oil, wipes, glass cleaning solution.

2.10. Image Analysis

1. Software for image processing and analysis.

   a. Free, public domain software: *NIH Image* (http://www.rsb.info.nih.gov/) is a public domain image processing and analysis program for the Macintosh. It was developed at the Research Services Branch (RSB) of the National Institute of Mental Health (NIMH), part of the National Institutes of Health (NIH). A free PC version of *Image*, *Scion Image for Windows*, is available from Scion Corporation (http://www.scioncorp.com/). *Image/J* (http://www.rsb.info.nih.gov/ij/) is a Java program inspired by *Image* that can be used with Windows or Macintosh computers. For system requirements, please refer to the respective software websites.

   The program developed by the Comet assay software project lab (*CASP*, http://www.casp.sourceforge.net/) automatically identifies Comet heads and tails, subtracts backgrounds, quantifies the signals from heads and tails, and tabulates the data as "% DNA in tails," "Olive moment," etc. This program chooses the area most intensely stained as the head of the Comet, thus it works very well when there is low damage and Comets have easily identifiable heads, but not for cells with more damage.

b. Commercial software packages include *Komet* from Andor (http://www.andor.com/software/komet/), *Comet Assay IV* from Perceptive Instruments (http://www.perceptive. co.uk/cometassay/), and others; for a complete list go to the Comet assay interest group at http://www.cometassay. com/.

2. Computer with appropriate spreadsheet, statistics, and graphics software.

3. Data storage device (CD, DVD, optical disc) reader.

3. Methods

3.1. Slide Preparation

1. Melt the 1% regular melting point "ultra pure" agarose in a microwave oven (see precautions above).

2. Pour agarose into a Coplin jar (or a 100 mL beaker) and equilibrate to 50°C in a water bath.

3. Dip the unfrosted end of the slides into the agarose up to the middle of the frosted area (see Note 4).

4. Wipe clean the backside of the slide with a paper towel; place the slide flat on a paper towel to dry at room temperature, then store at room temperature for 2–3 weeks before use (in high humidity climates, it might be necessary to dry the slides in an oven at 37°C for an empirically determined length of time).

3.2. Cell Preparation, DNA Damaging Treatment

Eukaryotic cells grown in suspension or as adherent monolayers should be seeded at appropriate densities to achieve the desired number of cells for Comets, for example, 2×10^5 cells per 10 cm dish typically yield enough cells for ~10 slides. If a DNA damaging agent is to be applied, the treatment should be carried out 1 day after seeding the cells (see Note 7).

1. Remove media from the cells by pipetting; save in a sterile bottle.

2. Wash cells with cold or warm PBS twice.

3. For UV irradiation, irradiate cells for the time necessary for a total dose of 0.10 J/m^2.

4. For chemical treatments, add the agent diluted in PBS, Hank's balanced buffered saline, or other appropriate isotonic buffer and incubate at the desired temperature (4°C to inhibit repair, 20 or 37°C to avoid thermal shock and to allow repair) for the desired length of time; then rinse the cells twice with PBS.

5. Proceed to cell harvest (below) or add saved media and incubate for the desired period(s) of time to allow repair or further induction of damage in the case of long-lasting chemicals.

3.3. Assay Preparation, Cell Harvest

1. Turn on UV lamp 20 min prior to use to achieve a stable spectrum, or prepare damaging agent in appropriate buffer.

2. Chill PBS at 4°C or warm to 37°C as appropriate; melt trypsin; prepare a box of ice.

3. For adherent cells, aspirate the medium or buffer covering the cells, rinse the cells with PBS twice, add 600 µL trypsin to the dish or flask, incubate at 37°C until the cells begin to detach (minimize exposure to trypsin, it might induce DNA strand breaks), and add 600 µL cold culture medium with 10% serum to inactivate the trypsin; resuspend the cells by gently pipetting up and down several times and transfer the suspension to a 1.5 mL Eppendorf tube, keep at 4°C until all the samples are ready.

4. For cells grown in suspension, spin down, aspirate supernatant, resuspend with 10 mL cold PBS, spin down, repeat wash with PBS, resuspend cells with 1.2 mL cold PBS, and keep at 4°C as above.

3.4. Cell Embedding and Lysis

1. Equilibrate a heat block for 1.5 mL Eppendorf tubes (or a water bath) at 37°C.

2. Melt 1.2% LMP agarose in a water bath at 70°C for 30 min; equilibrate 37°C.

3. Spin cells at low speed for 3 min at 4°C; wash with 600 µL cold PBS, repeat centrifugation and resuspension with PBS.

4. Place samples in the heat block at 37°C.

5. Mix 100 µL sample and 100 µL LMP agarose; keep at 37°C.

6. Using a micropipettor such as Pipetman, put 85 µL mixture (~15,000 cells) onto each slide; cover with a 24 × 60 mm coverslip leaving a 1 mm overhang to facilitate removal. If enough cells are available, prepare several slides from each sample. Label slides with a pen with alcohol-insoluble ink. Let gels harden at 4°C for 30 min; remove coverslips by lifting from the overhang with a fingernail.

7. Add 0.01 volumes Triton X-100 and 0.1 volumes DMSO to cold lysis solution.

8. Place slides in a box that will contain them in one layer, gently pour enough lysis solution to cover the slides, incubate at 4°C overnight. Solution changes should be performed slowly and gently to avoid detachment of the agarose from the slides.

3.5. Enzyme Treatment   To analyze the incidence and repair of DNA lesions that can be converted to single-strand nicks or to double-strand breaks, the agarose-embedded cells can be treated with glycosylases; for example, UV-induced CPD sites can be specifically nicked with T4 phage endonuclease V, 8-oxo-7,8-dihydroguanine with OGG1 or Fpg, etc.

1. For treatment with glycosylases, prepare 1 L of 1× reaction buffer using the appropriate formulation for each enzyme.

2. Discard the lysis solution; typically the slides stick to the bottom of the box, thus gently tipping the box works well to remove solutions.

3. Incubate the slides with reaction buffer for 5 min at room temperature; discard solution and repeat a total of three times.

4. Prepare appropriate enzyme dilution in 1× reaction buffer (usually 1:500–1:1,000; the optimal dilution must be determined for each cell type and batch of enzyme).

5. Place 50 μL enzyme dilution on appropriate slides and 50 μL 1× reaction buffer on control slides made with cells from the same sample; cover with 24×60 mm coverslips.

6. Incubate in a covered humidified chamber (place a folded paper towel in a plastic box, sprinkle with water to moisten) at 37°C for 45 min.

3.6. Electrophoresis   The electrophoresis should be carried out at 4°C to avoid melting the LMP agarose on the slides (see Note 5, Fig. 1).

1. Level the electrophoresis apparatus using a level device.

2. Remove coverslips; place slides in the apparatus, with frosted ends aligned and facing the (red) anode. Add additional empty slides if necessary to fill the width of the box (see Note 8).

3. Slowly add electrophoresis buffer until just covering the slides (caution: the high pH buffer is corrosive; wear gloves and protective gear); cover the tank. Incubate for 40 min.

4. Turn on the power supply to 0.75–1.00 V/cm (the distance between electrodes) constant voltage. The current should be approximately 300 mA; if it needs to be adjusted, turn off the power and add or remove electrophoresis buffer (observe cautions above).

5. Electrophorese for 30 min.

6. Carefully remove slides from electrophoresis apparatus (wearing gloves), touch edges on paper towels to drain, and place in a box for washes.

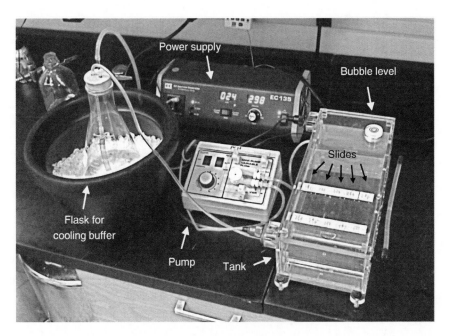

Fig. 1. Alkaline electrophoresis and buffer cooling system. The photograph shows the electrophoresis equipment used in our laboratory: the tank's buffer reservoir is connected to a pump that circulates the solution through a flask set in ice. A bubble level (shown on *top* of the tank) is used to ensure that the slides are horizontal.

7. Discard electrophoresis solution as recommended for corrosive materials.

8. Wash slides with 0.4 M Tris-HCl, pH 8 three times for 5 min.

9. Cover slides with cold 100% ethanol and incubate for 30 min at 4°C, remove ethanol.

10. Add 0.5 M NaOH for 25 min at room temperature.

11. Dehydrate slides by successive washes with 70, 85, and 95% ethanol in water for 5 min each at room temperature.

12. Let slides air-dry overnight in the dark; dehydrated slides can be kept for long periods of time.

3.7. Hybridization

1. Preincubate slides at 37°C for 30 min in a humidified chamber.

2. Prepare probe(s) according to appropriate instructions.

3. Prepare hybridization cocktail (hybridization solution, probe, and water if needed for 10 µL total volume) and keep on ice until ready to use.

4. Heat cocktail at 73°C for 5 min, quickly add to the center of the slide, and cover with a 22 × 22 mm coverslip. For multiple slides, probe mixtures should be kept covered and at 73°C between samplings.

5. Incubate slides at 37°C overnight in a humidified chamber.

6. Wash twice for 10 min with 50% formamide, 2× SSPE at 37°C.

7. Wash once for 10 min with 2× SSPE at 37°C.

8. Wash once for 5 min with 1× SSPE at 37°C.

9. If the signal needs to be developed by binding fluorescence-conjugated antibodies to the probes, the incubations should be carried out at this point in a moist box at 37°C for 30 min; between incubations the slides should be washed three times for 4 min in 4× SSPE, 0.05% Tween 20.

10. Drain slides on paper towels and allow to air-dry for at least 2 h.

3.8. Staining

1. Place a drop (~100 μL) of DAPI–VectaShield HardSet or the dye and antifade of choice onto each slide to counterstain the DNA.

2. Cover with a 24 × 60 mm coverslip and store at 4°C in a box protected from light until ready to view; time might be critical depending on the fluorescence half-life.

3.9. Microscopic Examination

1. Examine slides with an epifluorescence microscope coupled to a CCD camera and computer with image capture software; use appropriate filters to detect the fluorescent signal from the dyes used to label the DNA and the probes (see Note 9). An example of a Comet-FISH image of a cell damaged with UV light is depicted in Fig. 2.

3.10. Image Analysis

Comet analysis software programs are designed to distinguish Comet heads from tails, subtract backgrounds, and measure parameters including tail length, percent of total fluorescence in head and tail, and "tail moment," which represents the product of tail length and relative tail intensity. *Percent of DNA in tail* is linearly related to DNA break frequency when there is up to 80% of the DNA in the tail, and this defines the useful range of the assay. *Tail length* increases with dose at low levels of damage and thus its usefulness is limited. *Tail moment* combines the information of tail length and tail intensity.

1. When using CASP, the investigator must scan slides, find suitable Comets, draw a box around them, and adjust and position the background box above or below the Comet; the software automatically determines the area corresponding to the head, as discussed above. The datasets are exported to spreadsheets such as Excel.

2. For Image, ImageJ, or Scion, the user can decide the shape, size, and position of the areas occupied by Comet heads or tails, and can choose a dark area as background for each Comet. The raw data (signals from heads, tails, and background)

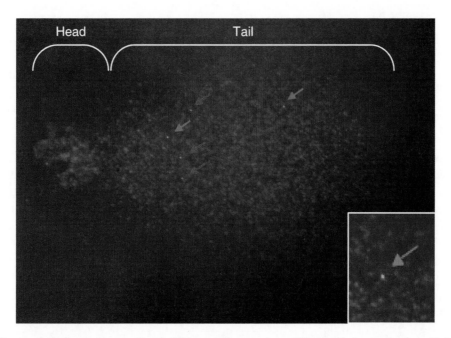

Fig. 2. Representative image from the Comet-FISH assay. Human primary skin fibroblasts were treated with 0.1 J/m^2 UV, harvested immediately, and processed for Comet-FISH as described in the text. The cells in this sample were treated with T4 endonuclease V to generate single-strand breaks at sites containing CPDs. The Comet's head and tail are indicated at the *top* of the image; the head shows more intense DAPI staining than the tail. The image was obtained with a Zeiss Axioplan microscope equipped with an AxioCam digital camera, DAPI, CY3 and FITC filters and AxioVision software. *Arrows* point to the positions of the p53 (*red*) and chromosome 7 centromere (*green*) domains. The Spectrum Orange-labeled 17p13.1 LSI p53 probe for a 145 kb region containing the *p53* gene, and the Spectrum Green-labeled CEP 7 probe for the alfoid satellite sequences in the centromeric region of chromosome 7 (purchased from Vysis– Abbott), were chosen to represent an actively transcribed domain and a transcriptionally silent domain, respectively. The signals from the FISH probes are shown as they were captured by the software, without enhancements. *Inset:* detail of a centromere spot.

can be entered into a spreadsheet designed to calculate the backgrounds for heads and tails proportionally to the sizes of the respective hand-drawn shapes, subtract them from the signals, calculate the percent DNA in tails, and finally the averages and standard errors.

3. As an alternative to computer-based analysis, visual scoring allows assignation to one of five classes of Comet, from class 0 (undamaged, no discernible tail) to class 4 (almost all the DNA in the tail, insignificant head). One hundred Comets should be selected at random from each slide and an overall score can be derived for each gel, ranging from 0 to 400 arbitrary units.

3.11. Calculations

Controls made with undamaged cells, treated or not with lesion-specific enzyme, should be included in each assay to assess the presence of endogenous or enzyme-induced unspecific DNA breaks, which can vary significantly between samples.

The UV-treated (or chemical-treated) controls (without enzyme treatment) provide an estimate of the background of DNA strand breaks that might result from the DNA-damaging agent. The damaging agent-treated, enzyme-treated gels reveal strand breaks and damaged bases or bulky lesions. Subtraction of background from the total gives a measure of damage caused by the agent of interest (see Note 10).

3.12. Storage and Reexamination

Place slides in a warm oven until the gel has dried. Slides can then be stored at room temperature. For reexamination, stain as above. The fluorescence in the probes lasts about 2 weeks for fluorescent probes; when using antibody cascades, the last antibody reaction may need duplication. In principle, the probes can be stripped and the gels can be rehybridized, but in our experience this resulted in high, grainy background fluorescence.

4. Notes

1. Efficient GGR of some types of lesions may mask the dedicated TCR pathway. TCR may then be revealed in mutant cells defective in global repair, such as cells deficient in the *XPC* gene product, which is not required for TCR of "bulky" adducts. Global repair can also be attenuated by interference RNA-mediated "knock-down" of relevant genes, such as *XPC* and *XPE* (21).

2. Selection of the type and dose of the damaging agent requires careful determination of the frequency of lesions induced. In practice, the analysis is limited to damage that would result in 20–80% of the DNA in Comet tails: too little or too much damage would render the tails or the heads undetectable, respectively, thus encumbering the quantification of lesions and assessments of repair.

3. "DNA repair" implies that the number of lesions should decrease after the time "0" when damage is induced; however, certain chemical treatments continue to exert their noxious effects after being removed from the media, other treatments might induce single-strand breaks that would hinder detection of the lesions of interest (if this is the case, allowing the cells to repair the breaks for 30 min usually solves the problem). Other factors that might interfere with measurements of repair include the incisions caused by repair processes (this can be used to one's advantage as mentioned above), the presence of alkali-sensitive replication forks, and DNA fragmentation due to apoptosis.

4. Occasionally, the researcher is plagued by the problem of agarose detaching from the slides. As stated in the Comet interest group FAQ, the problem might go away by itself. We found that the following measures usually remedied the situation: (a) use a fresh box of slides from a different batch; (b) make sure that the agarose comes up to the middle of the frosted area of the slides; (c) store freshly prepared slides for at least 2 weeks before using them; and (d) choose a day with low humidity for preparing slides.

5. Since the alkaline electrophoresis solution is not a buffer, the resistance generated by the electrical current is released as heat, which could melt the agarose on the slides. To avoid this problem, the electrophoresis can be carried out in a cold room, the electrophoresis tank can be placed in a large pan containing ice water, or a buffer recirculation system in which the solution passes through a flask set in ice can be assembled as shown in Fig. 1.

6. Selection of the genomic domain to be targeted by the probe must take into consideration the location of matrix-attachment sites, which might inhibit migration of neighboring sequences into Comet tails, as was found for the *DHFR* gene in Chinese hamster ovary cells (20). The intensity of the signal emitted by the probes is directly proportional to the size of the domain recognized by the probes; however, probes for large sequences should be fractionated by sonication, enzymatic digestion, or synthesized to be approximately 200 nt in length to allow their diffusion through the agarose gel.

7. Dividing cells migrate abnormally and replication forks are sensitive to alkali, thus replication might need to be inhibited for certain purposes; however, DNA synthesis inhibitors will also inhibit the synthesis step of excision repair processes, thus they should be removed prior to treatment or harvest of the cells (unless accumulation of repair-induced strand breaks is to be measured, as discussed above). The cells must be viable (DNA fragmentation caused by apoptosis or necrosis may confound the assay).

8. A typical electrophoresis tank can accommodate two rows of slides; the location of each slide within the tank may in some cases affect the migration of the DNA, thus it is recommended that a test trial should be carried out. The tank should be leveled to avoid abnormal migration patterns.

9. Avoid Comets on the edges of the gel, where anomalously high levels of damage are often seen. Since the cells are distributed in several planes within the agarose thickness, each microscopic field must be scanned vertically by adjusting the focus up and down. Once the Comet image is acquired, the

focus might need readjustment to capture the signals from FISH. Ideally, the software will integrate the monochromatic images obtained with the different filters and generate a compound image with all the colors superimposed.

10. The unique property of the Comet assay that permits observation of individual cells may result in significant heterogeneity and poor reproducibility, as mentioned above; cells within a slide may range from undamaged to highly damaged. In order to compare samples or to demonstrate a genotoxic effect, it might be necessary to employ statistical tools (22, 23).

Acknowledgments

The author would like to thank Alexia Chollat-Namy and Rachel A. Cox for their patience and perseverance in developing the assay, and Phil Hanawalt for believing in the project. This work was supported by a grant CA91456 from NIH.

References

1. Mellon, I. (2005) Transcription-coupled repair: a complex affair. *Mutat. Res.* **4**, 155–161.

2. Tornaletti, S., Reines, D., and Hanawalt, P. C. (1999) Structural characterization of RNA polymerase II complexes arrested by a cyclobutane pyrimidine dimer in the transcribed strand of template DNA. *J. Biol. Chem.* **274**, 24124–24130.

3. Scicchitano, D. A., Olesnicky, E. C., and Dimitri, A. (2004) Transcription and DNA adducts: what happens when the message gets cut off? *DNA Repair* **3**, 1537–1548.

4. Christians, F. C., and Hanawalt, P. C. (1992) Inhibition of transcription and strand-specific DNA repair by alpha-amanitin in Chinese hamster ovary cells. *Mutat. Res.* **274**, 93–101.

5. Spivak, G., and Hanawalt, P. C. (2006) Host cell reactivation of plasmids containing oxidative DNA lesions is defective in Cockayne syndrome but normal in UV-sensitive syndrome fibroblasts. *DNA Repair* **5**, 13–22.

6. Friedberg, E., Walker, G., Siede, W., Wood, R., Schultz, R., and Ellenberger, T. (2006) A brief history of the DNA repair field. *DNA Repair and Mutagenesis*, ASM Press, Washington DC.

7. Spivak, G., Pfeifer, G. P., and Hanawalt, P. C. (2006) In vivo assays for transcription-coupled repair, *in Methods Enzymol.: DNA Repair* (Campbell, J. C., and Modrich, P., Eds.), Vol. 408, pp. 223–246, Elsevier Inc., New York.

8. Kumaravel, T. S., and Bristow, R. G. (2005) Detection of genetic instability at HER-2/neu and p53 loci in breast cancer cells sing Comet-FISH. *Breast Cancer Res. Treat.* **91**, 89–93.

9. Escobar, P. A., Olivero, O. A., Wade, N. A., Abrams, E. J., Nesel, C. J., Ness, R. B., Day, R. D., Day, B. W., Meng, Q., O'Neill, J. P., Walker, D. M., Poirier, M. C., Walker, V. E., and Bigbee, W. L. (2007) Genotoxicity assessed by the comet and GPA assays following in vitro exposure of human lymphoblastoid cells (H9) or perinatal exposure of mother–child pairs to AZT or AZT-3TC. *Environ. Mol. Mutagen.* **48**, 330–343.

10. Glei, M., Schaeferhenrich, A., Claussen, U., Kuechler, A., Liehr, T., Weise, A., Marian, B., Sendt, W., and Pool-Zobel, B. L. (2007) Comet fluorescence in situ hybridization analysis for oxidative stress-induced DNA damage in colon cancer relevant genes. *Toxicol. Sci.* **96**, 279–284.

11. McKay, B. C., Chen, F., Clarke, S. T., Wiggin, H. E., Harley, L. M., and Ljungman, M.

(2001) UV light-induced degradation of RNA polymerase II is dependent on the Cockayne's syndrome A and B proteins but not p53 or MLH1. *Mutat. Res.* **485**, 93–105.

12. Spivak, G., Cox, R. A., and Hanawalt, P. C. (2009) New applications of the Comet assay: Comet-FISH and transcription-coupled DNA repair. *Mutat. Res.* **681**, 44–50.

13. Olive, P. L., and Banath, J. P. (2006) The comet assay: a method to measure DNA damage in individual cells. *Nat. Protoc.* **1**, 23–29.

14. Collins, A. R., Dobson, V. L., Dusinska, M., Kennedy, G., and Sttina, R. (1997) The comet assay: what can it really tell us? *Mutat. Res.* **375**, 183–193.

15. McKenna, D. J., Gallus, M., McKeown, S. R., Downes, C. S., and McKelvey-Martin, V. J. (2003) Modification of the alkaline Comet assay to allow simultaneous evaluation of mitomycin C-induced DNA cross-link damage and repair of specific DNA sequences in RT4 cells. *DNA Repair* **2**, 879–890.

16. Dhawan, A., Bajpayee, M., and Parmar, D. (2009) Comet assay: a reliable tool for the assessment of DNA damage in different models. *Cell Biol. Toxicol.* **25**, 5–32.

17. Rapp, A., Hausmann, M., and Greulich, K. O. (2005) The Comet-FISH technique: a tool for detection of specific DNA damage and repair. *Methods Mol. Biol.* **291**, 107–119.

18. Menke, M., Angelis, K. J., and Schubert, I. (2000) Detection of specific DNA lesions by a combination of comet assay and FISH in plants. *Environ. Mol. Mutagen.* **35**, 132–138.

19. Amendola, R., Basso, E., Pacifici, P. G., Piras, E., Giovanetti, A., Volpato, C., and Romeo, G. (2006) Ret, Abl1 (cAbl) and Trp53 gene fragmentations in comet-FISH assay act as in vivo biomarkers of radiation exposure in C57BL/6 and CBA/J mice. *Radiat. Res.* **165**, 553–561.

20. Horvâthovâ, E., Dusinskâ, M., Shaposhnikov, S., and Collins, A. R. (2004) DNA damage and repair measured in different genomic regions using the comet assay with fluorescent in situ hybridization. *Mutagenesis* **19**, 269–276.

21. Nouspikel, T. P., Hyka-Nouspikel, N., and Hanawalt, P. C. (2006) Transcription domain-associated repair in human cells. *Mol. Cell. Biol.* **26**, 8722–8730.

22. Duez, P., Dehon, G., Kumps, A., and Dubois, J. (2003) Statistics of the Comet assay: a key to discriminate between genotoxic effects. *Mutagenesis* **18**, 159–166.

23. Ejchart, A., and Sadlej-Sosnowska, N. (2003) Statistical evaluation and comparison of comet assay results. *Mutat. Res.* **534**, 85–92.

Chapter 10

Direct In Situ Hybridization with Oligonucleotide Functionalized Quantum Dot Probes

Laurent A. Bentolila

Abstract

Coming from the material sciences, fluorescent semiconductor nanocrystals, also known as quantum dots (QDs), have emerged as powerful fluorescent probes for a wide range of biological imaging applications. QDs have several advantages over organic dyes which include higher brightness, better resistance to photobleaching, and simplified multicolor target detection. In this chapter, we describe a rapid assay for the direct imaging of multiple repetitive subnuclear genetic sequences using QD-based FISH probes. Streptavidin-coated QDs (SAvQDs) are functionalized with short biotinylated oligonucleotides and used in a single hybridization/detection step. These QD-FISH probes penetrate both intact interphase nuclei and metaphase chromosomes and show good targeting of dense chromatin domains. Importantly, the broad absorption spectra of QDs allows two sequence specific QD-FISH probes of different colors to be simultaneously imaged with a single laser excitation wavelength. This method, which requires minimal custom conjugation, is easily expandable and offers the experimentalist a new alternative to increase flexibility in multicolor cytogenetic FISH applications of repetitive DNAs.

Key words: Biomaterial, Chromosome, Confocal laser scanning microscopy, Cytogenetic, FISH, Fluorescence, Hybridization, Imaging, Multiplex labeling, Nanotechnology, Oligonucleotide, Quantum dot, Repetitive DNA

1. Introduction

Quantum dot (QDs) nanocrystals are new fluorescent probes made of semiconductor materials that have become increasingly popular in virtually all aspects of fluorescence imaging, labeling, and sensing (reviewed in (1, 2)). Some of their unique optical properties offer substantial advantages over the use of conventional organic fluorophores in Fluorescence in situ Hybridization (FISH) applications (3). QDs are more photostable than organic

Joanna M. Bridger and Emanuela V. Volpi (eds.), *Fluorescence in situ Hybridization (FISH): Protocols and Applications*, Methods in Molecular Biology, vol. 659,
DOI 10.1007/978-1-60761-789-1_10, © Springer Science+Business Media, LLC 2010

fluorophores, which tend to undergo rapid and irreversible photobleaching upon viewing (4, 5). QDs can thus be observed over extended periods of time (6), and their superior photostability can be harnessed to improve quantification of the FISH signals (7). QDs also appear to be significantly brighter than their dye molecule counterparts primarily because at comparable quantum yield and similar emission saturation levels, QDs have larger molar absorption coefficients than those of most dyes (8, 9). The particular combination of improved signal brightness and resistance against photobleaching has firmly established QDs as the probes of choice for the observation of single-isolated molecules (reviewed in (10)).

FISH provides a direct correlation between the visualization of target nucleic acid sequences in situ and the morphology of the tissue sections, cells, or chromosomes. Such topological information is invaluable for both basic research and disease management in the clinic where FISH is used to detect and monitor chromosomal abnormalities associated with cancers (11). Simultaneous visualization of different targets in multiple, distinct colors is achieved by cohybridization of probes labeled with different fluorescent labels or label combinations. However, due to the spectral overlap of commonly used organic fluorophores, the number of probes identifiable on the basis of a unique fluorescence color is typically limited. Also, the multiple wavelengths required to excite polychromatic specimens may introduce chromatic aberrations that can reduce the reliability of multicolor fluorescent probes colocalization measurements. Here again, QD's optical properties which are size-dependent and governed by quantum confinement effects (12) can fundamentally overcome these limitations. The color emission of QDs can be fine-tuned anywhere from the UV to the infrared (IR) to match the experimentalist's needs. This is achieved by altering the QD's material composition and/ or their overall particle size during chemical synthesis. As the nanoparticles grow bigger, their emission shifts from blue to red in a continuum of colors. Also, QDs have symmetric and narrow emission bands (typically 25–30 nm at full-width-half-maximum) which enable multiple color combinations of QDs to be used with minimal spectral overlap. At the same time, QD's broad excitation spectrum makes it possible to use a single narrow-band excitation source to detect multiple QD colors simultaneously.

In this chapter, we describe how those unique QD's optical properties can be applied for the direct detection of repetitive genomic sequences by FISH. This protocol uses commercial streptavidin-coated QDs (SAvQDs). It details the preparation of novel QD-FISH probes based on the direct attachment of short biotinylated DNA oligonucleotides onto SAvQDs and their subsequent use in dual-color FISH analysis. The direct coupling of the oligonucleotide probes onto the QDs results in a faster, single-step FISH protocol that bypasses the need for lengthy secondary detection and washing steps. We also demonstrate that the same biotin–streptavidin interaction can be used to generate arrays of QD-oligonucleotide probes for multicolor FISH. Here we detail how two QD-oligonucleotide FISH probes with different emission spectra can be used in a one-step hybridization/detection experiment to visualize in situ two subchromosomal regions of highly condensed chromatin structure within the chromosome's centromeres. QD's broad excitation spectrum allows the two different color probes to be simultaneously excited by a single excitation wavelength and distinguished in a single exposure using standard far-field optics. These results demonstrate that while structurally larger than organic dye probes, QD-oligonucleotide probes are very effective in multicolor FISH applications. The direct visualization of repetitive DNA satellites by QD-oligonucleotide FISH provides a useful application for cytogenetic studies. It offers the opportunity to achieve high resolution, true-multicolor identification, and analysis of chromosomal structures since major structural chromosomal rearrangements frequently take place in heterochromatin regions rich in repetitive DNA (13).

2. Materials

2.1. Buffers, Reagents, General Supplies, and Equipment

1. Unless specified otherwise, all reagents are of the best analytical grade and all solutions should be prepared in Milli-Q water (resistivity of 18.2 MΩ/cm at 25°C).

2. 0.5 M EDTA: Add 18.6 g disodium EDTA to 80 mL H_2O. Stir and heat, and adjust pH 8.0 by adding about 2 g of sodium hydroxide pellets. Continue stirring until the solution clears. Make up to 100 mL and autoclave.

3. 10× TBE: Dissolve 108 g Tris Base (a.k.a. Trizma® base, Sigma–Aldrich), 55 g Boric Acid, 40 mL 0.5 M EDTA (disodium salt) in 700 mL H_2O in a 2 L flask. Stir to dissolve. Add water to bring up total volume to 1 L. Autoclave.

4. 5 M NaCl: Dissolve 146.1 g NaCl in 450 mL H_2O. Adjust volume to 0.5 L. Autoclave.

5. Ethyl alcohol 200 proof, absolute (Pharmco-AAPER).

6. Methanol.

7. Glacial acetic acid.

8. Water baths and heating blocks set for 37, 65, 70, and 85°C.

9. Boekel Inslide-out hybridization apparatus (Fisher Scientific).

10. Standard UV transilluminator.

11. Flat coverslip forceps.

2.2. Cell Culture

1. Mouse mast cell line P815 (American Type Culture Collection).

2. Dulbecco's modified Eagle's medium without phenol red supplemented with 5% (25 mL) inactivated fetal calf serum (Gemini Bio-products, Sacramento, CA), 5 mL of L-glutamine 200 mM, 5 mL of Penicillin–Streptomycin, and 4 μL of 2-βMercaptoethanol at 14.3 M.

3. Phosphate-buffered saline (D-PBS) without Ca^{2+} and Mg^{2+}.

4. 1× solution of 0.25% Trypsin–EDTA.

5. KaryoMAX® Colcemid™ liquid solution (10 μg/mL) in PBS (GIBCO® Invitrogen).

6. Vented 75 cm^2 – 250 mL Falcon tissue culture flasks.

7. 15 mL Falcon tubes with conical bottoms.

8. 50 mL Falcon tubes with conical bottoms.

9. Cell counter and phase hemacytometer (VWR international).

10. CO_2 cell culture incubator (SANYO) setup at 37°C.

11. Tissue culture hood biosafety cabinet (The Baker Company, SterilGARD III).

12. Phase-contrast microscope (Leica Microsystems Inc.).

2.3. Fixative and Sample Preparation

1. Carnoy's fixative: 3:1.5 (v/v) methanol/glacial acetic acid (water-free). The fixative solution should be prepared fresh before each use. Make 45 mL of fixative by mixing 30 mL of methanol and 15 mL of glacial acetic acid in a 50 mL Falcon tube, mix well, and store it at –20°C for at least 2 h prior to use.

2. Hypotonic solution: 0.4% KCl (w/v) prewarmed at 37°C.

3. Autoclaved polypropylene Coplin staining jars (VWR International).

2.4. Stock Solutions for Fluorescence In Situ Hybridization

1. 50× Denhart's solution: To 0.5 g Ficoll Type 400 (Sigma–Aldrich), 0.5 g polyvinylpyrolidone (Sigma –Aldrich), and 0.5 g BSA fraction V (Sigma–Aldrich), add 40 mL H_2O and stir vigorously overnight in a beaker. Make up to a final volume of 50 mL and filter sterilize through a 0.22 μm filter. Store in 10 mL aliquots at –20°C, where it is stable for several years.

2. Denatured, sheared salmon sperm DNA at 10 mg/mL: Denature the DNA (GIBCO® Invitrogen) by boiling the solution for 5 min and place immediately on ice to prevent renaturation. Store at –20°C. Stable for months to years.

3. Deionized formamide (Sigma–Aldrich, cat. no. F9037): Used as delivered. Keep in the dark and store at 4°C. Do not use formamide if it is yellowish in color. Formamide is a potential teratogen.

4. Formamide for molecular biology and in situ (Sigma–Aldrich, cat. no. 47671).

5. 1 M phosphate buffer for hybridization mix: Prepare 1 M solution of NaH_2PO_4 (H_2O) (dissolve 6.9 g in 50 mL H_2O, *solution #1* at about pH 4.0) and 1 M solution of Na_2HPO_4 (dissolve 7.1 g in 50 mL H_2O, *solution #2* at about pH 10.0). Mixing 5.8 mL *solution #2* (Na_2HPO_4) with 4.2 mL *solution #1* (NaH_2PO_4) gives a 1 M solution of phosphate buffer at about pH 7.0.

6. 20× SSC: Dissolve 175.3 g NaCl, 88.2 g Sodium Citrate in 800 mL H_2O. Adjust the pH to 7.0 with a few drops of 10 M NaOH or 14 N solution of HCl. Adjust the volume to 1 L with water. Sterilize by autoclaving.

7. 10× phosphate buffered saline (PBS): Prepare *solution #1* by dissolving 200 g NaCl, 5 g KCl in 1 L H_2O. Prepare *solution #2* by dissolving 72 g $Na_2HPO_4(2H_2O)$, 10 g KH_2PO_4 in 500 mL H_2O. Combine 400 mL of *solution #1*, 100 mL of *solution #2*, and 500 mL H_2O. Autoclave.

8. 10× Saline: Dissolve 85 g NaCl in 1 L H_2O (1.45 M final) and autoclave.

9. 2 M Tris–HCl: Dissolve 121.1 g Tris Base (a.k.a. Trizma® base, Sigma–Aldrich) in 300 mL H_2O and the pH down to 7.5 with concentrated HCl (about 64 mL will be needed). Bring final volume to 500 mL with water.

10. 100× pepsin stock: Dissolve 1 g of pepsin (Sigma–Aldrich) in 100 mL H_2O (1% final makes 100× stock solution). Stir, and filter through a 0.2 μm filter, make 0.5 mL aliquots, and store at −20°C.

11. Pepsin digestion buffer: Prepare a fresh solution of 0.01 M HCl (by adding 42 μL HCl to 50 mL H_2O) in a Coplin jar. Preheat to 37°C ahead of time. Add pepsin and mix (500 μL of 100× stock) just before treating the microscope slides with the cells.

12. Chromosome denaturation solution: 70% deionized formamide in 2× SSC (for 50 mL; 35 mL formamide, 5 mL 20× SSC, 10 mL H_2O, 10 μL 0.5 M EDTA pH 8.0).

13. Oligonucleotide hybridization buffer: 25% deionized formamide, 2× SSC, 200 ng/μL sheared salmon sperm DNA, 5× Denhardt's, 50 mM sodium phosphate, pH 7.0, and 1 mM EDTA. (For 10 mL; 2.5 mL deionized formamide, 1 mL 20× SSC, 200 μL salmon sperm DNA at 10 μg/μL;

1 mL 50× Denhart's solution, 0.5 mL 1 M phosphate buffer, 20 μL 0.5 M EDTA, 4.78 mL H₂O). Make 1 mL aliquots in Eppendorf tubes and store at –20°C.

14. Humidifier: 25% formamide in 2× SSC (for 5 mL; 1.25 mL formamide, 0.5 mL 20× SSC, 3.25 mL H₂O).

15. Tris-Saline (TS) posthybridization wash buffer: 0.1 M Tris–HCl pH 7.5, 0.145 M NaCl. Prepare 50 mL (2.5 mL 2 M Tris, 5 mL 10× Saline, 42.5 mL H₂O).

16. Tris-Saline-Tween (TST) posthybridization wash buffer: 0.1 M Tris–HCl pH 7.5, 0.145 M NaCl, 0.05% Tween 20. Prepare 2×50 mL (For 50 mL: 2.5 mL 2 M Tris, 5 mL 10× Saline, 42.5 mL H₂O, 25 μL Tween). Vortex well to dissolve Tween.

17. Mounting medium: 90% glycerol-10% PBS (v/v). Vortex well to mix and store at 4°C.

2.5. Quantum Dots Reagents and Oligonucleotide Sequences

1. Streptavidin-conjugated QDs (Qdot® 655) were purchased from Invitrogen while Qdot® 592 and 605 were kindly provided by Quantum Dot Corporation.

2. Low retention 1.7 mL micro-centrifuge tubes (VWR International).

3. Biotin (Pierce).

4. Hand-held long wave UV (365 nm) lamp (Entela, Model UVL-56).

5. Illustra MicroSpin™ S-300 HR Columns (GE Healthcare).

6. Biotinylated oligonucleotide consensus or degenerated probes for centromere repeat sequences were custom-synthesized on an Applied Biosystems 392 DNA/RNA synthesizer or obtained directly from a commercial vendor (Invitrogen). Mouse sequences used in this study are the following: Major γ-satellites: 5'-ATT TAG AAA TGT CCA CTG TAG GAC-3', 5'-CCT ᵀ/ₐCA GTG TGC ATT TCT CAT TTT TC-3'; Minor satellites: 5'-TGA TAT ACA CTG TTC TAC AAA TCC CG-3'. Oligonucleotides are reconstituted in water and spectrophotometrically quantified. Hundred nanograms per millilitre aliquots are kept at –20°C.

Important considerations when designing an oligonucleotide probe include adequate length (20–40 bases), GC content (35–55%), minimal self-complementary regions and mismatches (one to two can be tolerated if scattered throughout the sequence) (14). In practice, we first perform a multiple sequence alignment of all the sequences of interest using *ClustalW2* (http://www.ebi.ac.uk/Tools/clustalw2/index.html) (15) followed with a *BLAST* search (http://www.ebi.ac.uk/Tools/blast/) (16) to compare the subregion of interest with all known gene sequences in the database. The sequence is then adjusted and edited manually to optimize specificity.

7. SYBR® Green II RNA gel stain 10,000× concentrate in DMSO (Invitrogen).

8. 6× DNA gel-loading buffer (without Xilene Cyanol/ Bromophenol blue): 15% (w/v) Ficoll Type 400 (Sigma–Aldrich) in water. Store at room temperature.

2.6. Fluorescence Confocal Microscopy

1. Precleaned, Superfrost microscope slides, size 25 × 75 × 1 mm (Fisher Scientific) and cover slips, size 22 × 22 mm N1 (Fisher Scientific) and size 24 × 60 mm N1 (Fisher Scientific). Soak microscope slides and coverslips in 5 M HCl and wash for 1–2 h on a rocking platform (safety precautions for strong acids apply). Rinse thoroughly in running distilled water for a few hours until acidity is completely neutralized (check with pH paper). Store at room temperature in 100% ethanol in a sealed container. Air-dry before use.

2. Clear nail polish (Fisher Scientific).

3. Leica confocal inverted microscope DMIRE2 TCS SP2 AOBS (Leica Microsystems Inc.) equipped with a 405 nm violet diode laser 25 mW (Coherent Inc.) for excitation and a 63× oil-immersion objective (HCX PL APO, NA 1.40) for image acquisition.

3. Methods

The FISH protocol described in this section has been optimized for the detection of repetitive target sequences using consensus and degenerated short biotinylated oligonucleotide DNA probes directly conjugated to SAvQDs (3). This method, which requires minimal custom conjugation, has worked well in our laboratory. It should serve as a guide only. One should keep in mind that QDs are engineered materials that have a statistical size distribution (see Introduction) and, as such, are inherently heterogeneous in nature. Therefore, different batches of commercial QDs might be different both in terms of composition and performance in this assay (see Note 1). The following protocol uses two oligonucleotide-QD probes to visualize in situ, two subchromosomal regions of highly condensed chromatin structure within the centromere in a one-step hybridization/detection dual-color experiment. This method is easily expandable and offers the experimentalist a new alternative for increasing flexibility in multicolor FISH applications.

3.1. Cell Culture and Preparation of Metaphase Chromosome Spreads

1. Subculture cells for less than 24 h in order to achieve 70–80% confluence of actively growing cells.

2. Add Colcemid to the culture to a final concentration of 0.03 µg/mL and incubate for 60 min at 37°C (see Note 2).

3. Wash the cells with 10 mL of PBS.

4. Add 0.5 mL of trypsin to detach cells and incubate 4 min at 37°C. Monitor the enzymatic digestion with a light microscope.

5. Stop digestion by adding 10 mL of fresh complete medium. Transfer the cells into a 15 mL tube and centrifuge the solution at room temperature for 10 min at $130 \times g$.

6. Discard the supernatant by aspirating it off carefully with a Pasteur pipette. Leave 0.5 mL of supernatant above the cell pellet to avoid loss of material.

7. Carefully resuspend the pellet in 10 mL of KCl 0.4% (~0.05 M) prewarmed at 37°C while agitating gently (see Note 3).

8. Incubate 13 min at 37°C. This hypotonic treatment causes a swelling of the cells (see Note 4).

9. Add dropwise 0.5 mL of methanol/glacial acetic acid (3:1.5). This helps prevent clumping of the cells (see Note 5).

10. Centrifuge at $100 \times g$ for 5 min at room temperature (RT).

11. Discard the supernatant with a Pasteur pipette. Leave 0.5 mL above cell pellet. At this stage, the cell pellet should appear white and slightly larger than at the beginning.

12. Add 10 mL of methanol/glacial acetic acid (3:1.5) precooled at –20°C while agitating gently, add the first 5.0 mL dropwise, and add the rest faster.

13. Incubate on ice for 15 min.

14. Centrifuge at $100 \times g$ for 5 min at room temperature (RT).

15. Repeat the fixation procedure from step 11 three more times.

16. Resuspend the cell pellet in 200–1,000 µl of methanol/glacial acetic acid fixative according to the size of the pellet. Adjust final volume by examining under microscope the cell density on the slide (see also slide preparation below and Note 6).

17. Leftover cells can be kept in fixative for several months at 4°C or at –20°C for up to a year.

3.2. Slide Preparation

1. Lay the cleaned, acid-washed slides (see Subheading 2.6.1) at an angle with a pipette on a floating platform inside a 65°C steaming water bath then proceed immediately to next step (see Note 7).

2. Drop the cells (about 35 μL or three drops) from a distance of 15 cm onto the same spot of the moistened slides.

3. Immediately retrieve slides and blow hard on the drops to evenly spread the cells on the surface.

4. Optional: Delineate the cell area by marking the back of the slide with a diamond pencil.

5. When the fixative starts to evaporate (1–2 min) and a grainy surface begins to appear, add 1–2 drops of acetic acid to help disrupt membranes.

6. Place slides onto a 65°C heat block to air-dry.

7. Inspect the slides with a phase-contrast light microscope and select the best ones for hybridization (see Note 6).

3.3. Prehybridization Treatment

1. Equilibrate microscope slides with the cells in 2× SSC for 30 min at RT in a Coplin histology jar (see Note 8).

2. Digest with 0.01% pepsin in 0.01 M HCl for 5 min at 37°C (see Note 9).

3. Rinse in 2× SSC for 10 min at RT to get rid of the pepsin.

4. Dehydrate slides in successive 70, 90, and then 100% ethanol washes, 3 min each at RT.

5. Place slides in a clean rack to air-dry for about 20 min at RT. At this point, the slides can be stored in a box at RT for a couple of days.

6. Meanwhile, preheat the chromosome denaturation solution (70% deionized formamide in 2× SSC) to 70°C (see Note 10).

7. Prepare additional Coplin jars with 70, 90, and 100% ethanol and precool them at −20°C.

8. Incubate slides in 70% formamide in 2× SSC at 70°C for exactly 2 min (see Note 11).

9. Dehydrate slides in successive 70, 90, then 100% cold ethanol washes, 3 min each on ice.

10. Air-dry for about 20 min at RT before hybridizing with QD probe.

3.4. Titration and Preparation of QD-Oligonucleotide Complexes

The commercial SAvQDs have on average 5–10 streptavidin molecules covalently linked to an outer layer composed of an amphiphilic copolymer of octylamine and polyacrylic acid (5). Therefore, the relative stoichiometry of each new QD-oligonucleotide complexes needs to be titrated and characterized by electro-mobility shift assay on agarose gel. The goal is to saturate most of the streptavidin binding sites while minimizing the amount of unbound biotinylated oligonucleotides that will otherwise compete in the hybridization reaction.

We recommend performing this in-gel titration procedure when changing oligonucleotide sequence and/or starting with a new batch of SAvQDs.

1. Make twofold serial dilutions of QDs from 1 µM to 15 nM by diluting 1 µL of SAvQDs with 1 µL H$_2$O, vortex well, and spin. Prepare next dilution from previous one. Always mix thoroughly and centrifuge each intermediate dilution for consistency (see Note 12).

2. Dilute biotinylated oligonucleotides to 50 ng/µL in H$_2$O.

3. Add 1 µL of each SAvQD dilution to 1 µL of oligonucleotide (50 ng) and mix with a micropipette.

4. Incubate at RT for 30 min.

5. Optional: Block any remaining free streptavidin sites with an excess of biotin (1 µL at 25 ng/µL).

6. Add 3 µL of 0.5× TBE, 1 µL of gel loading buffer.

7. Run the QD-oligonucleotide complexes on a 2% agarose gel in 0.5× TBE along with the proper controls (oligonucleotide only and QD only lanes) for about 1 h at 110 volts.

8. Counterstain gel with SYBR® Green II (1:10,000-fold dilution as per manufacturer recommendations) 30 min at RT with agitation. Rinse in water before imaging to reveal free and bounded single-stranded oligonucleotide DNA.

9. Visualize red QDs and green DNA on a standard UV transilluminator to access binding efficiency/occupancy (see Note 13). SAvQD:oligonucleotide complexes will appear yellow (Fig. 1a).

10. Assess what is a reasonable dilution that gives good SAvQD:oligonucleotide binding and mobility shift while minimizing the amount of unbound biotinylated oligonucleotides. Make the appropriate dilution of QD and oligonucleotide as in steps 1–4 and prepare each FISH probe separately for hybridization.

11. Add 48 µL of hybridization buffer (25% deionized formamide, 2× SSC, 200 ng/µL sheared salmon sperm DNA, 5× Denhardt's, 50 mM sodium phosphate pH 7.0, and 1 mM EDTA) to the 2 µL of QD-oligonucleotide complexes (see Note 14).

12. Optional: Run QD-oligonucleotide complexes over a MicroSpin S-300 size exclusion column (according to the manufacturer's instructions) to remove small amounts of unbound ligands (streptavidin, biotin and oligonucleotides). The volume of the eluate collected is about 10% more than the starting volume (see Note 15).

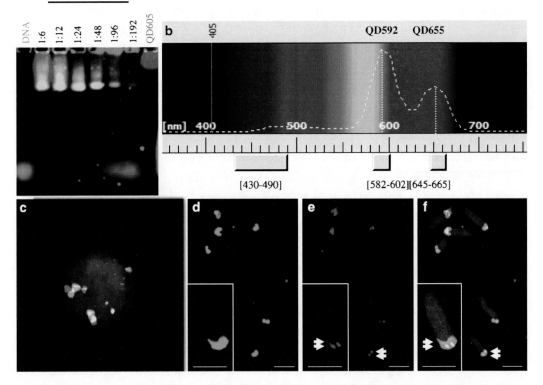

Fig. 1. Direct two color in situ hybridization with QD-oligonucleotide probes. (**a**) Characterization of the QD-oligonucleotide complexes used in FISH. Decreasing amounts of SAvQD (*red*) incubated with biotinylated DNA oligonucleotides (*green*) are electrophoresed on a 2% agarose gel in 0.5× TBE. Free oligonucleotide (*left lane*) migrates faster than QD-oligonucleotide conjugates (*middle lanes*) that migrate faster than QDs alone (*right lane*). Changes in mobility of the QD-oligonucleotide complexes correlate with yellow fluorescence color colocalization of the oligonucleotides (*green*) and the QDs (*red*). (**b**) Spectral analysis of the QD592- and QD655-FISH probe mixture recorded on the Leica DMIRE2 TCS SP2 AOBS confocal microscope using the 405 nm diode laser. The *grey boxes* at the bottom delineate the detection windows used for each PMT: QD592 (592 ± 10 nm); QD655 (655 ± 10 nm); Auto-fluorescence (460 ± 30 nm). (**c–f**) Dual color hybridization with γ- and minor satellite probes on murine nuclei (**c**) and chromosomes (**d–f**) using the same QD-FISH probes mixture and a single wavelength excitation at 405 nm. (**d**) γ- and (**e**) minor satellite probes (green QD592 nm and red QD655 nm respectively). (**f**) Merge signal. Multicolor images are acquired simultaneously. Scale bars, 4 μm. (Reproduced and adapted from ref. 3 with permission from Humana Press Inc., Springer Science and Business Media).

13. Incubate 10 min at 65°C to remove aggregates and melt secondary structures.

14. Spin briefly, check fluorescence with a hand-held long wave UV (365 nm) lamp or a UV transilluminator, then keep on ice until use.

15. The QD-FISH probe in hybridization buffer is used immediately or might be stored at −20°C for several weeks or at 4°C for a couple of days (see Note 16).

3.5. In Situ Hybridization

1. Lay microscope slides with cells flat and apply 14 μL of the QD-FISH probe hybridization mixture onto the cells.

2. Carefully cover with 22 × 22 mm acid-washed glass cover slips to avoid trapping air bubbles.

3. Bake for 3 min at 85°C.

4. Transfer slides in a sealed humidified chamber (humidifier is 25% formamide, 2× SSC) and incubate at 37°C overnight (see Note 17).

3.6. Posthybridization Washes and Mounting

1. Remove slides from humidified chamber with tweezers.

2. Dip the slides briefly in 2× SSC and lift them gently to let the coverslips slide off.

3. Wash slides three times in 2× SSC for 10 min each at 37°C without agitating (see Note 18).

4. Wash slides two times in TST for 5 min at RT without agitating.

5. Optional: Wash once in TS for 5 min to remove Tween.

6. After the last wash, remove excess liquid by blotting the slide but without totally drying it.

7. Apply 40 μL of 90% glycerol in PBS (v/v) in 2–3 drops spread along the stained slides.

8. Add carefully 24 × 60 mm cover slips with tweezers avoiding trapping air bubbles.

9. Seal with nail polish and let dry.

10. The slides are ready to be immediately examined with the confocal microscope or can be stored at 4°C for a couple of days (see Note 19).

3.7. Confocal Microscopy and Analysis

This protocol has been optimized for a Leica confocal inverted microscope DMIRE2 TCS SP2 (Leica Microsystems Inc.) equipped with an Acousto-Optical Beam Splitter (AOBS, programmable beam splitter). The AOBS, in combination with other tunable elements in both the excitation and emission paths, gives the user complete freedom to combine QDs for multicolor confocal analysis. Unless otherwise specified, the QD's fluorescence was detected in a 20-nm spectrally programmed window around their emission peaks (see Note 20). Analogous parameters should be used for other confocal microscopes. Custom filter sets for QDs are available from Chroma Technology or Omega Optical.

1. Examine the slides with a fluorescence confocal microscope using a single 405 nm diode laser for excitation (see Note 21).

2. Bring cells into focus using white light epi-illumination.

3. Switch to confocal mode.

4. Set up detection range of PMTs within 20 nm around the QD's respective emission maxima.

5. Set up desired image format (e.g., 512×512 pixels).

6. Perform laser scanning at a line frequency of 200 Hz averaging 2–4 times.

7. Acquire tri-color images in a simultaneous scan. Fig. 1 shows a representative dual-target detection of repeated sequences in murine nuclei and chromosomes, using two different color QD-FSH probes and a single wavelength excitation at 405 nm.

4. Notes

1. We have successfully tested a number of streptavidin-conjugated QDs (SAvQDs) from Invitrogen. SAvQDs with peak emission wavelengths of 565, 585, 592, 605, 611, and 655 nm performed equally well in that assay with the notable exception of Qdot® 525, which showed an irreversible spectral shift upon hybridization that precluded its subsequent use in FISH. SAvQDs are also available from Evident Technologies (Troy, NY, http://www.evidenttech.com) under the name EviFluors. QDs were sonicated prior to use for 5 min on ice to dissolve the aggregates.

2. The concentration of colcemid to use and the duration of treatment may vary for each cell line and subclone. It should be accessed empirically (suggested range: 0.03–0.1 µg/mL for 30–60 min). Colcemid is a potent microtubule inhibitor, and it should be handled with appropriate protection (i.e., laboratory coat and gloves).

3. The first 5 mL of KCl are added dropwise while agitating constantly at low speed on a vortex to avoid clumping of the cells. The rest is added faster. The concentration of KCl might be adjusted for other cell types (suggested range: 0.05–0.075 M).

4. The optimal time treatment varies from different cell types and must be determined empirically (suggested range: 13–20 min).

5. Methanol/glacial acetic acid fixative must be prepared fresh and added dropwise on the vortex. If some clumps appear, remove them with a Pasteur pipette before the next centrifugation step. We use a 3:1.5 (v/v) ratio that is optimized for the P815 cell line. When using a different cell line, we recommend starting with a 3:1 (v/v) ratio and then gradually

optimizing the ratio of acetic-acid according to the quality of the resulting nucleus/chromosome preparations.

6. If necessary, the cell concentration can be adjusted by adding fixative to obtain well spread chromosomes and nuclei. A good preparation should also contain minimal cellular debris.

7. This exposure to hot steam provides a uniform layer of moisture on the slide immediately prior to adding the cells which helps facilitate spreading of the cells over the glass surface.

8. Unless specified otherwise, we use autoclaved plastic Coplin jars for all solutions. In between each successive step, the slides are carefully blotted (but not dried) by both wiping the nonfrosted side as well as the edges of the slide with a folded tissue in order to remove the excess of liquid.

9. Concentration and duration of the pepsin treatment are important and should be optimized for each new batch of pepsin. Pepsin activity is optimum at 37°C around pH 1.5–2.0.

10. The temperature of the chromosome denaturation solution should be no less than 68°C and no more than 70°C. The denaturation solution must be used within 30 min while pH is stable to maintain optimal morphology of chromosomes and ensure efficient hybridization.

11. When handling more than two slides, we suggest leaving 30 s to 1 min intervals in between slides to ensure exact timing. The precise duration of this denaturation step is critical to ensure a successful hybridization.

12. We recommend using low retention tubes to minimize non-specific attachment of the QDs to the walls of the tubes.

13. One can spectrally characterize the electromobility shift and binding efficiency with a laser scanner imager. We use a MultiImager FX System (Bio-Rad, Hercules, CA) with the following filters: Qdot® 655 (532 nm green laser excitation, 640BP red emission) and SYBR Green DNA (488 nm blue laser excitation, 530BP green emission). The conjugation of the DNA on the SAvQDs is demonstrated by faster mobility of the complexes due to a net negative charge increase and by colocalization of both the red QDs and the green DNA fluorescent signals (yellow merge, Fig. 1a). The smearing of the QD-oligonucleotide complexes onto the agarose gel is likely to be the consequence of variations in the number of biotin sites (i.e., the number of streptavidin molecules) available on each SAvQD. This gives rise to particles with slightly different net charges after coating of the nanocrystals with the oligonucleotides. The size heterogeneity of the SAvQDs themselves does not contribute to that

phenomenon since bare SAvQDs (without DNA) migrate homogeneously as a single band in the agarose gel (Fig. 1a, left lane).

14. For dual labeling with two QD-FISH probes, adjust the volume of hybridization buffer accordingly up to a total volume of 50 μL which gives a final probe concentration of about 1 ng/μL for each oligonucleotide sequence.

15. This step is optional. Once the stoichiometry has been empirically accessed and optimized on gel (we have used a 1:12 molar ratio of QD-oligonucleotides for most experiments), the QD-oligonucleotide complexes diluted in hybridization buffer can be used as is. Any remaining unbound ligands (streptavidin, biotin, and oligonucleotides) that are smaller than the QD complexes can be removed by running a size exclusion spin column. Purification of the QD complexes is also recommended after storage since it has been reported that streptavidin sheds off of SAvQDs over time (17).

16. Upon retrieving from cold storage, always check if the sample fluoresces is using a hand-held long wave UV (365 nm) lamp or a UV transilluminator.

17. We use a micro-array oven (Boekel InSlide-Out hybridization apparatus) but a cheap alternative is to use any box with raised rails so that you can lay your slides flat on the rails. A plastic microscope slide storage box (e.g., VWR) works perfectly. Cover the bottom of the box with paper towels presoaked with the humidifying solution (not too much but just enough to wet the paper) and seal the box with tape to limit evaporation. Place the box in an oven at 37°C overnight.

18. These successive posthybridization washes are used to remove nonspecifically bound and unbound probe from the cell preparations. Very mild or no agitation should be applied so as not to lose the QDs from shear forces. Blot slides in between washes but do not let it dry. Temperature, number, and duration of washes can be optimized by periodically monitoring the quality of the FISH signal under the microscope.

19. QD's fluorescence is stable at 4°C in the dark but shows irreversible time-dependent decay after several weeks.

20. The fluorescence spectrum of each QD-FISH probe in hybridization buffer was first directly recorded in situ on the microscope and used to optimally set the detection range around the symmetric emission curve typical of each QD probe (Fig. 1b). The detection windows used for the photomultiplier tubes (PMTs) are: QD592 (592 ± 10 nm) and QD655 (655 ± 10 nm). The morphology of the nucleus was assessed through an auto-fluorescence detection window (460 ± 30 nm). Auto-fluorescence was used instead of the

more conventional DAPI counterstaining because we found DAPI to interfere with QD fluorescence upon 405 nm laser excitation.

21. We and others have observed that a brief period of illumination is sometimes required in order for the QD's fluorescence to emerge from the background (3, 18). To do so, simply place the FISH slides on a UV transilluminator or directly in the light excitation path of the confocal microscope for a few minutes prior to viewing. A 405 nm diode laser or the 488 nm line of an Argon laser can be used.

Acknowledgments

The author would like to thank Professor Shimon Weiss for advice and encouragement, Dr. Matthew Schibler for comments on this manuscript, and Tal Paley for her editorial assistance. Fluorescence microscopy was performed at the Advanced Light Microscopy/ Spectroscopy Shared Facility at the California NanoSystems Institute at UCLA. This work was funded in part by the National Institute of Health, Grant No. R01 EB000312-04.

References

1. Medintz, I. L., Uyeda, H. T., Goldman, E. R., and Mattoussi, H. (2005) Quantum dot bioconjugates for imaging, labeling and sensing. *Nat. Mater.* **4**, 435–446.

2. Michalet, X., Pinaud, F. F, Bentolila, L. A, Tsay, J. M, Doose, S., Li, J. J, Sundaresan, G., Wu, A. M, Gambhir, S. S, and Weiss, S. (2005) Quantum dots for live cells, *in vivo* imaging, and diagnostics. *Science* **307**, 538–544.

3. Bentolila, L. A. and Weiss, S. (2006) Single-step multicolor fluorescence *in situ* hybridization analysis using semiconductor quantum dot-DNA conjugates. *Cell Biochem. Biophys.* **4**, 59–70.

4. Bruchez, M., Moronne, M., Gin, P., Weiss, S., and Alivisatos, A. P. (1998) Semiconductor nanocrystals as fluorescent biological labels. *Science* **281**, 2013–2015.

5. Wu, X., Liu, H., Liu, J., Haley, K. N., Treadway, J. A., Larson, J. P., Ge, N., Peale, F., and Bruchez, M. P. (2003) Immunofluorescent labeling of cancer marker Her2 and other cellular targets with semiconductor quantum dots. *Nat. Biotechnol.* **21**, 41–46.

6. Jaiswal, J. K., Mattoussi, H., Mauro, J. M., and Simon, S. M. (2003) Long-term multiple color imaging of live cells using quantum dot bioconjugates. *Nat. Biotechnol.* **21**, 47–51.

7. Xiao, Y. and Barker P. E. (2004) Semiconductor nanocrystal probes for human metaphase chromosomes. *Nucleic Acids Res.* **32**, e28.

8. Doose, S., Tsay, J. M., Pinaud, F., and Weiss, S. (2005) Comparison of photophysical and colloidal properties of biocompatible semiconductor nanocrystals using fluorescence correlation spectroscopy. *Anal. Chem.* **77**, 2235–2242.

9. Yu, W. W., Qu, L., Guo, W., and Peng, X. (2003) Experimental determination of the extinction coefficient of CdTe, CdSe, and CdS nanocrystals. *Chem. Mater.* **15**, 2854–2860.

10. Bentolila, L. A., Michalet X., and Weiss S. (2008) Quantum optics: colloidal fluorescent semiconductor nanocrystals (quantum dots) in single-molecule detection and imaging. In: *Single-Molecules and Biotechnology*, (Rigler, R. and Vogel, E., Eds), Springer, Berlin, Heidelberg, pp. 53–81.

11. Nederlof P. M, van der Flier, S, Wiegant, J, Raap, A. K, Tanke, H. J, Ploem, J. S, and van der Ploeg, M. (1990) Multiple fluorescence *in situ* hybridization. *Cytometry* **11**, 126–131.

12. Alivisatos, A. P. (1996) Semiconductor clusters, nanocrystals, and quantum dots. *Science* **271**, 933–937.

13. Lukusaa, T. and Fryns, J. P. (2008) Human chromosome fragility. *Biochim. Biophys. Acta.* **1779**, 3–16.

14. Rattray, M. and Michael, G. J. (1998) Oligonucleotide probes for *in situ* hybridization. In: *In Situ* Hybridization (Wilkinson, D. G., Ed.), Oxford University Press, Oxford, pp. 23–67.

15. Higgins, D. G., Thompson, J. D., and Gibson, T. J. (1996) Using CLUSTAL for multiple sequence alignments. *Methods Enzymol.* **266**, 383–402.

16. Altschul, S. F., Gish, W., Miller, W., Myers, E. W., and Lipman, D. J. (1990) Basic local alignment search tool. *J. Mol. Biol.* **215**, 403–410.

17. Lidke, D. S, Nagy, P, Heintzmann, R, Arndt-Jovin, D. J, Post, J. N, Grecco, H. E, Jares-Erijman, E. A, and Jovin, T. M (2004) Quantum dot ligands provide new insights into erbB/HER receptor-mediated signal transduction. *Nat. Biotechnol.* **22**, 198–203.

18. Silver, J. and Ou, W. (2005) Photoactivation of quantum dot fluorescence following endocytosis. *Nano Lett.* **5**, 1445–1449.

Chapter 11

LNA-FISH for Detection of MicroRNAs in Frozen Sections

Asli N. Silahtaroglu

Abstract

MicroRNAs (miRNAs) are small (~22 nt) noncoding RNA molecules that regulate the expression of protein coding genes either by cleavage or translational repression. miRNAs comprise one of the most abundant classes of gene regulatory molecules in multicellular organisms. Yet, the function of miRNAs at the tissue, cell, and subcellular levels is still to be explored. Especially, determining spatial and temporal expression of miRNAs has been a challenge due to their short size and low expression. This protocol describes a fast and effective method for detection of miRNAs in frozen tissue sections using fluorescence in situ hybridization. The method employs the unique recognition power of locked nucleic acids as probes together with enhanced detection power of the tyramide signal amplification system for detection of miRNAs in frozen tissues of human and animal origin within a single day.

Key words: Locked nucleic acids, MicroRNA, Frozen section, Cryosections, FISH, LNA-FISH, In situ hybridization, Tyramide signal amplification

1. Introduction

The key principles and most of the technical details regarding LNA-FISH method for detection of microRNAs (miRNAs) in frozen sections have been described before (1). The method is based on the use of locked nucleic acids (LNA) and the tyramide signal amplification (TSA) system. The chemistry of both these processes is already known to increase the sensitivity of the in situ hybridization experiments (2, 3). A combination of LNA probes with the TSA detection system reduces the experimental time drastically, making the analysis possible in a clinical environment. Methods for detection of miRNAs that employ different methodologies have been published before (4–6), but none of the methods are as fast as the one presented here.

Joanna M. Bridger and Emanuela V. Volpi (eds.), *Fluorescence in situ Hybridization (FISH):*
Protocols and Applications, Methods in Molecular Biology, vol. 659,
DOI 10.1007/978-1-60761-789-1_11, © Springer Science+Business Media, LLC 2010

This fluorescence in situ hybridization (FISH) protocol for frozen sections employs LNA-modified oligonucleotides complementary to the mature miRNA. LNA is an RNA analog with a high hybridization affinity for complementary DNA and RNA (7). Hybridization with LNA probes takes only 1 h. The in situ signal is detected by incubation with horseradish peroxidase-conjugated antidigoxigenin (anti-DIG) or antifluorescein isothiocyanate (anti-FITC) antibodies followed by amplification of the signal with FITC- or DIG-conjugated tyramide. This results in the detection of gene expression at a single cell level. The whole protocol, from the fixation of the frozen sections to microscopy, takes around 6 h, which makes it possible to obtain results within 1 day.

The protocol has been successfully used in a variety of different biological settings. It has been applied to detect miRNAs in cancer tissues from human colon and bladder (8, 9) as well as brain sections from the developing and adult mouse to investigate the miRNAs role in normal brain development (10) and in neurological disorders (11) (Fig. 1).

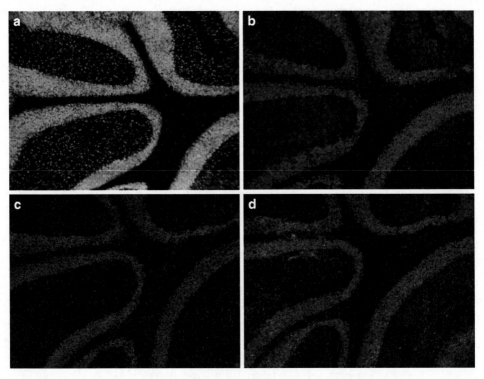

Fig. 1. Specificity of the LNA-FISH for detecting microRNAs in situ. Positive signals are seen as *green* and the nuclei are *stained blue*. Since miR-124a is known to be highly expressed in the brain (a) it is used as a positive control to show the specificity of the technique. (b) Hybridization with an oligonucleotide probe for miR-124a without any LNA modifications is used to show the power of LNA in detecting the small noncoding RNAs (c) To show the importance of tyramide amplification, an FITC-labeled miR-124a probe has been hybridized, washed, and mounted without the amplification step. No signals were obtained. (d) A no probe control has been through the whole procedure. One can see a slight background staining in slides that have been through the amplification process. One should decrease the incubation with tyramide fluorescein and wash more stringently after incubation.

2. Materials

The laboratory should be equipped with a fume hood, a minimum of one hybridizer, an incubator, and a shaker. All the solutions used prior to the hybridization step are prepared using DEPC-treated water. The glassware should be washed thoroughly including an acid wash after use; yet, no special treatment is necessary before use.

2.1. Probes

1. LNA probes (Mi Curry) (see Note 1).
2. DIG oligonucleotide tailing kit (Roche Applied Sciences), if needed (see Note 1).

2.2. Pretreatment of the Frozen Sections

1. Glass Coplin jars.
2. Encircling pen (Dako).
3. 4% (w/v) PFA in PBS, pH 7.6 (Bie & Berntsen). Aliquot and keep at –20°C and thaw before fixation. After thawing use within 24 h.
4. 10× PBS: 137 mM NaCl, 2.7 mM KCl, 4.3 mM Na_2HPO_4, and 1.47 mM KH_2PO_4. Adjust pH to 7.4. Autoclave and store at room temperature (RT).
5. DEPC-treated H_2O: Prepare 0.1% diethylpyrocarbonate (DEPC; Fluka) in Milli-Q water. Mix and leave overnight at RT in a fume hood. Autoclave solutions the next day. Note that solutions that contain Tris base cannot be treated with DEPC.
6. PBS: 100 mL of 10× PBS is mixed with 900 mL of DEPC-treated water.
7. Acetylation buffer: Final concentration 0.6% acetic anhydride (Sigma-Aldrich), 1.3% triethanolamine (Fluka), and 0.6 N HCl (Merck). Prepare just before use by adding 500 µL of 6 N HCl and 670 µL of triethanolamine into 48.5 mL Milli-Q water, dissolve and add 300 µL of acetic anhydride.

2.3. Hybridization and Postwashes

1. Hybridizer (Dako).
2. Hybridization buffer: 50% (v/v) formamide (F-5786, Sigma-Aldrich), 5× SSC, 500 µg/µL yeast transfer RNA (Invitrogen), 1× Denhardt's solution (D-2532, Sigma-Aldrich), and DEPC-treated water (see Note 2).
3. 20× SSC: Dissolve 175.3 g NaCl and 88.2 g sodium citrate–$2H_2O$ in 800 mL H_2O. Adjust pH to 7.0 and complete to 1,000 mL.
4. NescoFilm (Fisons Scientific Apparatus).
5. TNT buffer: 0.1 M Tris–HCl, pH 7.5, 0.15 M NaCl, 0.3% Triton X-100.

6. TN buffer: 0.1 M Tris–HCl, pH 7.5, 0.15 M NaCl.

7. Shaker (The Belly Dancer® Laboratory Shaker, Stovall Life Science).

2.4. Signal Amplification and Detection

1. Moist chamber (VWR).

2. 3% (v/v) H_2O_2, prepared from 30% H_2O_2 (Perhydrol®; Merck KGaA) in PBS. Aliquoted in 10-mL tubes and kept at –20°C until the day of use.

3. Blocking buffer: 0.1 M Tris–HCl, pH 7.5, 0.15 M NaCl, 0.5% blocking reagent (Roche Applied Sciences), 0.5% bovine serum albumin (BSA) (see Note 3).

4. Sheep anti-DIG, Fab fragments, POD conjugated (Roche Applied Sciences).

5. Sheep anti-FITC-POD, Fab fragments (Roche Applied Sciences).

6. TSA plus fluorescence system containing fluorochrome tyramide and amplification buffer (Perkin Elmer) (see Note 4).

7. Prolong gold antifade reagent with DAPI (Invitrogen-Molecular Probes).

8. Olympus MVX110 macroview microscope equipped with Olympus F-10 camera and Olympus Cell P software (Olympus A/S).

   FITC filter, excitation 475–505/emission 497–540.

   DAPI filter, excitation 315–380/emission 415–507.

   TRITC filter, excitation 525–565/emission 555–600.

3. Methods

3.1. Preparation of the Probe Mixture

1. Label your LNA probes if unlabeled (see Note 1).

2. Use 60–100 μL of 30 nM of probe mixture in hybridization buffer per slide according to the size of the section.

3.2. Preparation of the Tissue Sections

The in situ hybridization technique described here has been tried on frozen sections prepared by different histology/pathology laboratories with great success. Though snap-frozen sections worked the best, the system works well with all frozen sections. Sections between 8- and12-μm thicknesses give good results. We have generally used 10-μm sections in our lab.

3.3. Pretreatment of the Tissue Sections

1. Remove sections from the –80°C freezer and thaw them for 10 min at RT.

2. Make a circle on the glass around the tissue with Dako pen.

3. Dry the slides for an additional 10 min at 55°C.

4. Fix the sections with 4% PFA in a fume hood for 10 min (see Note 5).

5. Wash the slides with DEPC-PBS three times for 3 min using agitation.

6. Make fresh acetylation buffer and treat the slides for 5–10 min.

7. Wash the slides three times for 3 min with DEPC-PBS on a shaker.

3.4. In Situ Hybridization and Posthybridization

1. Add 60–100 μL of hybridization mixture to the section and cover with a piece of NescoFilm (see Note 6).

2. Put the slides in the hybridizer and prehybridize for 30 min at the hybridization temperature (see Note 7).

3. Remove the NescoFilm carefully, shake the liquid off the slide, and dry the area around the section with a tissue paper.

4. Add 60–100 μL of 30 nM of probe mixture on the slides, cover with a new piece of NescoFilm, and incubate the slides for 1 h at 22–25°C below the predicted T_m value of the LNA probe used.

5. Remove NescoFilm and wash the slides three times for 10 min per wash, with agitation in 0.1× SSC at 4–8°C above the hybridization temperature.

6. Wash with 2× SSC for 10 min at RT.

3.5. Signal Amplification and Detection

1. Treat the sections with 3% (v/v) H_2O_2 for 10 min horizontally in the incubation chamber. Peroxidase must be freshly prepared before use. Store 30% hydrogen peroxide solution in the refrigerator.

2. Wash slides three times for 3 min with TN buffer on the shaker.

3. Incubate slides with blocking buffer for 30 min at RT.

4. Dilute primary antibody (anti-DIG-POD or anti-FITC-POD, Roche Biosciences) 1:25–1:1,000 in blocking buffer.

5. Incubate sections with 100–150 μL of antibody solution for 30 min at RT.

6. Wash slides three times for 5 min with TNT buffer at RT on the shaker.

7. Dilute FITC-tyramide 1:50–1:75 in the amplification buffer, add 100 μL on the sections, and incubate for 10 min at RT in dark.

8. Wash slides three times for 5 min with TNT buffer in dark with shaking.

9. Dry the slides for 2 min.

10. Mount the slides with 25 μL of "Prolong Gold" containing DAPI and cover with a glass coverslip.

4. Notes

1. In most of the cases LNA probes are purchased labeled at the 5′ end either with FITC or digoxigenin (DIG). If the probes are also meant to be used for other purposes, i.e. northern blot, they are purchased unlabeled and labeled afterwards using DIG oligonucleotide tailing kit (Roche Applied Sciences) according to instructions for short tailing. Probes are kept at –20°C.

2. Formamide is both mutagenic and terotagenic. Work in a fume hood when working with formamide.

3. It is possible to exchange 0.5% (w/v) BSA with 10% (v/v) fetal calf serum within the blocking buffer.

4. Choice of TSA plus fluorescence system depends on the filter set available on the user's microscope. No difference has been observed in the results when using TSA plus fluorescein or tetramethylrhodamine (TMR) systems.

5. Efficient fixation of the tissue sections is very crucial for the success of the experiment. If the sections are thicker, the fixation step can be prolonged.

6. Instead of NescoFilm, coverslips can also be used during prehybridization and hybridization.

7. Hybridization temperature should be around 22–25°C below the T_m value of the probes used.

Acknowledgments

The author acknowledges the financial support from the Lundbeck Foundation, the Danish Research Agency and the Dr Sofus Carl Emil Friis and wife Olga Doris Friis Foundation. Wilhelm Johannsen Centre for Functional Genome Research is established by the Danish National Research Foundation.

References

1. Silahtaroglu AN, Nolting D, Dyrskjøt L, Berezikov E, Møller M, Tommerup N, Kauppinen S (2007). Detection of microR-NAs in frozen tissue sections by fluorescence in situ hybridization using Locked Nucleic Acid probes and tyramide signal amplification. *Nat Protoc.* **200**, 2520–2528.

2. Silahtaroglu AN, Tommerup N, Vissing H (2003). FISHing with locked nucleic acids (LNA): evaluation of different LNA/DNA mixmers. *Mol Cell Probes.* **17**, 165–169.

3. Kerstens HM, Poddighe PJ, Hanselaar AG (1995). A novel *in situ* hybridization signal amplification method based on the deposition of biotinylated tyramide. *J Histochem Cytochem.* **43**, 347–352.

4. Thompson RC, Deo M, Turner DL (2007). Analysis of microRNA expression by in situ hybridization with RNA oligonucleotide probes. *Methods.* **43**, 153–161.

5. Obernosterer G, Martinez J, Alenius M (2007). Locked nucleic acid-based *in situ* detection of microRNAs in mouse tissue sections. *Nat Protoc.* **2**, 1508–1514.

6. Kloosterman WP, Wienholds E, de Bruijn E, Kauppinen S, Plasterk RH (2006). In situ detection of miRNAs in animal embryos using LNA-modified oligonucleotide probes. *Nat Methods.* **3**, 27–29.

7. Koshkin AA, Nielsen P, Meldgaard M, Rajwanshi VK, Singh SK, Wengel J (1998). LNA (locked nucleic acid): An RNA mimic forming exceedingly stable LNA: LNA duplexes. *J Am Chem Soc.* **120**, 13252–13253.

8. Stenvang J, Silahtaroglu AN, Lindow M, Elmen J, Kauppinen S (2008). The utility of LNA in microRNA-based cancer diagnostics and therapeutics. *Semin Cancer Biol.* **18**, 89–102.

9. Schepeler T, Reinert JT, Ostenfeld MS, Christensen LL, Silahtaroglu AN, Dyrskjøt L, Wiuf C, Sørensen FJ, Kruhøffer M, Laurberg S, Kauppinen S, Ørntoft TF, Andersen CL (2008). Diagnostic and prognostic microR-NAs in stage II colon cancer. *Cancer Res.* **68**, 6416–6424.

10. Bak M, Silahtaroglu A, Møller M, Christensen M, Rath M, Skryabin B, Tommerup N, Kauppinen S (2008). MicroRNA expression in the adult mouse central nervous system. *RNA.* **14**, 432–444.

11. Hébert SS, Horré K, Nicolaï L, Papadopoulou AS, Mandemakers W, Silahtaroglu A, Kauppinen S, Delacourte A, De Strooper B (2008). Loss of microRNA cluster miR-29a/b-1 in sporadic Alzheimer's disease correlates with increased BACE1/beta-secretase expression. *Proc Natl Acad Sci U S A.* **105**, 6415–6420.

Chapter 12

Chromosome Orientation Fluorescence In Situ Hybridization or Strand-Specific FISH

Susan M. Bailey, Eli S. Williams, Michael N. Cornforth, and Edwin H. Goodwin

Abstract

Chromosome Orientation FISH (CO-FISH) is a technique that can be used to extend the information obtainable from standard FISH to include the relative orientation of two or more DNA sequences within a chromosome. CO-FISH can determine the absolute 5′-to-3′ direction of a DNA sequence relative to the short arm-to-long arm axis of the chromosome, and so was originally termed "COD-FISH" (Chromosome Orientation and Direction FISH). CO-FISH has been employed to detect chromosomal inversions associated with isochromosome formation, various pericentric inversions, and to confirm the origin of lateral asymmetry. More recent and sophisticated applications of CO-FISH include distinction between telomeres produced via leading- vs. lagging-strand DNA synthesis, identification of interstitial blocks of telomere sequence that result from inappropriate fusion to double-strand breaks (telomere-DSB fusion), discovery of elevated rates of mitotic recombination at chromosomal termini and sister chromatid exchange within telomeric DNA (T-SCE), establishing replication timing of mammalian telomeres throughout S-phase (ReD-FISH) and to identify chromosomes, in combination with spectral karyotyping (SKY-CO-FISH).

Key words: CO-FISH, ReD-FISH, SKY-CO-FISH, Telomeres

1. Introduction

Fluorescence in situ hybridization (FISH) is a powerful tool for exploring genomes at the chromosomal level. The procedure can be used to identify individual chromosomes, rearrangements between chromosomes, and the location within a chromosome of specific DNA sequences such as centromeres, telomeres, and even individual genes. CO-FISH extends the information obtainable from standard FISH to include the relative orientation of two or

Joanna M. Bridger and Emanuela V. Volpi (eds.), *Fluorescence in situ Hybridization (FISH): Protocols and Applications*, Methods in Molecular Biology, vol. 659,
DOI 10.1007/978-1-60761-789-1_12, © Springer Science+Business Media, LLC 2010

more DNA sequences within a chromosome (1). With a suitable reference probe, CO-FISH can also determine the absolute 5′-to-3′ direction of a DNA sequence relative to the short arm-to-long arm axis of the chromosome. This variation of CO-FISH was originally termed "COD-FISH" (Chromosome Orientation and Direction FISH) to reflect this fact (2). A probe to telomeric DNA is typically used for COD-FISH because G-rich telomere strands, i.e., the $5′-(TTAGGG)_n-3′$ strands, are known to be located at chromosome ends and oriented such that the 3′ ends of the strands are at chromosomal termini.

In the beginning, CO-FISH was used to detect obligate chromosomal inversions associated with isochromosome formation (3), various pericentric inversions (4), and to confirm the origin of lateral asymmetry (5). More recent and sophisticated applications of CO-FISH include distinction between telomeres produced via leading- vs. lagging-strand DNA synthesis (6), identification of interstitial blocks of telomere sequence that result from inappropriate fusion to double-strand breaks (telomere–DSB fusion) (7), discovery of elevated rates of mitotic recombination at chromosomal termini (8) and sister chromatid exchange within telomeric DNA (T-SCE) (9), establishing replication timing of mammalian telomeres throughout S-phase (Replicative Detargeting FISH (ReD-FISH)) (10, 11), and in combination with spectral karyotyping (SKY-CO-FISH) (12). For more information, the reader is referred to several reviews (13–15).

CO-FISH differs from standard FISH in its ability to make hybridizations strand-specific. As shown in Fig. 1, the procedure works by culturing cells for a single round of replication in the presence of the thymidine analog 5′-bromo-2′-deoxyuridine (BrdU). During semiconservative DNA synthesis, BrdU incorporates into

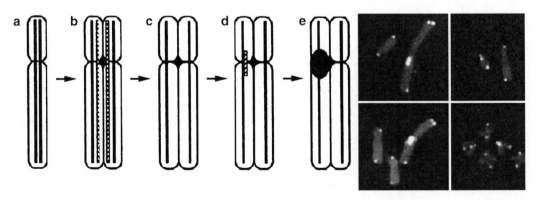

Fig. 1. CO-FISH steps. (**a**) G_1 chromosome prior to replication; (**b**) incorporation of BrdU; (**c**) selective removal of newly replicated strands leaving the target parental strands oriented in opposite directions on the two chromatids; (**d**) probe hybridization; (**e**) probe detection. The photographs illustrate single-sided signals following hybridization of single-stranded probes to repetitive sequences found at telomeres, the centromeres of chromosomes 1 and 18, and the 1p36 region on the short arm of chromosome 1 (adapted from (4) and (2)).

replicating DNA strands opposite to the original parental strands. Cell-cycle progression is blocked in metaphase with a microtubule inhibitor, typically colcemid, and cells are fixed using standard cytogenetic techniques. The fixed cells are dropped onto microscope slides causing the mitotic cells to spread out and display their condensed chromosomes. The slides are then stained with Hoechst 33258 and exposed to long-wave UV light, a process that preferentially nicks bromo-substituted DNA. Next cells are treated with exonuclease III, an enzyme that catalytically degrades one strand of the double helix starting at nicks. Because the newly replicated BrdU-substituted strands are heavily nicked, Exo III selectively and effectively removes these strands. The mitotic chromosomes are rendered single stranded such that each chromatid contains one of the original parental strands. The single-stranded chromosomal DNA is an ideal target for probe hybridization. Moreover, because the original parental strands are complementary to one another, a single-stranded probe hybridizes only to its complementary chromosomal target on just one chromatid. Overall, the effect is to produce single-sided signals.

The relative orientation of two chromosomal DNA sequences is revealed by unique hybridization patterns. If the sequences are oriented in parallel, probes to both sequences hybridize to, and produce visible signals on, the same chromatid; signals on opposite chromatids reveal an antiparallel orientation. Repetitive sequences are a special case in which a single probe binds to multiple repeats. Several probes were hybridized to repetitive sequences with results shown in the photographs in Fig. 1 (2). In each case, and as we have found to be generally true of tandem repeats (3–5), these repetitive sequences are oriented head to tail (i.e., in parallel) as determined from the CO-FISH pattern (single-sided signals). In contrast, dispersed repeats, such as Alu, appear to have no preferential orientation, at least not at the level of resolution of light microscopy.

The CO-FISH concept has been used to provide information not easily obtainable through other means. For example, it has been modified for use in determining the replication timing of specific mammalian telomeres during S-phase, a technique termed ReD-FISH (11). Replication of a particular probed locus or sequence in the presence of BrdU results in a "switch" from a normal FISH (two-sided; one on each chromatid) signal to a CO-FISH (one-sided; only one chromatid involved) signal as the replication fork passes the region in question. When monitoring such a change during the ensuing metaphase, Fig. 2 shows a smooth sigmoid curve in the ratio of CO-FISH (single-sided) to normal FISH (double-sided) signals, consistent with the proposition that telomeres in normal human cells replicate throughout S-phase. It is tempting to speculate that protracted telomere replication may assist mammalian cells in preventing

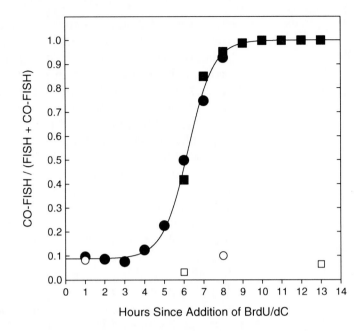

Fig. 2. ReD-FISH in human fibroblasts. Two overlapping experiments (*squares* and *circles*). The sigmoid shape of the curve suggests that telomeres replicate throughout S-phase. *Open symbols* are time-matched controls that were not allowed to incorporate BrdU.

chromosome end fusions by minimizing the number of "open" DNA ends at any given time that are vulnerable to inappropriate nonhomologous end joining during S-phase. Interestingly, the mammalian situation is in stark contrast to yeast, which replicate their telomeres coordinately in late S-phase. Yeast have little nonhomologous end-joining activity other than in G_1, perhaps limiting their vulnerability to chromosome end fusions during telomere replication.

In addition to the applications described above, CO-FISH studies are providing unique insights into the genetic basis of telomere function. As the significance of strand specificity has been recognized and better appreciated, it was realized that probes made from G- and C-rich telomeric strands could be used to distinguish between telomeres based on their mode of replication. Because replication forks proceed from the most distal origin of replication toward the chromosome ends, the G- and C-rich probes hybridize, respectively, to telomeres that were replicated either through leading-strand or lagging-strand DNA synthesis (6). The detection of elevated SCE-like recombination within telomeric DNA (T-SCE) (9, 16, 17), as well as its regulation and its implications for replicative senescence (9, 18), has also been recognized and is currently enjoying widespread application. Most studies of telomeres now use probes made of peptide nucleic acid (PNA) rather than DNA. Prelabeled PNA telomere probes are commercially available and generally give excellent results. Hybridization conditions for PNA probes are substantially different from those commonly used for DNA probes. Detailed methods for both types of telomere probe are given below.

2. Materials

2.1. Reagents

1. 10^{-3} M bromodeoxyuridine stock solution in water (BrdU; Sigma-Aldrich).
2. Trypsin–EDTA (Gibco).
3. Colcemid (Gibco).
4. 75 mM KCl.
5. 3:1 methanol: acetic acid fixative (always make up fresh).
6. RNAse A.
7. 1× phosphate-buffered saline (PBS).
8. 37% formaldehyde.
9. 75, 85, and 100% ethanol.
10. Hoechst 33258.
11. 2× SSC prepared from 20× SSC stock (1× SSC is 0.15 M NaCl, 0.015 M sodium citrate).
12. Exonuclease III (and buffer provided by manufacturer).
13. 100% formamide.
14. Labeled PNA telomere probe (commercially available).
15. PN buffer (Phosphate NP-40 buffer).
16. Vectashield antifade with DAPI (4′,6-diamidine-2-phenylindole dihydrochloride) (1.5 μg/mL; Vector Laboratories).
17. Anti-BrdU antibody (Molecular Probes).

2.2. Equipment

1. Moist chamber or a sealed Tupperware container lined with moist paper towels and pipettes to raise the slides off the towels is perfectly suitable. Hybridization chambers are also commercially available.
2. High-quality glass microscope slides, very clean.
3. Glass coverslips (22 mm × 50 mm, No.1).
4. Coplin jars with lids.
5. 37°C waterbath, preferably shaking.
6. 75°C waterbath.

7. 37°C oven (or incubator).

8. Ultraviolet light source, ~365 nm (e.g., Stratalinker 1800 UV Irradiator, Stratagene).

9. Fluorescence microscope equipped with appropriate filter sets and objectives.

3. Methods

3.1. PNA Telomere CO-FISH: Outline of the Procedure

1. Culture cells for one round of replication in BrdU.

2. Chromosome preparations.

3. Air-dry slides (store desiccated at RT).

4. Treat slides with RNAse A at 37°C for 10 min.

5. Rinse in PBS.

6. Postfixation in 4% formaldehyde at RT for 10 min.

7. Rinse in PBS.

8. Dehydrate slides in ice-cold 75, 85, and 100% ethanol.

9. Air-dry.

10. Remove newly synthesized, BrdU-incorporated strand (CO-FISH).

11. Stain slides in 0.5 μg/mL Hoechst 33258 in 2× SSC at RT for 15 min.

12. Flood slide with 2× SSC and coverslip.

13. Expose to 365 nm UV light for 30 min.

14. Rinse in distilled water and air-dry.

15. Place 50 μL exonuclease III solution onto slide, add coverslip, and incubate at RT for 10 min.

16. Rinse in distilled water and air-dry.

17. Denature chromosomes in 70% formamide, 2× SSC at 75°C for 2 min.

18. Dehydrate slides in ice-cold 75, 85, and 100% ethanol.

19. Air-dry.

20. Denature hybridization cocktail at 75°C for 5 min.

21. Apply 15–25 μL of hybridization cocktail to each slide and add coverslip.

22. Hybridization at 37°C for 3 h.

23. Posthybridization washes in 70% formamide/2× SSC at 32°C for 15 min (twice), then PN buffer at RT for 5 min.

24. Counterstain and mount in DAPI plus antifade.

25. Anti-BrdU controls/tests.

3.2. Incorporation of BrdU, Cell Collection, and Slide Preparation

Near confluent cell cultures are subcultured into fresh medium containing 5′-bromo-2′-deoxyuridine at a total final concentration of 1×10^{-5} M (cell type dependent; see Notes) and collected one cell cycle later (cell type dependent; ~24 h for human cells).

Colcemid (0.2 µg/ml, mouse; 0.1 µg/ml human) is added for the final 4 h to accumulate mitotic cells. Cells, a majority of which are now singly substituted with BrdU, are dislodged with trypsin and suspended in 75 mM KCl hypotonic solution at 37°C for 15 min before fixation in 3:1 methanol/acetic acid. Fixed cells are dropped onto cold, wet glass microscope slides and allowed to dry slowly in a humid environment. Slides are aged overnight in a drying oven or several days at room temperature.

Step by step:

1. Add BrdU to cell culture to achieve a final concentration of 10 µM.

2. Incubate the cells for one doubling time.

3. Add colcemid for the last 4 h of culture.

4. Collect and pellet cells (centrifuge at $210 \times g$ for 4–5 min).

5. Aspirate, break up pellet by the flicking tube and resuspend in 75 mM KCl (hypotonic); and leave at RT for 15 min.

6. Prefix the cells by adding 1 ml of 3:1 methanol:acetic acid and mix by flicking and/or inverting the tube.

7. Repellet the cells, aspirate supernatant, and fix cells in 3:1 methanol:acetic acid added dropwise with mixing.

8. Fix at RT for 10 min. Pellets can then be stored at –20°C for years.

9. Re-pellet cells, aspirate supernatant, and re-suspend in fresh 3:1 methanol:acetic acid at RT for 5 min. This step should be repeated at least twice.

10. Drop onto clean (cold and wet) microscope slides and allow to air-dry. Variation in slide drying time and cell density has a significant effect on the quality of the metaphase spreads and conditions must be optimized for each case. High-quality metaphase spreads are essential for robust, specific hybridization of the PNA telomere probe.

3.3. Slide Pretreatment

We find these steps, though not strictly required for adequate signal, significantly improve the signal strength and reduce non-specific probe binding. Prior to hybridization of the single-stranded PNA (or DNA, see Notes) telomere probe, slides are treated with 100 µg/ml RNase A for 10 min at 37°C (19), fixed in 4% formaldehyde at RT for 10 min, rinsed in PBS, and dehydrated through a cold ethanol series.

Step by step:

1. Incubate slides in RNAse A (100 µg/ml) at 37°C for 10 min, followed by 1× PBS rinse.

2. "Postfix" chromosomes in 4% formaldehyde at RT for 10 min, followed by one PBS rinse.

3. Dehydrate slides in 75, 85, and 100% ice-cold ethanol for 2 min each; allow slides to air-dry.

3.4. Preparation of Single-Stranded DNA/ Degradation of Newly Synthesized DNA Strands (CO-FISH)

Degradation of newly synthesized, BrdU-substituted strands is the foundation of the CO-FISH procedure. Staining with Hoechst 33258 sensitizes BrdU-substituted strands to induction of DNA single-strand breaks from exposure to UV light. Acting at the sites of these "nicks", enzymatic digestion with Exonuclease III preferentially removes the BrdU-substituted DNA strands.

Slides are stained with 0.5 µg/ml Hoechst 33258 in 2× SSC for 15 min at room temperature. Slides are then exposed to 365-nm UV light (Stratalinker 1800 UV irradiator) for 25–30 min. Enzymatic digestion of the BrdU-substituted DNA strands with 3 U/µL of Exonuclease III in buffer supplied by the manufacturer (50 mM Tris–HCl, 5 mM $MgCl_2$, and 5 mM dithiothreitol, pH 8.0) is allowed to proceed for 10 min at room temperature. An additional denaturation in 70% formamide, 30% 2× SSC at 70°C for 1–2 min can be performed, followed by dehydration in a cold ethanol series (70, 85, and 100%); allow slides to air-dry.

Step by step:

1. Stain slides in Hoechst 33258 at RT for 15 min. This step should be performed in the dark due to the light sensitivity of the Hoechst dye. Briefly rinse the slides in deionized distilled water and allow to air-dry.

2. Flood the slide with ~50 µl 2× SSC and apply coverslip. Expose slides to long-wave (~365 nm) UV light for 25–30 min (do not allow to dry out). Remove the coverslip, briefly rinse slides in distilled water and air-dry.

3. Place 50 µl of exonuclease III solution (3 U/µl) onto the slide and apply coverslip. Incubate at RT for 10 min. Remove the coverslip, briefly rinse slides in distilled water, and air-dry. Care should be taken at all times to avoid scratching slide.

3.5. Denaturation of Chromosomes and Hybridization

Denaturation of chromosomes is not strictly required because the DNA is single stranded. However, the addition of a denaturation step may be useful to compensate for any incomplete preparation of single-stranded DNA. Denaturation should be performed in 70% formamide/2× SSC at 70°C for 1–2 min, followed by dehydration in a cold ethanol series (70, 85, and 100%); allow slides to air-dry.

Step by step:

1. Denature chromosomes in 70% formamide, 30% 2× SSC solution for 1–2 min at 72°C. Note that adding one slide to a solution in a Coplin jar temporarily decreases the temperature by 1°C, adjust the waterbath temperature accordingly. Immediately dehydrate slides in an ice-cold ethanol series (75, 85, and 100% ethanol, 2 min each) and allow slides to air-dry.

2. Prepare hybridization cocktail consisting of 0.2 μg/ml fluorochrome-labeled PNA probe, 70% (v/v) formamide, 12 mM Tris–HCl pH 7.2, 5 mM KCl, 1 mM MgCl$_2$. Excessive light should be avoided from this point forward to prevent bleaching of probe fluorescence. Denature probe for 5 min at 72°C, and move immediately to ice for 5 min.

3. Add 25 μl of the denatured hybridization cocktail to each slide. Apply coverslip to slide, remove any bubbles that may have formed by applying slight pressure to the coverslip, and seal the edges with rubber cement. Place the slides in a dark, moist chamber for 3 h at 37°C.

3.6. Posthybridization Washes and Counterstaining

1. After hybridization, carefully remove rubber cement and coverslip. Wash slides in 70% formamide, 2× SSC solution for 15 min at 32°C with shaking, followed by a second wash in 2× SSC for an additional 15 min at 32°C with shaking.

2. Transfer slide to PN buffer for 5 min at RT.

3. Counterstain chromosomes by adding 18 μl of antifade with DAPI (Vectashield; 1.5 μg/mL; Vector Laboratories).

4. Apply coverslip and remove air bubbles through gentle pressure.

5. Immediate analysis of the slides is recommended as the probe signal and slide quality will deteriorate with time. However, signal should remain stable for a few weeks if stored in the dark at 4°C.

3.7. Anti-BrdU Controls/Tests

To confirm the one cell cycle in BrdU criterion has been met, and to confirm CO-FISH is working properly, we routinely perform and highly recommend the following anti-BrdU controls/tests.

3.7.1. Slides Not Processed for CO-FISH

1. Denature slides in 70% formamide, 2× SSC at 70°C for 2 min.

2. Dehydrate through ethanol series (75, 85, and 100%).

3. Hybridize ~40 μl of 1:100 dilution of anti-BrdU antibody in PN buffer; coverslip and incubate at 37°C for 30–45 min in moist chamber.

4. Rinse two times in PN buffer for 3–5 min each at RT.

5. Mount in Vectashield containing DAPI.

6. Cells that have been through a complete S-phase in BrdU will be very brightly stained on both chromatids, whereas a banding pattern indicates cells have replicated for a partial S-phase in BrdU and are not suitable for CO-FISH analysis.

3.7.2. Slides Processed for CO-FISH

1. After 2× SSC rinses (CO-FISH washes, above), apply anti-BrdU antibody as above. Do NOT let slides dry.

2. Rinse in PN buffer; mount in Vectashield + DAPI.

3. Dimly fluorescing chromosomes indicate that cells are in the first cycle and that the newly replicated DNA strands have been effectively removed. Second-cycle cells will have characteristic harlequin staining and are not suitable for CO-FISH analysis.

---

4. Notes

1. A probe to telomeric DNA can be prepared by synthesizing an oligomer having either the sequence $(TTAGGG)_7$ or $(CCCTAA)_7$ and labeled by terminal deoxynucleotidal transferase tailing (Boehringer Mannheim) with, for example, Cy3-dCTP according to the manufacturer's instructions (Amersham). A hybridization mixture containing 0.4 μg/ml probe DNA in 30% formamide and 2× SSC is applied to slides that have been prepared for CO-FISH. Following an overnight hybridization at 37°C in a moist chamber, slides are washed in 2× SSC at 42°C (five times, 15 min each), placed in PN buffer (Phosphate NP-40) at room temperature for 5 min and mounted in a glycerol solution containing 1 mg/ml of the antifade compound *p*-phenylenediamine HCl and 0.1 μg/ml DAPI (Vectashield; Vector Laboratories). Cells are examined with a microscope outfitted for fluorescence and image analysis.

2. The protocols presented above were developed for specific telomere probes. Other probes may have different melting temperatures and will require adjusting hybridization and wash temperatures accordingly.

3. It may be necessary to increase BrdU concentration to obtain satisfactory results in some transformed cell lines having defects in nucleotide metabolism. For example, when working with TK6 cells that have a mutation in the thymidine kinase gene, BrdU concentration had to be increased by tenfold.

4. Other than for ReD-FISH, the necessity for a single round of replication in BrdU is a strict requirement. When BrdU is

added to exponentially growing cultures, this requirement will not be met for all cells, thus the importance of the anti-BrdU tests. Whenever possible, we suggest allowing a culture to become confluent so the cells are synchronized in a G_0 state, then subculturing into fresh medium containing BrdU and blocking cell-cycle progression in metaphase with colcemid.

5. For irradiation studies, confluent cultures are held for 24 h after exposure before subculturing to ensure that cellular DNA is not bromo-substituted either at the time of irradiation or during the subsequent time interval when most DNA damage was repaired.

References

1. Goodwin, E. and Meyne, J. (1993) Strand-specific FISH reveals orientation of chromosome 18 alphoid DNA. *Cytogenet Cell Genet*, 63, 126–127.

2. Meyne, J. and Goodwin, E.H. (1995) Direction of DNA-sequences within chromatids determined using strand-specific fish. *Chromosome Res*, 3, 375–378.

3. Bailey, S.M., Goodwin, E.H., Meyne, J. and Cornforth, M.N. (1996) CO-FISH reveals inversions associated with isochromosome formation. *Mutagenesis*, 11, 139–144.

4. Bailey, S.M., Meyne, J., Cornforth, M.N., McConnell, T.S. and Goodwin, E.H. (1996) A new method for detecting pericentric inversions using COD-FISH. *Cytogenet Cell Genet*, 75, 248–253.

5. Goodwin, E.H., Meyne, J., Bailey, S.M. and Quigley, D. (1996) On the origin of lateral asymmetry. *Chromosoma*, 104, 345–347.

6. Bailey, S.M., Cornforth, M.N., Kurimasa, A., Chen, D.J. and Goodwin, E.H. (2001) Strand-specific postreplicative processing of mammalian telomeres. *Science*, 293, 2462–2465.

7. Bailey, S.M., Cornforth, M.N., Ullrich, R.L. and Goodwin, E.H. (2004) Dysfunctional mammalian telomeres join with DNA double-strand breaks. *DNA Repair (Amst)*, 3, 349–357.

8. Cornforth, M.N. and Eberle, R.L. (2001) Termini of human chromosomes display elevated rates of mitotic recombination. *Mutagenesis*, 16(1), 85–89.

9. Bailey, S.M., Brenneman, M.A. and Goodwin, E.H. (2004) Frequent recombination in telomeric DNA may extend the proliferative life of telomerase-negative cells. *Nucleic Acids Res*, 32, 3743–3751.

10. Cornforth, M.N., Eberle, R.L., Loucas, B.D., Fox, M.H. and Bailey, S.M. (2003) *Cold Spring Harbor Symposium: Telomeres and Telomerase*, Cold Spring Harbor, NY.

11. Zou, Y., Gryaznov, S.M., Shay, J.W., Wright, W.E. and Cornforth, M.N. (2004) Asynchronous replication timing of telomeres at opposite arms of mammalian chromosomes. *Proc Natl Acad Sci U S A*, 101, 12928–12933.

12. Williams, E.S., Klingler, R., Ponnaiya, B., Hardt, T., Schrock, E., Lees-Miller, S.P., Meek, K., Ullrich, R.L. and Bailey, S.M. (2009) Telomere dysfunction and DNA-PKcs deficiency: characterization and consequence. *Cancer Res*, 69, 2100–2107

13. Bailey, S.M., Goodwin, E.H. and Cornforth, M.N. (2004) Strand-specific fluorescence in situ hybridization: the CO-FISH family. *Cytogenet Genome Res*, 107, 14–17.

14. Bailey, S.M. and Cornforth, M.N. (2007) Telomeres and DNA double-strand breaks: ever the twain shall meet? *Cell Mol Life Sci*, 64, 2956–2964

15. Bailey, S.M. (2008) Telomeres and double-strand breaks – All's well that "Ends" well. *Radiat Res*, 169, 1–7.

16. Bechter, O.E., Zou, Y., Walker, W., Wright, W.E. and Shay, J.W. (2004) Telomeric recombination in mismatch repair deficient human colon cancer cells after telomerase inhibition. *Cancer Res*, 64, 3444–3451.

17. Londono-Vallejo, J.A., Der-Sarkissian, H., Cazes, L., Bacchetti, S. and Reddel, R.R. (2004) Alternative lengthening of telomeres is characterized by high rates of telomeric exchange. *Cancer Res*, 64, 2324–2327.

18. Blagoev, K.B. and Goodwin, E.H. (2008) Telomere exchange and asymmetric segregation of chromosomes can account for the unlimited proliferative potential of ALT cell populations. *DNA Repair (Amst)*, 7, 199–204.

19. Hayata, I. (1993) Removal of stainable cytoplasmic substances from cytogenetic slide preparations. *Biotechnic & Histochemistry*, 68, 150–152.

Chapter 13

Combinatorial Oligo FISH: Directed Labeling of Specific Genome Domains in Differentially Fixed Cell Material and Live Cells

Eberhard Schmitt, Jutta Schwarz-Finsterle, Stefan Stein,
Carmen Boxler, Patrick Müller, Andriy Mokhir, Roland Krämer,
Christoph Cremer, and Michael Hausmann

Abstract

With the improvement and completeness of genome databases, it has become possible to develop a novel fluorescence in situ hybridization (FISH) technique called COMBinatorial Oligo FISH (COMBO-FISH). In contrast to other (standard) FISH applications, COMBO-FISH makes use of a bioinformatic approach for probe set design. By means of computer genome database search, oligonucleotide stretches of typical lengths of 15–30 nucleotides are selected in such a way that they all colocalize within a given genome (gene) target. Typically, probe sets of about 20–40 stretches are designed within 50–250 kb, which is enough to get an increased fluorescence signal specifically highlighting the target from the background. Although "specific colocalization" is the only necessary condition for probe selection, i.e. the probes of different lengths can be composed of purines and pyrimidines, we additionally refined the design strategy restricting the probe sets to homopurine or homopyrimidine oligonucleotides so that depending on the probe orientation either double (requiring denaturation of the target double strand) or triple (omitting denaturation of the target strand) strand bonding of the probes is possible. The probes used for the protocols described below are DNA or PNA oligonucleotides, which can be synthesized by established automatized techniques. We describe different protocols that were successfully applied to label gene targets via double- or triple-strand bonding in fixed lymphocyte cell cultures, bone marrow smears, and formalin-fixed, paraffin-wax embedded tissue sections. In addition, we present a procedure of probe microinjection in living cells resulting in specific labeling when microscopically detected after fixation.

Key words: COMBO-FISH, Combinatorial oligo fluorescence in situ hybridization, FISH, Computer-based probe selection, DNA/PNA oligonucleotides, Lymphocyte hybridization, Tissue hybridization, Bone marrow smear hybridization, Microinjection of oligonucleotides, Microscopic detection of genome targets

Joanna M. Bridger and Emanuela V. Volpi (eds.), *Fluorescence in situ Hybridization (FISH):*
Protocols and Applications, Methods in Molecular Biology, vol. 659,
DOI 10.1007/978-1-60761-789-1_13, © Springer Science+Business Media, LLC 2010

1. Introduction

Fluorescence in situ hybridization (FISH) is an essential tool in many fields of biological and biomedical research and has become indispensable in routine medical diagnostics. FISH offers attractive possibilities to label DNA sequences from whole chromosomes to centromeres or telomeres and other subchromosomal regions to individual genes with high specificity and sensitivity for (multicolor) microscopic visualization. Specific FISH kits for the detection of numerical (e.g., gene copy number changes) or structural (e.g., translocations) chromosome or gene alterations are available for medical diagnostics, which can be applied to isolated cells in suspension, blood or bone marrow cell smears, and tissue sections (fresh, frozen, fixed, or paraffin-wax embedded). Usually, the DNA probes are derived from BAC-, Cosmid-, or YAC-clones, which are amplified by standard techniques. They have to be constructed by means of molecular biology, e.g., isolation, enzymatic cutting, and cloning of DNA from sorted metaphase chromosomes.

Although these sophisticated techniques of molecular biology have made nearly any genome target accessible as DNA probes, some shortcomings still exist. More precisely: (a) for medical diagnostics, procedures are required that can be easily standardized and automatized (1, 2); (b) many probes do not exactly map to their given genome target regions of diagnostic interest. Thus, for instance, small genes or translocation breakpoint regions of only a few kilo basepairs lengths cannot be labeled without simultaneous probing of neighboring target sites; (c) the design of probes directed to a certain genome target often requires complex molecular biology procedures; and (d) all current FISH protocols require a denaturation step of the native DNA double strand in the cell nucleus to bind the single-stranded DNA probe to the complementary target sequence. This denaturation step is performed by heat treatment (typically over 70°C) accompanied by an extensive use of chaotropic agents such as formamide (3, 4). Due to the requirement for denaturation, standard FISH procedures cannot be used in living cells (5).

On the basis of low temperature FISH procedures (6) omitting chaotropic agents and heat treatment for target denaturation, COMBinatorial Oligo FISH (COMBO-FISH) has been developed (7, 8) to specifically label genome targets exactly using by a beginning and an end nucleotide in the human DNA sequence database (in principle the procedure works for all other species for which a DNA database exists such as mouse). A set of 20–40 distinct oligonucleotide probes of about 15–30 bases, each localizing to a defined genomic region, is selected by computer search and can be synthesized as PNA or DNA probes according to automatic synthesis procedures. In this respect, the technique is easy to standardize and automatize (8, 9). It has also been shown that SMART probes (probes with a stem/loop conformation allowing fluorescence only by an open loop after probe to target binding) are appropriate for COMBO-FISH (10).

COMBO-FISH takes advantage of homopurine/homopyrimidine oligonucleotides that, depending on their 3′–5′ orientation, form double-strand (via the so-called Watson–Crick bonding) as well as triple-strand helices (via the so-called Hoogsteen bonding) with complementary homopurine/homopyrimidine sequences in intact genomic DNA. We will therefore distinguish between double-strand COMBO-FISH (DSCF) and triple-strand COMBO-FISH (TSCF). Each oligonucleotide probe is carrying at least one fluorochrome at one end. Due to the optical diffraction of a microscope lens, the fluorescence signals of the computer-designed oligonucleotide probe set merge into a homogeneous FISH "spot". If some of the oligonucleotide probes have additional binding sites forming clusters with other probes at targets elsewhere in the genome, they are excluded from a probe set. According to geometric imaging parameter (11), intensity barycenters of standard FISH labels and COMBO-FISH labels were found at comparable target positions (12).

COMBO-FISH works without the use of chaotropic agents, which results in a procedure that is less destructive for the chromatin morphology. In the case of Hoogsteen bonding, it can be performed without prior thermal denaturation of the DNA target sequence, too. This offers the principle possibility for DNA labeling in living cells with high specificity and without any modification of gene activity and expression (13). COMBO-FISH has successfully been applied to the analysis of the micro- and nanoarchitecture of the cell nucleus. Probe sets for *ABL* on chromosome 9 and *BCR* on chromosome 22 were applied to archival specimens from patients with hematological diseases such as chronic myelogenous leukemia (CML) (14). Improved measurements of the nanoarchitecture of the genome by spatially modulated illumination (SMI) microscopy (15) have become feasible using COMBO-FISH (16). So far, these experiments are more like feasibility studies. They, however, demonstrate the potential of the technique and may be extended to other applications.

2. Materials

In the following, we list all materials required for the different protocols of DSCF and TSCF presented here.

2.1. Standard Materials for All Protocols

1. 20× SSC – sodium saline citrate (Sigma-Aldrich).
2. HCl – hydrochloric acid (J.T. Baker).
3. 10× PBS – phosphate-buffered saline (Sigma-Aldrich).
4. Ethanol (Riedel De Haen) – store at –20°C for ice-cold application.
5. Formamide (Merck).
6. Vectashield antifade solution (Vector Laboratories).
7. Fixogum rubber cement (Marabu).
8. DNA or PNA oligonucleotide probes (see Subheading 3.1.2).

2.2. Materials for TSCF and DSCF with DNA Oligonucleotides (On Fixed Blood Cells)

1. Chromosome medium B (Biochrom).
2. Colcemid solution – concentration 10 µL/mL (Sigma-Aldrich).
3. KCl – potassium chloride (Merck).
4. Methanol/acetic acid (3:1) (Merck) – prepare fresh for each experiment.
5. 0.7% Triton X-100 (Merck).
6. 0.1% Saponin (Serva) – prepare freshly for each experiment.
7. RNase A (Sigma-Aldrich) – dilute 10 mg/mL in water with 10 mM Tris–HCl, pH 8.0 to 100 µg/mL; store at –20°C.
8. Pepsin (100 mg/mL ddH$_2$O) (Sigma-Aldrich) – prepare a stock solution consisting of 10% pepsin in 0.01 N HCl, pH 2.3; store at –20°C.
9. Hybridization buffer, pH 5.0, containing: 3 M sodium acetate (J.T. Baker), 3 M NaCl (J.T. Baker), and 3 M MgCl$_2$·6H$_2$O (Sigma-Aldrich).
10. Counterstain: TOPRO-3 (Invitrogen) or Yo-Yo (Invitrogen).

2.3. Additional Materials for TSCF with DNA Oligonucleotides (On Bone Marrow Smears)

1. Formaldehyde – prepare solution freshly from paraformaldehyde, dilute in highly purified water, pH 8.0.
2. Nonidet P40 substitute (Fluka).

2.4. Additional Materials for DSCF with PNA Oligonucleotides (On Formalin-Fixed, Paraffin-Wax Embedded Tissues)

1. Xylol (Merck).
2. Isopropanol (J.T. Baker).

2.5. Additional Materials for COMBO-FISH with PNA Nucleotides After Microinjection (Vital Lymphocytes)

1. Poly-L-lysine 0.01% (Sigma-Aldrich).

2. CD3-magnetic beads for specific cell separation (Miltenyi Biotec).

3. Cleaning solution containing: 50% H_2O/25% ammonia/25% isopropanol.

4. HEPES (20 mM) culture medium solution (Carl Roth).

5. Merckofix (Merck).

3. Methods

Like most standard FISH techniques, COMBO-FISH can be applied to any type of cells obtained from blood, tissues, bone marrow, or established permanent cell lines. In the next paragraph, we first describe the design and preparation of the probes, which is the same procedure for all cell types and species.

The following protocols differ in some steps depending whether the probes bind via Watson–Crick bonding or via Hoogsteen bonding. Here, we show protocols for fixed cells in suspension, which is exemplified by blood cells (routine material used in research). Such protocols may be also applied to other cell suspensions. Special attention is drawn on the applications that are useful in pathology either for research or diagnostic purposes, i.e. DSCF of air-dried bone marrow smears and tissue sections embedded in paraffin-wax after formalin fixation according to routine laboratory standards in pathology. Finally, we present an initial approach to label living lymphocytes by probe microinjection. This technique is so far still under further development but shows the potential of COMBO-FISH as a FISH technique for living cells.

3.1. COMBO-FISH Probes

3.1.1. DNA Database Search for TSCF and DSCF Probes

The design of sets of COMBO-FISH probes, which colocalize at a given locus within the genome, is accomplished in two principal parts (see Note 1).

The first part, namely the creation of a database with special properties for an efficient search for each probe set, has to be carried out only once for each genome or each new sequence release. For each contigue of the sequenced genome, two data files are created: an "annotation file" and a "PY – sequence file", which will be the basis for an effective search for suitable probe sets for any given genetic element. The annotation file contains all annotations such as gene names and the locations of reading frames within the nucleotide sequence of the respective contigue and also some additional information such as mRNA locations, coding regions, SNPs, etc. The PY – sequence file contains all sequences of consecutive homopurines or homopyrimidines of a total length of 15 nucleotides or more and their locations within the contigue. To create the annotation and PY – sequence files for all contigues of a genome, the gene bank files from NCBI (http://www.ncbi.nlm.nih.gov) are downloaded and processed. For the human genome e.g., there are 24 chromosomes described in files <hs_ref_chr1.gbk> through <hs_ref_chrY.gbk> (ftp://ftp.ncbi.nih.gov/genomes/H_sapiens). For each chromosome, the information in the gene bank file is analyzed by our program Crte_PoPY, and the contigue annotation and PY – sequence files are created between 1 (for chromosome 14) and 51 (for chromosome 1) files of different lengths. The automatic program needs several hours up to 1 day to perform this task on a PC, mainly depending on the input and output performance of the computer. The annotation files contain the same data as the annotation parts of the gene bank files. The PY – sequence files contain the homopurine sequences in the same FASTA – format, whereas the additional information, such as location within the contigue, is given as a sequence of numbers. The overall storage requirements are between 1 and 3% of the original sequence and annotation information of the gene bank file, depending on the specific organism.

The actual search for a set of homopurine (or homopyrimidine) probes, specifically colocalizing at a defined genetic element, is performed on the basis of the data files created in Part I and has to be performed for each genetic element separately. As purines are paired with corresponding pyrimidines on the reverse strand, the search for homopyrimidine oligonucleotides is equivalent to a search for homopurine oligonucleotides. Furthermore, from the structure of (reverse) Hoogsteen conformations, it is clear that triple-helix binding probes have the same sequence as the double-helical pairing partners, but in the reverse direction. Therefore, the search is performed for homopurines only, but both strand directions are considered. The search consists of four steps:

1. The program Where_Genes extracts the location within the genome of the respective genetic element from all contigue annotation files (see Note 2).

2. The program Basic_PoPY extracts the basic set of possible homopurine probes of a minimal length of 15 nucleotides from the PY – sequence file for the respective location (see Note 3).

3. The program Clean_PoPY searches for the occurrences of the probes in the basic set of the largest three chromosomes. Those probes, which occur more than 50 times on these chromosomes, are deleted from the basic set (see Notes 4 and 5).

4. The program Final_CFst outputs an optimal set of probes, which colocalize exclusively at the given location. The program Final_CFst searches for all occurrences in the genome of the reduced basic set and transfers all clusters of more than five probes within 250 kb into memory. In this step, one can distinguish between double- and triple-helical binding and use one or both for a selection criterion. In an iterative process, single probes that contribute to clusters are removed from the reduced basic set, and clusters are reinspected for their multiplicity of probes. This operation proceeds until no clusters with more than five stretches within 250 kb remain (see Note 6).

5. In fact, the program Final_CFst outputs all possible four lists of final COMBO-FISH sets: double- and triple-helical binding probe sets of homopurines and homopyrimidines (see Note 7).

3.1.2. Preparation of Oligonucleotide Probes

Although different probe types have been tested and successfully applied for COMBO-FISH, the best results concerning probe handling and hybridization efficiency were obtained by DNA or PNA oligonucleotides, which are commercially available or can be prepared according to standard techniques (see Notes 8 and 9). In the case of repetitive elements, it is often enough to prepare one specific oligonucleotide as a probe that binds at a high repetition rate. Typically, these probes carry a dye molecule at the 3' end of the sequence. The amount of fluorescence dye on the target is thus equal to the number of annealing probes.

In the case of small targets such as genes or breakpoint regions, this may amount only to some 20–40 dye molecules. Therefore, it is recommended to order such probes with a label at both ends 3'and 5' to further increase the fluorescence signal on the target.

3.2. Protocol for TSCF and DSCF with DNA Oligonucleotides (On Fixed Blood Cells)

The standard applications of COMBO-FISH are TSCF and DSCF with appropriately designed homopurine or homopyrimidine DNA oligonucleotide probes. In Fig. 1a–c, some typical results obtained for blood cell nuclei are presented. The *BCR* breakpoint region on chromosome 22 was labeled by TSCF (Fig. 1a, b) using a probe set of 31 stretches (14) (see Note 10), whereas the *TBX* gene region on chromosome 22 was labeled by DSCF (Fig. 1c) using a probe set of 30 stretches (10) (see Note 11). We also labeled the centromere of chromosome 9 using the repetitive but specific oligonucleotide AATCAACCCGAGTGCAAT (10), which contains both purines and pyrimidines, and therefore only allows the application of the DSCF protocol.

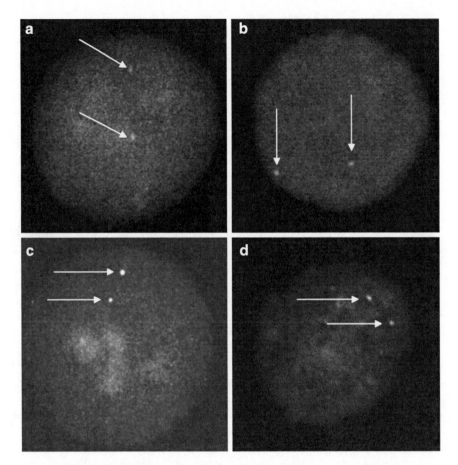

Fig. 1. (**a**, **b**) TSCF of the bcr breakpoint region (chromosome 22) in methanol/acetic acid fixed lymphocytes (see also ref. 14). The bcr region is labeled by a DNA oligonucleotide probe set of 31 probes carrying two fluorescence dye molecules each (TAMRA, *red*). The nuclei were counterstained by Yo-Yo. (**c**) DSCF of the gene region TBX (chromosome 22) in methanol/acetic acid fixed lymphocytes (see also ref. 10). The TBX region is labeled by a DNA oligonucleotide probe set of 30 stretches carrying two fluorescence dye molecules each (OregonGreen 488, *green*). The nuclei were not counterstained. (**d**) TSCF of the breakpoint region abl (chromosome 9) in a granulocyte of a bone marrow smear. The abl region (see also ref. 12) is labeled by a DNA oligonucleotide probe set of 31 probes carrying two fluorescence dye molecules each (OregonGreen 488, *green*). The nuclei were counterstained by Yo-Yo. In all figures (**a–d**), the labeled sites are highlighted by *arrows*.

3.2.1. Culturing and Pretreatment of Cells Isolated from Peripheral Blood

1. Defrost chromosome medium B and preheat to 37°C.

2. Isolate lymphocytes or all blood cells from of the fresh heparinized blood and prepare according to standard conditions.

3. Place 50 mL of preheated chromosome medium B in three culture flasks.

4. Place 8 mL of lymphocytes/blood cells into the three culture flasks.

5. Incubate culture flasks at 37°C for 72 h (turn the cap once so that it remains loose).

6. Add 200 µL of demecolchin solution (Colcemid; concentration 10 µg/mL) to each flask after 70-h incubation time.

7. Incubate flasks further at 37°C for 2 h.

8. Preheat 100 mL of 75 mM KCl solution to 37°C.

9. Place the cells in 4 × 50 mL centrifuge tubes.

10. Centrifuge at 200 × g for 10 min at room temperature (RT).

11. Reduce supernatant to 4–5 mL.

12. Agitate the pellet, but do not vortex the pellet.

13. Add 25 mL of preheated KCl solution.

14. Incubate at 37°C for 25 min.

15. Centrifuge at 200 × g for 10 min at RT.

16. Reduce the supernatant to 4–5 mL.

17. Agitate the pellet, but do not vortex the pellet.

18. Carefully drop (drop by drop) ice-cold methanol/acetic acid with the Pasteur pipette, agitate slowly during the process, and fill up to 30 mL.

19. Incubate for 1 h at RT.

20. Centrifuge at 170 × g for 10 min at RT.

21. Reduce the supernatant to 4–5 mL.

22. Agitate the pellet well, but do not vortex the pellet.

23. Add freshly prepared methanol/acetic acid under agitation with the Pasteur pipette up to 20–25 mL.

24. Incubate for 1 h at RT.

25. Centrifuge at 170 × g for 10 min at RT.

26. Reduce the supernatant to 4–5 mL.

27. Agitate the pellet well, but do not vortex the pellet.

28. Add freshly prepared methanol/acetic acid under agitation with the Pasteur pipette up to 20 mL.

29. Incubate for 1 h at RT.

30. Centrifuge at 170 × g for 10 min at RT.

31. Reduce the supernatant to 4–5 mL (pellet should be white or light colored).

32. Wash with 10 mL freshly prepared methanol/acetic acid three times.

33. Centrifuge at 170 × g for 10 min at RT.

34. Resuspend the pellet in 2–5 mL methanol/acetic acid.

35. Store cell suspension at 4°C until required.

3.2.2. Specimen Preparation on Slides

1. Drop cell suspensions of lymphocytes or blood cells on to slides.

2. Incubate the slides in 0.7% Triton X-100, 0.1% Saponin in 2× SSC for 30 min at RT for permeabilization of the nuclear membrane.

3. Rinse in 2× SSC for 5 min twice.

4. Incubate in 200 µL RNase A at 37°C for 1 h (see Note 12).

5. Wash three times in 2× SSC for 3 min each.

6. Equilibrate the slides in 1× PBS for 5 min.

7. Treat with 15 µL pepsin diluted in 50 mL 0.01 N HCl at 37°C for 1–3 min (see Note 12).

8. Stop the reaction with 1× PBS for 5 min.

9. Dehydrate using an ethanol series (70, 80, and 95%) for 3 min each.

10. Air-dry.

11. Only for DSCF:

    (a) Denaturate at 75°C in 70% formamide/2× SSC (pH 7.0–7.2) for 5 min.

    (b) Wash using an ethanol series (ice cold; 70, 90, and 100%) for 3 min each.

3.2.3. In Situ Hybridization and Washing

1. Add 30 µL hybridization mixture consisting of 188 ng probe DNA (8 µL) in equal amounts of all oligonucleotide probes.

2. Drop 22 µL hybridization buffer on to the slide, cover with a coverslip, and seal with Fixogum rubber cement.

3. Incubate in a humidified chamber at 37–42°C for 24 h.

4. Wash in 2× SSC at 37°C for 5 min (see Note 13).

5. Only in the case of counterstaining:

    (a) Dilute TOPRO-3 or Yo-Yo 1:1,000.

    (b) Incubate the slides in the dye solution for 30–45 s.

6. Mount the slides in 15 µL Vectashield antifade solution, cover with a coverslip, and seal with Fixogum rubber cement.

3.3. Protocol for TSCF with DNA Oligonucleotides (On Bone Marrow Smears)

We used standard bone marrow smears air-dried on to slides for routine pathology. In Fig. 1d, an example of such a cell is shown.

3.3.1. Slide Preparation

1. Fix bone marrow smear before starting with the hybridization process with 1% formaldehyde (freshly prepared from paraformaldehyde) for 10 min at RT.

2. Incubate in 1% Nonidet/2× SSC for 30 min at RT for permeabilization of the nuclear membrane.

3. Rinse in 2× SSC for 5 min twice.

4. Incubate in 100 µL RNase A at 37°C for 30 min (see Note 12).

5. Wash three times in 2× SSC for 3 min each.

6. Expose to 10 mM HCl for 5 min.

7. Treat with 15 μL pepsin diluted in 50 mL 0.01 N HCl at 37°C for 1–3 min (see Note 12).

8. Stop the reaction with 1× PBS for 5 min.

9. Wash in 2× SSC for 5 min.

10. Fix again with 4% formaldehyde (freshly prepared from para-formaldehyde) for 10 min at RT.

11. Wash twice in 2× SSC for 5 min each.

12. Dehydrate by an ethanol series (70, 80, and 95%) for 2 min each.

13. Air-dry.

3.3.2. In Situ Hybridization and Washing

1. Add 30 μL hybridization mixture consisting of 188 ng probe DNA (8 μL) in equal amounts of all oligonucleotide probes.

2. Drop 22 μL hybridization buffer on to the slide, cover with a coverslip, and seal with Fixogum rubber cement.

3. Incubate in a humidified chamber at 37°C for 24 h.

4. Wash in 4× SSC and 200 μL Tween 20 for 5 min.

5. Mount the slides in 15 μL Vectashield antifade solution, cover with a coverslip, and seal with Fixogum rubber cement.

3.4. Protocol for DSCF with PNA Oligonucleotides (On Formalin-Fixed, Paraffin-Wax Embedded Tissues)

In the case of formalin-fixed paraffin-wax embedded tissues, the application of DNA probes appeared to be highly loaded by background of nonspecifically attached probe material in the tissue. This could be overcome by the application of PNA oligonucleotide probes. Figure 2 shows an example where the repetitive centromere 9 oligonucleotide probe was applied.

3.4.1. Pretreatment of Paraffin-Wax Embedded Tissue Sections

1. Remove paraffin by incubation in 100% Xylol for 15 min twice at RT.

2. Wash with an isopropanol ethanol series (100% Isopropanol, 96% EtOH, and 70% EtOH) for 5 min each at RT.

3. Wash in 1× PBS for 5 min at RT.

4. Incubation in 100 mL citrate buffer in a microwave oven: 1 min preheating, 3–5 min 290 W, 1–3 min 465 W (see Note 14).

5. Wash in 1× PBS for 2 min at RT.

6. Denature in 50% formamide/2× SSC for 1 min at RT.

7. Denature in 70% formamide/2× SSC for 15 min at 75°C (see Note 15).

8. Wash in 1× PBS for 5 min at RT.

9. Dehydrate by an ethanol series (70, 80, and 95%) for 3 min each.

10. Air-dry.

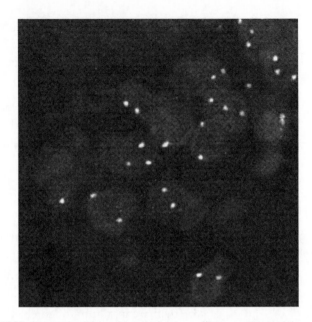

Fig. 2. DSCF of centromere 9 in a formalin-fixed, paraffin-wax embedded tissue (breast carcinoma) section of 8 µm thickness. The centromere is labeled by the repetition of the PNA oligonucleotide probe AATCAACCCGAGTGCAAT carrying a fluorescent dye molecule (TAMRA) at the N-terminus.

3.4.2. In Situ Hybridization and Washing

1. Add 0.5 µL PNA probe (~300 ng) and 29.5 µL hybridization buffer.
2. Cover with a coverslip, seal with Fixogum rubber cement, and incubate in a humidified chamber at 37°C for 24 h.
3. Wash in 2× SSC (pH 7.0) for 5 min at RT.
4. Wash in 1× SSC for 5 min at 65°C.
5. Wash in 1× PBS for 5 min.
6. Mount the slides in 15 µL Vectashield antifade solution, cover with a coverslip, and seal with Fixogum rubber cement.

3.5. Protocol for COMBO-FISH with PNA Nucleotides After Microinjection (Live Lymphocytes)

As a proof of principle of vital COMBO-FISH in lymphocytes, a PNA oligonucleotide probe was microinjected that uniquely exists as a repetitive sequence in the centromere region of chromosome 9. In Fig. 3, a typical result is shown. The nuclei were not counterstained. Very little background is visible although a high gain of the detection system has been chosen to detect the nuclear boundaries. Although the microscopic detection has been done under fixed cell conditions to detect the low fluorescence and the cell nucleus during long time image acquisition, it should be emphasized that neither denaturation nor harsh chemical treatment has been used.

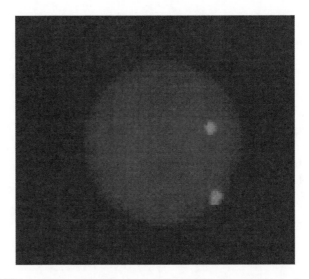

Fig. 3. DSCF of centromere 9 after microinjection of PNA oligonucleotide probes in vital lymphocytes and Merckofix fixation after 24 h further cell cultivation The repetitive probes AATCAACCCGAGTGCAAT are carrying fluorescent dye molecules (TAMRA) at the N-terminus. The cell nuclei were not counterstained.

3.5.1. Prepreparation of Coverslips for Cell Growth and Microinjection

1. Agitate the coverslips in the cleaning solution for 5 min.
2. Remove the cleaning solution by washing five times with deionized and autoclaved H_2O for 3 min each.
3. Sterilize coverslips at 200°C for 2 h.
4. Continue working under laminar flow:
   (a) Put the coverslips into Petri dishes.
   (b) Add 100 μL of 0.01% poly-L-lysine solution and incubate for 3 h at RT.
   (c) Take out and dry coverslips (see Note 16).

3.5.2. Lymphocyte Preparation (Under Laminar Flow)

1. Separate and purify T-lymphocytes from peripheral blood according to a standard protocol (see Note 17).
2. Loosen lymphocytes by tapping the culture flask.
3. Transfer cell suspension to a 15-mL Falcon tube.
4. Centrifuge at $200 \times g$ for 10 min at RT.
5. Reduce the supernatant of the pellet.
6. Resuspend the pellet in 10 mL 1× PBS (preheated to 37°C).
7. Incubate for 5 min.
8. Centrifuge at $200 \times g$ for 10 min at RT.
9. Remove the supernatant.
10. Resuspend the pellet into 20 mM HEPES – culture medium to a final concentration of 500,000 cells/mL.

11. Drop 0.1 mL of the cell suspension on the poly-L-lysine-treated coverslips in Petri dishes.

12. Incubate coverslips for 3 h at 37°C under 5% CO_2 atmosphere.

13. Add 4 mL of 20 mM HEPES – culture medium.

3.5.3. Microinjection of PNA Oligonucleotide Probes and Postmicroinjection Treatment

1. Dilute the PNA oligonucleotide probe in buffer for microinjection to a concentration of 67 µM.

2. Inject probe solution with 50 hPa for 0.5 s in each cell (compensation pressure about 30 hPa) (see Notes 18 and 19).

3. Replace the HEPES – culture medium by standard (HEPES free) cell culture medium (see Note 20).

4. Incubate the cells for 24 h at 37°C under 5% CO_2 atmosphere in the dark and identify dead (apoptotic) cells by standard procedures.

5. Remove the culture medium.

6. Wash with 1× PBS for 30 s.

7. Fix the cells with Merckofix for 3–10 min.

8. Wash twice with 1× PBS for 7.5 min each.

9. Wash with 2× SSC for 1 min.

10. Mount the slides in 15 µL Vectashield antifade solution, cover with a coverslip, and seal with Fixogum rubber cement.

4. Notes

1. Since the data search is of principal importance for the technique, it is described in detail. The applicant, however, who is interested in specific probe sets, but does not have the bioinformatics background for database search, may contact the authors.

2. In step 1, all annotation files are scanned for information concerning the genetic element in question. It should be noted that in special cases different names of genes are used and one has to specify several variants. The location of the genetic element in its specific contigue is written to a file and used for input to step 2, which can follow automatically or after inspection and eventual change of the location, e.g. by narrowing or extending the range. Usually, at least about 50 kb is necessary to allow a suitable number of homopurine probes to be found. It should be noted that location information can also be written to a file without referring to Where_Genes, if this information is available from other sources and if this file has the same format as the Where_Genes output file.

3. There is an option to select all possible probes or only those which are more favorable concerning their binding kinetics. The file containing this set of basic homopurine probes is used by program Clean_PoPY in step 3.

4. In our experience, the frequent probes do not only contribute to an increased background noise as they bind to many locations in the genome, but they also form large clusters of repetitive or pseudorepetitive sequences. Additional information on the binding locations of the most frequent probes is given in an extra output file and can be utilized in editing the files for the basic set and the reduced basic set of probes so that further candidate sequences can be deleted or introduced.

5. At this stage of the search, it may also be convenient to inspect the positions of these probes within the targeted region to analyze its local structural properties for clusters and repetitions or to locate the probes on exons or introns, etc. Reruns of programs Basic_PoPY and Clean_PoPY can be combined with manual editing until a satisfying reduced basic set has been defined.

6. This iterative removal is based on several heuristics. In a last step, all removed probes are reinspected and added to the remaining set again one by one to see, if their inclusion would again produce clusters. If not, they are included to the final set.

7. It can take several hours to complete the searches, since the final cluster is normally user inspected for the total binding frequency of its probes within the whole genome. For additional clusters within longer nucleotide sequences, or further criteria for an optimal probe set. Therefore, step 4 can be repeated with changed reduced basic sets as input, until a satisfying final set has been constructed. However, the programs can also be run completely automatically in series by a suitable shell script. In our experience, user interference does not improve the resulting outcome over automatical generation, unless additional requirements like observation of binding properties as mentioned above are introduced into the searching criteria.

8. DNA oligonucleotides are commercially available in high quality from several companies. In our experiments, we used the probes manufactured by IBA GmbH, Göttingen.

9. PNA sequences were prepared by solid phase synthesis (17) with an Expedite 8909 PNA/DNA synthesizer according to the manufacturer's protocol for 2 μM synthesis using commercially available PNA building blocks (Applied Biosystems). The fluorophores (TAMRA or Coumarin 343) were linked to

the N-terminus of the strand. After filtering and drying, HPLC purification was performed at 49°C on a Shimadzu liquid chromatograph.

10. *BCR* as well as *ABL* are famous breakpoint regions involved in the formation of the Philadelphia chromosome (a reciprocal translocation t(9;22)), a small chromosome that is strongly correlated to the induction of leukemia like CML. The Philadelphia chromosome is the result of a molecular rearrangement between the abl proto-oncogene on chromosome 9 and the *BCR* breakpoint cluster region on chromosome 22 leading to a fusion region that is responsible for the production of 220 kDa protein.

11. The gene *TBX-1* is involved in different tumors, where it shows changes in the gene copy number, especially deletions (18). It is also involved in cardiac diseases such as the DiGeorge syndrome.

12. RNase and pepsin treatment should be tested on a separate slide, because the concentration and incubation time critically depend on the specimen.

13. Unspecific binding can be reduced by lower concentrations of salt and higher temperature.

14. Treatment in the microwave oven has to be done under continuous observation. The exposure time is usually depending on the position of the slide in the oven and the quality of the tissue.

15. Optional pronase treatment is possible. The concentration and incubation time is depending on the specimen. We had good experience with Pronase E 0.05% in 1× PBS at 37°C for 1 min.

16. Closed Petri dishes with poly-L-lysine can be stored for a few days at RT.

17. We recommend the cell separation with magnetic beads coupled to CD3 antibodies directly. The procedure can be done according to the manufacturer's description.

18. Microinjection can be performed at RT without CO_2 atmosphere using a standard microinjection setup mounted on an inverted microscope. To avoid cell death, the whole procedure should not take more than 30 min for one coverslip.

19. Microinjection was performed with an AIS I system, a completely automated microinjection system for adherent cells. It was possible to inject approximately 200 cells in 30 min.

20. HEPES – culture medium is only used to stabilize the pH.

Acknowledgment

The financial support of the German Federal Minister of Education and Research (BMBF) is gratefully acknowledged (grant no. 01IG07015G (Services@MediGrid) and grant no. 13N8350).

References

1. Hausmann, M. and Cremer, C. (2003) Standardisation of FISH-procedures: summary of the first discussion workshop. *Anal. Cell. Pathol.* **25**, 201–205.

2. Hausmann, M., Cremer, C., Linares-Cruz, G., Nebe, T.C., Peters, K., Plesch, A., Tham, J., Vetter, M. and Werner, M. (2004) Standardisation of FISH-procedures: summary of the second discussion workshop. *Cell. Oncol.* **26**, 119–124.

3. Rauch, J., Wolf, D., Hausmann, M. and Cremer, C. (2000) The influence of formamide on thermal denaturation profiles of DNA and metaphase chromosomes in suspensions. *Z. Naturforsch. C* **55**, 737–746.

4. Solovei, I., Walter, J., Cremer, M., Habermann, F., Schermelleh, L. and Cremer, T. (2001). FISH on three-dimensionally preserved nuclei, in *FISH: a practical approach* (Squire, J., Beatty, B. and Mai, S., eds.) Oxford University Press, Oxford.

5. Tanke, H.J., Dirks, R.W. and Raap, T. (2005) FISH and immunocytochemistry: toward visualising single target molecules in living cells. *Curr. Opin. Biotechnol.* **16**, 49–54.

6. Winkler, R., Perner, B., Rapp, A., Durm, M., Cremer, C., Greulich, K.O. and Hausmann, M. (2003) Labelling quality and chromosome morphology after low temperature FISH analysed by scanning far-field and scanning near-field optical microscopy. *J. Microsc.* **209**, 23–33.

7. Hausmann, M., Winkler, R., Hildenbrand, G., Finsterle, J., Weisel, A., Rapp, A., Schmitt, E., Janz, S. and Cremer, C. (2003) COMBO-FISH: specific labelling of nondenatured chromatin targets by computer-selected DNA oligonucleotide probe combinations. *Biotechniques* **35**, 564–577.

8. Cremer, C., Hausmann, M. and Cremer, T. Markierung von Nukleinsäuren mit speziellen Probengemischen. Patent DE 198 06 962.6 Deutsches Patentamt München (5.8.2004).

9. Schmitt, E., Finsterle, J., Stein, S. and Hausmann, M. (2005) Focussed COMBO-FISH for selected nanosized genomic regions. *Cell Prolif.* **38**, 182.

10. Nolte, O., Müller, M., Häfner, B., Knemeyer, J.-P., Stöhr, K., Wolfrum, J., Hakenbeck, R., Denapaite, D., Schwarz-Finsterle, J., Stein, S., Schmitt, E., Cremer, C., Herten, D.-P., Hausmann, M. and Sauer, M. (2006) Novel singly labelled probes for identification of microorganisms, detection of antibiotic resistance genes and mutations, and tumor diagnosis (SMART PROBES), in *Biophotonics* (Popp, J. and Strehle, M., eds.) Wiley-VCH, Weinheim, pp. 167–230.

11. Wiech, T., Timme, S., Riede, F., Stein, S., Schuricke, M., Cremer, C., Werner, M., Hausmann, M. and Walch, A. (2005) Archival tissues provide a valuable source for the analyis of spatial genome organisation. *Histochem. Cell. Biol.* **123**, 229–238.

12. Schwarz-Finsterle, J., Stein, S., Großmann, C., Schmitt, E., Trakhtenbrot, L., Rechavi, G., Amariglio, N., Cremer, C. and Hausmann, M. (2006/2007) Comparison of triplehelical COMBO-FISH and standard FISH by means of quantitative microscopic image analysis of abl/bcr genome organisation. *J. Biochem. Biophys. Methods* **70**, 397–406.

13. Hausmann, M., Stein, S., Kaya, Z., Finsterle, J., Schmitt, E., Krämer, R. and Cremer, C. (2005) COMBO-FISH of living cells. *Cell Prolif.* **38**, 182.

14. Schwarz-Finsterle, J., Stein, S., Großmann, C., Schmitt, E., Schneider, H., Trakhtenbrot, L., Rechavi, G., Amariglio, N., Cremer, C. and Hausmann, M. (2005) COMBO-FISH for focussed fluorescence labelling of gene domains: 3D-analysis of the genome architecture of abl and bcr in human blood cells. *Cell Biol. Int.* **29**, 1038–1046.

15. Hildenbrand, G., Rapp, A., Spöri, U., Wagner, C., Cremer, C. and Hausmann, M. (2005)

Nano-sizing of specific gene domains in intact human cell nuclei by Spatially Modulated Illumination (SMI) light microscopy. *Biophys. J.* **88**, 4312–4318.

16. Hausmann, M., Hildenbrand, G., Schwarz-Finsterle, J., Spöri, U., Schneider, H., Schmitt, E. and Cremer, C. (2005) New technologies measure genome domains – high resolution microscopy and novel labeling procedures enable 3-D studies of the functional architecture of gene domains in cell nuclei. *Biophotonics Int.* **12**(**10**), 34–37.

17. Mayfield, L.D. and Corey D. R. (1999) Automated synthesis of peptide nucleic acids and peptide nucleic acid – peptide conjugates. *Anal. Biochem.* **268**, 401–404.

18. Albrecht, B., Hausmann, M., Zitzelsberger, H., Stein, H., Siewert, J.R., Hopt, U., Langer, R., Höfler, H., Werner, M. and Walch, A. (2004) Array-based comparative genomic hybridization for the detection of DNA sequence copy number changes in Barrett's adenocarcinoma. *J. Pathol.* **203**, 780–788.

Chapter 14

Simultaneous Visualization of FISH Signals and Bromo-deoxyuridine Incorporation by Formamide-Free DNA Denaturation

Daniela Moralli and Zoia L. Monaco

Abstract

The replication timing of different DNA sequences in the mammalian cell nucleus is a tightly regulated system, which affects important cellular processes such as genes expression, chromatin epigenetic marking, and maintenance of chromosome structure. For this reason, it is important to study the replication properties of specific sequences, to determine for example, if the replication timing varies in different tissues, or in the presence of specific reagents, such as hormones, or other biologically active molecules. In this chapter, we present a technique, which allows identification of specific DNA sequences by fluorescence in situ hybridization (FISH) and simultaneously analyses the incorporation of a thymidine analogue, 5-bromo-2-deoxyuridine (BrdU), to mark DNA replication. First, tissue culture cells are synchronized at the beginning of the S-phase. BrdU is then added, either at specific time-points during S-phase or during the whole of the cell cycle. After harvesting the cells, the chromosomal DNA is hybridized to FISH probes that identify specific DNA sequences; this is performed without the teratogen formamide normally used in FISH. Finally, the cell preparations are analysed with an epifluorescence microscope to determine if the sequence of interest incorporates BrdU and in which point of the S-phase.

Key words: BrdU, DNA replication, FISH

1. Introduction

The replication of the DNA in mammalian cells does not occur in a linear timeline along the whole genome length. Gene-rich regions are generally replicated earlier than gene poor areas (1), and differences in tissue specific gene expression are mirrored in differential replication timing of the corresponding sequences (2). Chromatin packaging affects differential replication, by

Joanna M. Bridger and Emanuela V. Volpi (eds.), *Fluorescence in situ Hybridization (FISH): Protocols and Applications*, Methods in Molecular Biology, vol. 659,
DOI 10.1007/978-1-60761-789-1_14, © Springer Science+Business Media, LLC 2010

making specific DNA sequences more easily accessible to the DNA replication machinery. Thus, epigenetic factors, such as histone modifications, can act as regulators of this process (3). Moreover, the resolution of problems in DNA replication, such as stalled replication forks, has been linked to the formation of gross chromosomal rearrangements, especially in the presence of inverted repeat sequences (4). Several human diseases are thought to be linked to the incorrect replication timing of specific genes (5, 6).

Therefore, the analysis of DNA replication timing on cytological preparations can provide an important tool for the characterization of complex cellular processes that can have clinical relevance. Such approaches are based on the incorporation in tissue cultured cells of 5-bromo-2-deoxyuridine (BrdU), an analogue of thymidine, for short periods of time (pulses), so that it will only be incorporated into the DNA that is replicating during that time. Thus, using specific FISH probes, the replication timing of specific loci of interest can be determined. Other cellular processes, such as the formation of sister chromatid exchanges (SCE) by mitotic recombination, can be analysed by incorporating BrdU for two more complete replication cycles, so that the two chromatids will be labelled differentially (7).

The protocol presented in this chapter details how to detect simultaneously the BrdU and the FISH signals. Since the technique does not involve formamide for denaturation of the target DNA and in post-hybridization washes, it represents a safer alternative to classical FISH techniques. In our experience, it proves to be reliable and fast when compared to alternative methods (8).

2. Materials

2.1. Cell Culture and Cell Harvesting

2.1.1. Cell Lines

The protocol is suitable for cells that grow in adhesion, such as human fibrosarcoma cell line HT1080 or murine cell lines LA-9 and STO.

2.1.2. Equipment and Consumables

1. Laminar flow cabinet for sterile tissue culture.
2. CO_2 thermostatic incubator.
3. Inverted microscope, equipped for phase contrast microscopy.
4. Bench top centrifuge, with swing-out rotor.
5. Electric pipet-aid filler/dispenser.
6. Sterile tissue culture dishes, flasks, and serological pipettes (Nunc).
7. Micropipettes and sterile micropipette tips (Gilson).
8. Burker cell counter.
9. Super premium microscope slides (VWR International).

<table>
<tr><td>

2.1.3. Tissue Culture Media and Solutions

</td><td>

1. Phosphate buffer saline (PBS) without calcium and magnesium (KCl 2 mM, KH_2PO_4 1.5 mM, NaCl 137 mM, Na_2HPO_4 8 mM), dissolved in distilled water, and sterilized by autoclaving. Store at room temperature.

2. Dulbecco's Modified Eagle Medium, with high glucose, GlutaMAX™, and Sodium Pyruvate (DMEM-Glutamax, Invitrogen Ltd), supplemented with 10% foetal bovine serum (FBS) (Sigma-Aldrich) and penicillin–streptomycin solution (Sigma-Aldrich) 1× (corresponding to 1,000 U/mL penicillin and 0.1 mg/mL streptomycin). From here, we refer to this solution as "complete medium". Store at 4°C.

3. Trypsin with Versene (BioWhittaker Lonza) 10×. Store at –20°C. To prepare the working solution, dilute to 1× (Trypsin 0.25%) in sterile PBS. Store at 4°C.

4. KaryoMAX Colcemid Solution in PBS (Invitrogen Ltd). Store at 4°C.

5. Aphidicolin (Sigma-Aldrich) dissolved in dimethyl sulphoxide (DMSO) (Sigma-Aldrich) at 1 mg/mL (see Note 1). Store at –20°C.

6. BrdU (Sigma-Aldrich) is dissolved in PBS at a final concentration of 3.22 M (corresponding to 1 mg/mL), sterilized by filtration through a 0.2-μm filter, and stored at –20°C (see Note 2).

7. Hypotonic solution: KCl 75 mM. Store at room temperature.

8. Fixative solution: 25% glacial acetic acid, 75% methanol (see Note 3). Cool it down to –20°C.

9. Silica gel granules, self-indicating (VWR International).

</td></tr>
</table>

2.2. DNA Probe Labelling

2.2.1. Equipment and Consumables

1. Microcentrifuge.
2. Micropipettes and sterile micropipette tips (Gilson).
3. Refrigerated thermostatic waterbath (Grant Instruments) (see Note 4).

2.2.2. Buffers and Solutions

1. DNA, 1 μg (see Note 5).
2. 0.5-mL microcentrifuge tubes (various suppliers, e.g. Eppendorf).
3. Nick Translation System (Invitrogen Ltd).
4. Biotin-16-2′-deoxyuridine-5′-triphosphate (Bio-16-dUTP) (Roche Diagnostic Ltd).
5. Digoxigenin-11-2′-deoxyuridine-5′-triphosphate (Dig-11-dUTP), alkali-stable (Roche Diagnostic Ltd).
6. $CH_3CH_2NH_4$ (ammonium acetate) 7.5 M.

7. Ethanol.

8. Human C_0t-1 DNA (Roche Diagnostic Ltd).

9. Dextran blue (Sigma-Aldrich), 1.8% in distilled water. Sterilize by filtration and store at 4°C.

10. Sheared Salmon sperm DNA (Sigma-Aldrich).

11. Formamide, especially purified for biochemistry (BDH, VWR International) (see Note 6). Store at −20°C.

12. Dextran sulphate 50% in distilled water. Sterilize by filtration. Store at room temperature.

13. 100× Denhardt's solution: Dissolve in distilled water 1% bovine serum albumin (BSA) fraction V, 1% Ficoll 4000000, 1% polyvinylpyrrolidone 40000. Store at −20°C.

14. 20% sodium dodecyl sulphate (SDS) in water (see Note 7).

15. 20× SSC buffer: 3 M NaCl, 0.3 M HOC(COONa) $(CH_2COONa)_2 \cdot 2H_2O$ (trisodium citrate, dihydrate) in distilled water. Adjust the pH to 7.0 with 1 M citric acid or HCl (see Note 8). Store at room temperature.

2.3. FISH and Signal Detection

2.3.1. Equipment and Consumables

1. Microcentrifuge.

2. Micropipettes and sterile micropipette tips.

3. PCR Thermocycler.

4. Thermostated incubator.

5. Microscope, equipped for epifluorescence.

6. Wheaton 8-slide glass Coplin jars.

7. Airtight plastic box.

8. 24 mm × 50 mm glass coverslips (VWR International).

9. Parafilm (VWR International).

2.3.2. Buffers and Solutions

1. Denaturation buffer: 10 mM Tris–HCl pH 8.0, 50 mM KCl, 5% glycerol. Prepare fresh when needed.

2. Post-hybridization washing buffer: 0.1× SSC. Store at room temperature.

3. Sheep antidigoxigenin, rhodamine conjugated (Roche Diagnostic Ltd) (see Note 9).

4. Avidin, Alexa Fluor® 488 conjugate (Invitrogen Ltd) (see Note 10).

5. Sheep anti-BrdU antibody (ab1893 Abcam) (see Note 9).

6. Mouse anti-BrdU antibody clone BU-33 (Sigma-Aldrich) (see Note 9).

7. Anti-sheep antibody, rhodamine conjugated (Jackson Immunoresearch, Stratech).

8. Biotinylated anti-avidin D antibody (Vector Laboratories).

9. Anti-mouse antibody, Alexa Fluor® 488 conjugate (Invitrogen Ltd) (see Note 10).

10. Diluent buffer: 4× SSC, 0.1% Tween 20, 3% BSA.

11. Wash buffer A: 4× SSC, 0.1% Tween 20.

12. Wash buffer B: 4× SSC.

13. Counterstaining solution: 4′,6-diamidino-2-phenylindole dihydrochloride (DAPI) (Sigma-Aldrich) 10 ng/mL in 4× SSC. Protect from light (see Note 11).

14. Antifade solution: Dissolve 0.233 g of 1,4-diazabicyclo(2.2.2) octane (DABCO) (Sigma-Aldrich) in 1 mL of 10× PBS. Add 9 mL of glycerol, molecular biology grade (Sigma-Aldrich) and resuspend thoroughly. Store at 4°C, protected from light (see Note 12).

3. Methods

3.1. Cell Synchronization by Aphidicolin Block

Unless otherwise stated, the following manipulation should be carried out in a laminar flow cabinet in sterile conditions.

For the study of DNA replication, it is useful to synchronize the cell population (see Note 13).

1. The day before the start of the experiment, check the cell density in the tissue culture dish, using an inverted microscope. If the cell population is fully confluent, remove the spent medium by careful aspiration. Add 5 mL of sterile PBS, warmed at 37°C, to wash away any leftover medium, and remove. Add 5 mL of sterile trypsin solution 1×, warmed at 37°C (see Notes 14 and 15). Leave it for 20 s, gently swirling the solution around the dish, then remove completely. Put the dish in a 37°C incubator for 5 min. Using an inverted microscope, check that all the cells have detached from the tissue culture dish (see Note 15), then resuspend in 10 mL of complete medium, warmed at 37°C, and move to a 15-mL conical centrifuge tube.

2. Count the cell number using a Burker cell counter (see Note 16). Centrifuge the cell suspension at $300 \times g$ for 10 min and resuspend in an appropriate volume of complete medium to have approximately 10^6 cells/mL (see Note 17).

3. Plate 2×10^6 cells in a 10-cm tissue culture dish, so that the cells are at approximately 60% confluence (see Note 17), and place in the CO_2 thermostatic incubator for at least 2 h to allow the cells to reattach themselves.

4. Add aphidicolin at a final concentration of 3 μg/mL (see Note 18) and incubate for 24 h in the CO_2 thermostatic incubator (see Note 19).

3.2. Block Release and BrdU Incorporation

Unless otherwise stated, the following manipulation should be carried out in a laminar flow cabinet, in sterile conditions.

1. To release the cells from the DNA synthesis block, remove the medium containing aphidicolin (see Note 20) and replace it with 10 mL of complete medium, warmed to 37°C. Gently swish the medium around the dish, then remove it, and replace with 10 mL of complete medium. Repeat twice, for a total of three washes, then add complete medium.

2. To analyse a specific time-point in the S-phase (Fig. 1), place the cells back in the CO_2 thermostatic incubator. After 3 h (early S), 6 h (mid-S), or 8 h (late S), add BrdU to the cells at a final concentration of 40 μM (see Note 21). Incubate for 15 min in the CO_2 incubator at 37°C, then harvest the cells.

3. To analyse the replication of the whole genome (Fig. 1), add BrdU to the cells, at a final concentration of 40 μM, and place back in the CO_2 incubator at 37°C. Leave for 24–48 h before harvesting (see Note 22).

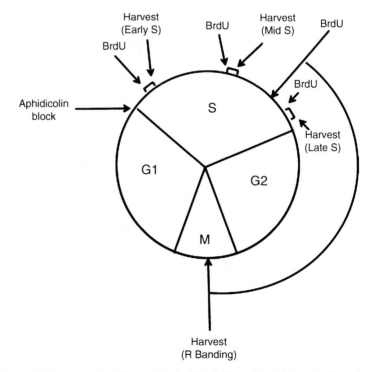

Fig. 1. BrdU incorporation: Time-points at which BrdU can be added, post-release from synchronization.

4. Two hours before harvesting, add Colcemid at 0.03 μg/mL and place the cells back in the incubator.

3.3. Cell Harvesting

1. In sterile conditions, aspirate the medium from the treated cells (see Note 23) and detach the cells by gentle trypsinization. Collect the cells in complete medium in a 15-mL conical centrifuge tube.

2. Pellet the cells by centrifugation at $300 \times g$ for 10 min.

3. Using a disposable Pasteur pipette, carefully remove all the supernatant (see Note 24) and then resuspend the cell pellet by flicking the tube.

4. From here onwards, it is no longer necessary to work under sterile conditions. Add 10 mL of hypotonic solution and gently mix the solution by pipetting up and down. Incubate at room temperature for 6–10 min (see Note 25).

5. Centrifuge the cells and remove the supernatant as indicated above.

6. Gently resuspend the cell pellet and quickly add about 500 μL of cold fixative solution (see Note 26). Mix by pipetting up and down, then take the total volume up to 10 mL with fresh cold fixative. Incubate for 15 min at room temperature.

7. Centrifuge as above and replace the supernatant with 10 mL of fresh fixative. Incubate for 15 min.

8. Centrifuge as before, then remove all the supernatant, and replace with 500 μL to 1 mL of fresh fixative.

9. Drop 50–100 μL of cell suspension onto a slide, cleaned in absolute ethanol, and allow to air-dry (see Note 27).

10. Check the slide quality using a phase contrast microscope. The nuclei or chromosomes should appear dark grey and not reflective. No cytoplasm should be visible (see Note 28).

11. Add to the remaining cell suspension 500 μL of fresh fixative and store at −20°C (see Note 29).

3.4. Probe Labelling

1. Mix the following reagents in a 0.5-mL microcentrifuge tube: 1-μg DNA, 5-μL dNTP mix minus dTTP, 5-μL DNAseI/polymerase mix (see Note 30), either 1-μL Dig-11-dUTP (Roche), corresponding to a final concentration of 20 μM, or 1-μL Bio-16-dUTP (Roche), corresponding to a final concentration of 20 μM (see Note 31), sterile distilled water to make up the final volume to 50 μL (provided by the kit).

2. Incubate at 15°C in a waterbath (see Note 4) for a minimum of 2 h up to 16 h.

3. Add 5 μL of Stop buffer, 5 μL of dextran blue solution (see Note 32) and 10× carrier DNA (see Note 33).

4. Precipitate the labelled DNA by adding half a volume ((reaction mix + Stop buffer + dextran blue + carrier DNA)/2) of ammonium acetate 7.5 M, mix well by inverting the tube several times, and two volumes of ethanol ((reaction mix + Stop buffer + dextran blue + carrier DNA + ammonium acetate) × 2).

5. Incubate at room temperature for 15 min, then centrifuge at $10,000 \times g$ for 10 min. Meanwhile, prepare the hybridization mix: 50% formamide, 10% dextran sulphate, 1× Denhardt's, 0.1% SDS, 2× SSC (see Note 34).

6. Discard the supernatant and resuspend in hybridization mix at a final probe concentration of 10 ng/μL (see Note 35). Vigorous vortexing will ensure that the pellet is fully resuspended.

3.5. FISH and Signal Detection

1. Set the thermocycler at 95°C.

2. Place the slides, cells side up, on the thermocycler plate. Leave 20 s, then add 150 μL of denaturation mix, and cover with a coverslip. Close the thermocycler lid and incubate for 10 min at 95°C (see Note 36).

3. Meanwhile, denature the labelled probe by incubating at 80°C for 10 min. Place the probe on ice once the denaturation is completed.

4. Remove the slides from the thermocycler. Allow to cool down for a few seconds, then dip them in 0.1× SSC, in a Coplin jar, to remove the coverslips.

5. Dehydrate the cells by incubating the slides in 70, 90, and 100% ethanol in Coplin jars at room temperature for 2 min each. Finally, air-dry them.

6. Place the slides flat on an even surface, cells side up. Add 15 μL of denatured probe in the centre of each slide, then cover with a coverslip, trying to avoid trapping any air bubbles.

7. Keeping the slides flat, put them in an airtight box, with absorbent paper drenched in water at the bottom (see Note 37).

8. Incubate overnight at 37–42°C (see Note 38).

9. Place three Coplin jars containing the post-hybridization wash buffer in a waterbath and heat to 65°C (see Note 39).

10. Remove the slides from the box and place in the first Coplin jar (see Note 40). Shake gently the slides, so that all coverslips are washed off. Remove the coverslip from the jar. Incubate for 5 min after all coverslips have fallen off.

11. Transfer the slides to the second Coplin jar, and then to the third one, and incubate for 5 min each at 65°C (see Note 40).

12. Transfer the slides to a jar containing wash buffer A at room temperature. Incubate for 5 min.

13. Dilute the primary antibodies: For detection of BrdU and biotinylated probes, sheep anti-BrdU 1:100+avidin, Alexa Fluor® 488 1:200 in diluent buffer; for detection of BrdU and digoxigenin probes, mouse anti-BrdU 1:100+sheep antidigoxigenin rhodaminated 1:50 in diluent buffer. Keep on ice, and prepare fresh each time.

14. Keeping the slides flat, put them in an airtight box, with absorbent paper drenched in water on the bottom. Apply 60–100 μL of primary antibody dilution, and cover with Parafilm cut to the same size of the slide (see Note 41), and incubate for 30 min at 37°C (see Note 42), protected from light.

15. Remove the Parafilm with tweezers, and place the slides in wash buffer A, at 42°C for 5 min (see Note 43). Repeat twice for a total of three washes, keeping the slides protected from light.

16. Dilute the secondary antibodies: For detection of BrdU and biotinylated probes, anti-sheep antibody, rhodamine conjugated 1:100+biotinylated anti-avidin D antibody 1:100 in diluent buffer; for detection of BrdU and digoxigenin probes, anti-mouse antibody, Alexa Fluor® 488 conjugate 1:100+anti-sheep antibody, rhodamine conjugated 1:100 in diluent buffer. Keep on ice, and prepare fresh each time.

17. Apply the antibodies and incubate as indicated above in step 14.

18. Wash the slides as indicated in step 16. Then, for detection of BrdU and biotinylated probes go the next step, whereas for detection of BrdU and digoxigenin probes go to step 22.

19. For detection of BrdU and biotinylated probes, dilute avidin, Alexa Fluor® 488 1:200 in diluent buffer.

20. Apply the avidin solution and incubate as indicated above in step 14.

21. Wash the slides as indicated in step 16.

22. Transfer the slides to wash buffer B, and incubate for 5 min at room temperature, protected from light (see Note 44).

23. Transfer the slides to the staining solution, and incubate at room temperature for 6 min, protected from light.

24. Wash the slides in wash buffer B for a few seconds.

25. Place the slides flat on an even surface, cells side up. Add 20 μL DABCO mounting medium then cover with a coverslip, trying to avoid trapping any air bubbles (see Note 45).

26. Put the slides in a cardboard slide holder, and keep at 4°C, protected from light.

3.6. Slide Analysis   The cells are analysed with a microscope equipped for epifluorescence to determine if the sequence of interest have incorporated BrdU by checking if there is colocalization between the BrdU and FISH signals.

4. Notes

1. Aphidicolin is a powerful inhibitor of DNA polymerase and suitable protective measures should be used. It is not soluble in water, but it can be dissolved in methanol (at 10 mg/mL) or ethanol (1 mg/mL). In our experience, however, it is best to dissolve aphidicolin in DMSO, because the solution is stable for longer at –20°C, especially if frequent freezing and thawing is avoided. It is not necessary to filter-sterilize the solution. Keep protected from light by wrapping the tube in foil.

2. BrdU is a powerful mutagenic agent. Use protective clothing and gloves when manipulating the solution. It is best to buy small amounts of powder, and resuspend the whole of it, to avoid weighing it. Keep protected from light by wrapping the tube in foil.

3. The fixative solution works by replacing the water in the cells. To guarantee optimal fixation, it is important that both acetic acid and methanol are anhydrous. Keep the stock bottles well closed to prevent them from absorbing humidity from the environment. Prepare the fixative solution immediately before use, and discard any amount not used at the end of the procedure. Methanol is toxic, so suitable protective measures should be used.

4. If a refrigerated thermostatic waterbath is not available, a standard waterbath kept in the cold room (4–8°C) at the minimum setting will maintain the necessary temperature.

5. DNA can be prepared from a variety of sources with a variety of methods. For labelling purposes, it is important that the DNA is not contaminated by proteins or organic agents, such as phenol or chloroform, which will reduce the efficiency of the nick translation reaction. This can be ascertained by reading the DNA OD (optical density) with a spectrophotometer at 230, 260, and 280 nm. High-quality DNA should have an OD_{260}/OD_{280} ratio close to 1.8. Lower values indicate that the DNA sample is contaminated by proteins. The OD_{260}/OD_{230} ratio should be close to 2. If lower, it indicates the presence of contaminating organic reagents.

6. It is advisable to use formamide especially purified for biochemistry. Standard formamide solutions contain a mixture of highly reactive molecules, which can damage the chromosome structure. Formamide is a known teratogen, and is a toxic substance, so it should be used in a chemical hood, and suitable protective clothing should be worn.

7. To help dissolve SDS use a magnetic hotplate stirrer, set at around 40°C. SDS easily precipitates when the temperature drops below 20°C, and in high salt concentration. It can be dissolved again when the solution is heated.

8. It is not necessary to sterilize 20× SSC. The high salt content will prevent most bacteria and moulds from growing. Moreover, after autoclaving, it can sometime form a precipitate. Discard the solution if this occurs.

9. The antidigoxigenin antibody is only available from Roche, either in unconjugated, POD (horseradish peroxidase) fluorescein or rhodamine conjugated forms, and is raised in sheep. This can generate problems when it is necessary to detect other molecules, with antibodies raised in the same host. Such is the case of BrdU: in our experience, the best available antibody is Abcam sheep anti-BrdU antibody (ab1893). It reliably gives a strong, good signal. However, alternatives are available, such as the mouse anti-BrdU antibody clone BU-33, from Sigma-Aldrich.

10. In our hands, Alexa Fluor conjugates generally give stronger signals, with improved background.

11. DAPI binds to DNA, and as such it is a potential mutagenic compound. Care should be taken when counterstaining. It does not easily dissolve in water. To help dissolve the powder, add methanol to the solution at 20% final concentration.

12. Fluorochromes are sensitive to the quenching action of oxidative processes. To preserve their fluorescence for the analysis, slides are mounted in antifading solution, which usually contains antioxidant reagents, such as DABCO and large percentages of glycerol, to reduce the effect of atmospheric oxygen. The DABCO antifading solution should be clear. If it has a straw yellow colour, then it should be discarded.

13. In any population of actively growing cells, the single cells will be in different stages of the cell cycle. Hence, when it is necessary to analyse complex processes, such as DNA replication, it is useful to synchronize the cells by increasing the number of cells in the cell cycle phase of interest. There are a variety of cell synchronization protocols, and we here present a technique that is efficient and fast.

14. To passage a cell line that is adherent, it is necessary to remove the protein bridges that keep the cells attached to each other and to the tissue culture dish. This can be done by treatment with trypsin, a broad-spectrum protease.

15. The time necessary to detach cells by trypsin treatment varies depending on the cell line. For example, primary fibroblast cells need to be exposed to the protease for a longer period. However, avoid incubations of more than 15 min, as this will results in a high proportion of cell death.

16. A Burker cell counter consists of a glass slide with an etched grid, whose dimensions are known. A small volume of cell suspension (10–15 μL) is applied to the Burker slide and then covered with a coverslip. This forms a small chamber, whose volume generally corresponds to $1/10^4$ of a millilitre (different commercially available models may however differ). Thus, by counting the number of cells in the grid and multiplying by 10^4, we can then estimate how many cells are present on average in 1 mL of cell suspension. Hence, to correctly evaluate the total cell number, it is important that the cells are well resuspended, and no cell clumps are present. It is also important to count the cells as soon as possible after the trypsin treatment. If left too long after resuspension in complete medium, the cells will tend to attach to each other, forming cell clumps.

17. The total number of cells per dish depends, of course, on the cell line employed. When setting up a synchronization experiment, it is important to seed an appropriate number of cells. If the starting number is too low, there will not be enough cells to analyse a statistically significant sample. However, if the plating cell density is too high, the cell population will become confluent, and stop dividing due to contact inhibition. It might be necessary to set up titration experiments for different cell lines.

18. We found this dosage to be optimal for cell lines such as human HT1080 and murine LA-9 or STO. Different cell lines might require higher or lower concentrations of aphidicolin.

19. In the presence of aphidicolin, an inhibitor of DNA polymerase, the cells enter the S-phase but are not able to complete DNA replication. So all the cycling cells will eventually stop at the beginning of the S-phase. In our experience, 24-h incubation is sufficient to guarantee that the majority of the cells are synchronized. However, other cell lines might require a longer exposure. Moreover, the synchronization procedure is never fully efficient. Even in a synchronized population, there will always be a fraction of cells, which is not in the same cell

cycle stage as the rest. Finally, the synchronization will be maintained for a short period of time only. Generally, after two cell divisions, the cells return to an unsynchronized status.

20. Aphidicolin is a toxic substance. Check your local regulations on how to dispose of the aphidicolin-containing medium.

21. Usually, when characterizing the replication timing of specific DNA sequences, it is desirable to analyse specific time-points during the S-phase. Note that it is necessary to prepare a dish of synchronized cells for each of the time-points required. To this aim, BrdU is incorporated for a short time (usually referred to as a pulse) at regular time post-release, and the cells are then immediately harvested. To correctly identify each of the S-phase stages (early S, mid S, and late S), it is necessary to know how long the whole S-phase lasts for the specific cell line analysed. The times indicated here can be used as starting point, but they should be determined empirically for each new cell type.

22. It might be of interest to incorporate BrdU in the whole cellular DNA, for one or two replication cycles. This is the case when we want to visualize SCEs. BrdU is added immediately after release, and the cells are allowed to keep growing until they have replicated twice. To obtain a chromosome banding R pattern, the BrdU is added towards the end of the S-phase (Fig. 1). The cells are then allowed to complete the S- and G2-phases, and Colcemid is added to arrest the cells at metaphase. Then the cells are harvested as indicated.

23. BrdU is a mutagenic compound. Check your local regulations on how to dispose of the BrdU-containing medium.

24. Never discard the supernatant by pouring it off the tube. Doing so usually leads to the loss of the top layer of the cell pellet, which contains mostly cells in mitosis/metaphase.

25. In a hypotonic solution, the cells will absorb water, thus leading to the cell swelling. When swollen, the cells are quite fragile and can burst if left for too long in hypotonic solution. The length of incubation required varies depending on the cell line.

26. The fixative solution denatures the cellular proteins and replaces the water present in the cytoplasm. If too much fixative is added to the cell pellet, it will lead to the formation of cell aggregates that are difficult to break apart. The aggregates are usually impervious to DNA denaturation and so cannot be analysed. The formation of aggregates can be reduced by adding a small amount of fixative and then immediately resuspending the pellet.

27. To obtain optimal fixation, the slides should be dried off as quickly as possible. This can be achieved by placing them on a dry block heater kept at 37°C. Avoid higher temperatures as this can sometime lead to the corrugation of the cell layer on top of the slide.

28. The presence of visible cytoplasm surrounding the nuclei is usually an indication of poor fixation. This can be a result of an insufficient incubation time in fixative solution or due to some remaining hypotonic solution when the fixative solution was added for the first time. The cell suspension can be slightly improved by incubating for a further 15 min in fresh fixative, but the results of the successive experiments are generally sub-optimal. Good quality slides can be stored up to 2 weeks at room temperature, protected from dust and light, if the cells contain BrdU. For longer storage, place the slides at 4°C in a box with hygroscopic salts (silica gel, self-indicating) and seal the box with tape.

29. The fixed cell suspension is stable at –20°C for years, and new slides can be prepared as need occurs. Over very long periods (more than 5 years), the acetic acid in the fixative will depurinate the DNA, thus making it unrecognizable to DNA probes.

30. For reliability, we label our probes by nick translation using a commercial kit (Invitrogen Nick Translation System) following the manufacturer's instruction. Several other kits are available from various suppliers.

31. In our experience, the DNA labelling efficiency is the same with either Dig-11d-UTP or Bio-16-dUTP.

32. Dextran blue is an inert sugar that will precipitate along with the DNA, thus making the pellet blue and easily identifiable. It can interfere with enzymatic reactions, so it should not be used when the DNA probe has to be further manipulated.

33. To enhance the precipitation of the labelled DNA (whose size is around 300 bp), it is necessary to add an unlabelled carrier DNA. We normally use sheared Salmon sperm DNA. However, if the labelled DNA contains repetitive elements that could cross-hybridize with specific targets, it is advisable to add C_0t-1 DNA as carrier. This corresponds to the highly repetitive fraction of the genomic DNA (e.g. satellite sequences, interspersed repeats, etc.), and once the probe is denatured, will rehybridize to the corresponding sequences, making them unavailable for hybridization to the target DNA. Note that C_0t-1 DNA is species-specific, and so for example, murine C_0t-1 should be used when hybridizing murine probes to murine targets.

34. The hybridization mix can be prepared in large batches. It is stable for years, when aliquoted and stored at –20°C.

35. Once the probe is resuspended in hybridization mix, it is stable at –20°C for about a year. For longer storage, it is advisable to resuspend the probe in TE buffer (10 mM Tris–HCl, 1 mM EDTA, pH 8.0).

36. This denaturation mix makes no use of formamide. As such, it is safer to use. Moreover, it reliably gives a good FISH signal even when the quality of cells preparations on the slides is poor for the presence of leftover cytoplasm. The chromosome morphology is usually well preserved, but if it is necessary to karyotype the cells, it is advisable to reduce the denaturation time to 5 min.

37. The paper drenched in water is necessary to keep a high humidity level in the box to prevent the slides from drying out. However, the slides should not be placed directly on top of the paper, but rather on stands (such as glass lid of Coplin jars), to prevent water being transferred from the paper to the slides by capillarity phenomena.

38. A hybridization temperature of 37°C is generally ideal for the majority of probes. When it is necessary to increase the stringency, to ensure high specific binding of the probes, the hybridization temperature should be of 42°C.

39. Glass jar are sensitive to brusque temperature changes. Place the jars with the buffer in the cold waterbath and then increase the temperature. Doing otherwise might crack the jars.

40. These washes are at high stringency. The very low salt content of the post-hybridization wash buffer and the high temperature ensure that all of the probe not bound to a specific site is fully removed. Thus, it is important that the buffer temperature stays at 65°C upon addition of the slides, or the post-hybridization washes might not be stringent enough. This is why it is preferable to have three independent jars with buffer at the right temperature to move the slides sequentially from one to another.

41. A minimum of 60 µL is required to guarantee that the whole surface of the slide is homogenously covered by the primary antibody mix. Using Parafilm instead of glass coverslips reduces the formation of air bubble. Parafilm cannot be used for the hybridization step because of the presence of formamide in the hybridization buffer.

42. Incubation times longer than 30 min at 37°C increase the overall background staining of the cells. For more specific binding of the antibodies, an overnight incubation at 4°C might be useful. In our experience, however, it is not necessary with the antibodies we use.

43. Wash buffer A has a very high NaCl content. This ensures simultaneously that no DNA probe is removed from its target and that the antibody binding is highly specific.

44. This step is necessary to remove any traces of the Tween 20 detergent from the slides, as it interferes with the mounting medium.

45. It is possible to add DAPI directly to the DABCO antifading solution at 2 ng/μL. This makes steps 23 and 24 redundant.

References

1. Folle GA. (2008) Nuclear architecture, chromosome domains and genetic damage. *Mutat. Res.* **658**, 172–183.

2. Watanabe Y, Shibata K, Ikemura T, Maekawa M. (2008) Replication timing of extremely large genes on human chromosomes 11q and 21q. *Gene* **421**, 74–80.

3. Kemp MG, Ghosh M, Liu G, Leffak M. (2005). The histone deacetylase inhibitor trichostatin A alters the pattern of DNA replication origin activity in human cells. *Nucleic Acid Res.* **33**, 325–336.

4. Voineagu I, Narayanan V, Lobachev KS, Mirkin SM. (2008). Replication stalling at unstable inverted repeats: Interplay between DNA hairpins and fork stabilizing proteins. Proc Natl Acad Sci USA **105**, 9936–9941.

5. Barbosa AC, Otto PA, Vianna-Morgante AM. (2000) Replication timing of homologous alpha satellite DNA in Roberts syndrome. *Chromosome Res.* **8**, 645–650.

6. D'Antoni S, Mattina T, Di Mare P, et al. (2004) Altered replication timing of the HIRS/Tuple1 locus in the DiGeorge and Velocardiofacial syndromes. *Gene* **333**, 111–119.

7. Shiraishi Y. (1990) Nature and role of high sister chromatid exchanges in Bloom syndrome cells. Some cytogenetic and immunological aspects. *Cancer Genet. Cytogenet.* **50**, 175–187.

8. Moralli D, Monaco ZL. Simultaneous detection of FISH signals and bromo-deoxyuridine incorporation in fixed tissue cultured cells. *PLOS ONE* 2009, 4(2): e4483. doi:10.1371/journal.pone.0004483

Chapter 15

CryoFISH: Fluorescence In Situ Hybridization on Ultrathin Cryosections

Sheila Q. Xie, Liron-Mark Lavitas, and Ana Pombo

Abstract

The visualization of cellular structures and components has become an invaluable tool in biological and medical sciences. Imaging subcellular compartments and single molecules within a cell has prompted the development of a wide range of sample preparation techniques as well as various microscope devices to obtain images with increased spatial resolution. Here, we present cryoFISH, a method for fluorescence in situ hybridization (FISH) on thin (~150 nm thick) cryosections from sucrose-embedded fixed cells or tissues. CryoFISH can be used in combination with immunodetection (IF) of other cellular components. The main advantages of cryoFISH and cryoIF over whole-cell labeling methods are increased spatial resolution with confocal microscopy, greater sensitivity of detection due to increased probe accessibility, and better image contrast. CryoFISH and cryoIF methods typically used on samples fixed in conditions that preserve ultrastructure, are compatible with the labeling of cells in their tissue context and are ideal for correlative studies that compare fluorescence with electron microscopy.

Key words: Imaging, Immunofluorescence, Fluorescence in-situ hybridization, Interchromosomal interactions, Tokuyasu cryosections, RNA polymerase II

1. Introduction

Immunofluorescence (IF) and Fluorescence in situ hybridization (FISH) are powerful tools to visualize the spatial relationships between proteins and nucleic acids within subcellular compartments. Conventional laser scanning confocal microscopes provide at best a resolution of ~200 nm in the x- and y-axes and $\geq$500 nm in the z-axis (1). Subcellular structures often have smaller dimensions than the resolution afforded by confocal microscopy and appear larger (at least 200 nm) and more elongated (>500 nm) in the resulting images than their real dimensions, decreasing the

Joanna M. Bridger and Emanuela V. Volpi (eds.), *Fluorescence in situ Hybridization (FISH):*
Protocols and Applications, Methods in Molecular Biology, vol. 659,
DOI 10.1007/978-1-60761-789-1_15, © Springer Science+Business Media, LLC 2010

contrast between objects in the same optical sections (2, 3). For example, nucleosomes are 10 nm, ribosomes are ~20 nm, and transcription factories are ~45–75 nm, as determined by electron microscopy. To take advantage of the ease and speed of labeling and data collection by confocal microscopy while achieving higher spatial resolution, image contrast, and ultrastructural preservation, we have developed labeling strategies to be used on thin (i.e., 100–200 nm) physical sections prepared according to the Tokuyasu procedure (4, 5). In brief, cells are optimally fixed (as for electron microscopy), embedded in a cryoprotectant (high concentration sucrose), frozen in liquid nitrogen, before thin sections are cut on dry glass or diamond knives, collected on a sucrose drop, transferred to a coverslip, and the sucrose is washed off prior to labeling. Combining confocal microscopy with cryosectioning provides a significant increase in resolution, such that structures close to each other can be resolved as discrete bright sites against a higher contrast background, improving overall quantitative analyses (6). Besides better spatial resolution, cryosectioning offers additional advantages. Whereas intact three-dimensional (3D) cells fixed under robust conditions often resist penetration of DNA probes, whole antibody molecules, and colloidal gold conjugates, cryosections are highly accessible to large probes (e.g., IgM antibodies), especially when treated with a mild detergent prior to IF/FISH (6). Imaging thin cryosections by fluorescence microscopy minimizes chromatic aberration (different focusing positions of emitted light of different wavelengths), as optimal z positions can be set for each channel yielding images of highest contrast at optimal focusing positions (6). CryoIF and cryoFISH are also ideal approaches for correlative studies between light and electron microscopy (6–10).

CryoFISH was first used in our laboratory to investigate the extent of intermingling between chromosome territories in human cells and the positioning of single loci or transcription factories relative to chromosome territories (7). With cryoFISH, we could measure a significant amount of chromosome intermingling and found that transcription factories are present to an equal extent at the interior and interface between chromosomes. The more recent application of cryoFISH to study the association of specific genes with transcription factories in combination with labeling of CTs has also allowed three-way correlations between the positioning of a locus relative to its CT and to transcription factories, at high spatial resolution and nuclear preservation (10). Recently, cryoFISH was also successfully applied to measure, at the single-cell level, the frequency of long-range chromatin interactions derived from studies using 4C (chromosome conformation capture–on-chip) (11, 12). Whereas 4C measures DNA interactions in whole cell populations, high-resolution FISH represents an essential complementary approach to this biochemical assay for the assessment of long-range DNA interactions on a single-cell basis. In comparison with 3D-FISH, cryoFISH offers a higher spatial resolution and improved ultrastructural preservation, minimizing the risk of capturing false-positive chromatin interactions, and providing a more unbiased quantitative approach to measure chromatin interactions, as only closely positioned loci will colocalize within thin sections.

In this chapter, we describe representative protocols for cryoFISH in combination with antigen immunodetection. We combine labeling of sites containing the active form of RNA polymerase II marked with an IgM antibody against phosphorylated Serine2 residues (S2p pol II), with labeling of genomic regions, both single alleles (CD2 locus), or whole chromosomes within the cell nucleus (Fig. 1).

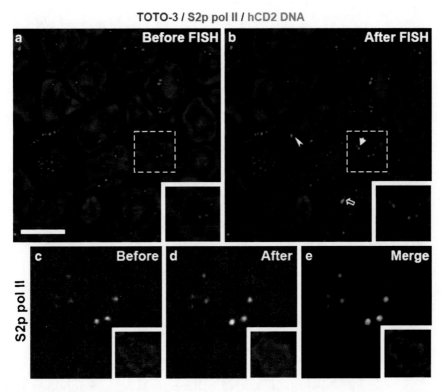

Fig. 1. Localization of hCD2 transgenes and transcription factories in thymic sections from hCD2 transgenic mice bearing ~15 tandem transgene copies. Ultrathin cryosections (~150 nm thick) from murine thymus were immunolabelled with H5 antibody against phosphorylated Serine2 residues to mark active RNA polymerase II (S2p pol II) complexes in transcription factories and images collected on a confocal microscope [*red*; (**a**)]. After imaging, the same samples were hybridized with DIG-labeled hCD2 probe that recognizes hCD2 transgenes [*green*; (**b**)], counterstaining DNA with TOTO-3 (*blue*) and new images of S2p pol II sites collected [*red*; (**b**)]. Images of active pol II sites collected before (**a**, **c**) and after (**b**, **d**) FISH yields identical patterns of S2p pol II distribution [*merge*; (**e**)]. *Arrow* indicates hCD2 transgenes associated with transcription factories, *open arrow* and *arrowhead* show hCD2 loci adjacent to and separate from factories, respectively (**b**). Bar; 2 μm.

2. Materials

2.1. Fixation of Cells Grown in Suspension, Attached to Plastic or Whole Tissues

1. Phosphate buffered saline (PBS) is 10 mM Na_2HPO_4/NaH_2PO_4, 137 mM NaCl, 2.7 mM KCl at pH 7.4 and stored at 4°C.

2. 0.5 M HEPES–NaOH is made from HEPES (Sigma) dissolved in distilled water (VWR), and pH is adjusted to 7.6.

3. Paraformaldehyde (PFA, Sigma-Aldrich) is dissolved in distilled water at 16% (w/v), depolymerized, pH is adjusted to 7.6 with 1M NaOH, and stored in aliquots at –20°C (see Note 1).

4. Petri dish (VWR), cell scrapers (Fisher), and scalpel.

2.2. Embedding and Cryosectioning

1. Sucrose (Molecular Biology Grade) is dissolved with PBS at 2.1 M (can be stored at –20°C) for embedding to prevent biological ice-crystal damage (see Note 2).

2. 1.5 mL nonstick hydrophobic Eppendorf tube (Fisher); wooden cocktail sticks; centrifuge; light dissecting microscope (Zeiss); liquid nitrogen, sample stud (Leica) to support cell pellet for tissue; forceps and hardened filter paper (Whatmann).

3. Leica EM FCS Ultracryomicrotome (Leica Microsystems).

4. Glass knife is made by LEICA EMKMR2 knife maker (Leica) or Diatome 45° dry cryo-diamond knife (DiATOME Ltd.).

5. Glass coverslips (Agar; 10-mm diameter, thickness 1.5) for supporting Tokuyasu cryosections.

6. Loop for collecting cryosections. A small (2 mm) diameter metal bacteriological loop (Agar) or homemade using copper wire.

2.3. Cryoimmunolabeling

1. Permeabilization solution: 0.1% Triton X-100 in PBS.

2. Quenching solution: 20 mM glycine in PBS.

3. Blocking solution: Prepare 10% BSA, 0.5% casein, and 10% fish skin gelatin stock solutions in PBS (stocks are kept frozen). The working solution is 1% BSA, 0.1% casein, and 0.1% gelatin in PBS (adjust to pH 8.0 with 1 M NaOH).

4. Antibodies: RNA polymerase II primary antibody (clone H5, mouse IgM; 1:2,000; Covance); AlexaFluor555 donkey anti-mouse IgG (1:2,000; Invitrogen).

5. Nuclei counterstain: 2 µM TOTO-3 (Invitrogen) and 300 nM DAPI (Sigma) in PBS/0.05% Tween-20.

6. Mounting medium: Antifade solution VectaShield (Vector Laboratories).

7. Confocal laser scanning microscope (Leica TCS SP5; 100× objective, NA 1.4), equipped with Diode (405 nm), Argon (488 nm), HeNe (594 nm), and HeNe (633 nm) lasers. Images are collected with pinhole equivalent to one Airy disk, without signal saturation or thresholding.

2.4. Probe Preparation

1. Human ~30 kb CD2-cos1 cosmid (hCD2; kind gift from Dimitris Kioussis) (13).

2. Probe labeling: Digoxigenin-translation kit (Roche).

3. 0.5 M EDTA (pH 8.0) (Sigma) to stop the nick translation reaction.

4. Heating block, Shaker.

5. Agarose minigel including 1% agarose in 0.5× TBE running buffer, loading buffer (Promega), and 100 bp DNA molecular weight marker (Promega), for checking the size of probe.

6. MicroBioSpin P-30 chromatography columns (BioRad) for removal of unincorporated nucleotides.

7. 1× TE buffer (Promega).

2.5. CryoFISH and Immuno-cryoFISH

1. 8% PFA in PBS fixative is prepared by dilution of 1/2 of 16% PFA in distilled water stock solution with 2× PBS (for immuno-cryoFISH).

2. Washing solutions: 2× SSC; 0.1× SSC; 4× SSC with 0.05% Tween-20; (20× SSC stock, Sigma, Molecular Biology Grade).

3. 250 μg/mL RNase A in 2× SSC (prepared freshly from 30 mg/mL stock from Sigma).

4. 0.1 M HCl in distilled water (prepared freshly from 1 M concentrated stock; VWR).

5. Ethanol series: 30, 50, 70, 90, and 100% Ethanol.

6. Deionized formamide (Sigma, Molecular Biology Grade).

7. Homemade 2× hybridization mixture: it contains 10% dextran sulphate (from autoclaved 50% solution), 4× SSC, 2 mM EDTA, 20 mM Tris–HCl (pH 7.2).

8. For 6 μL of final probe volume, coprecipitate 1 μg salmon sperm DNA (Sigma) carrier and 5 μg human C_0t1 DNA (Roche) with the nick-translation labeled probe to block repetitive DNA sequences.

9. Denaturation buffer: it contains 70% deionized formamide, 2× SSC, and 50 mM phosphate buffer (pH 7.0).

10. Hybrislips (Invitrogen, 25 mm × 25 mm).

11. Rubber cement (Qbiogene).

12. Omnislide In Situ Hybridization System (ThermoHybaid, UK).

13. Antibodies for amplification of probe signal: sheep Fab fragments anti-digoxigenin (1:100; Roche) and AlexaFluor488 donkey anti-sheep IgG (1:1,000; Invitrogen).

3. Methods

The labeling of chromosome territories in intact nuclei often requires harsh treatments that compromise the three-dimensional structure, or which rely on critical conditions that minimize extraction procedures while providing sufficient access of probes to their cellular targets (14, 15). The combination of 3D-FISH with immunodetection of transcription factories is however more complex, as RNA polymerase II is prone to aggregation in 3D cells when using fixation procedures compatible with 3D-FISH (16). In contrast, sucrose-embedded Tokuyasu cryosections (~150 nm thickness) are ideal samples for IF and FISH as their preparation does not depend on treatments with organic solvents, resins, or detergent extractions prior to fixation. Tokuyasu cryosections are compatible with labeling of target sites throughout the thickness of sections (17), resulting in excellent signal to noise ratios compared with other sectioning approaches that rely on surface labeling. Combining Tokuyasu cryosectioning with FISH and IF procedures has allowed us to achieve good probe accessibility and spatial resolution under conditions that preserve cellular ultrastructure and the distribution of nuclear components. Here, we provide an example of the use of cryoFISH in combination with immunolabeling of transcription factories. Sites containing the active form of RNA polymerase II were labeled with an IgM antibody specific for Serine2 phosphorylated form (S2p pol II), together with FISH detection of hCD2 transgene loci in sections from the thymus of transgenic mice bearing 15 tandem copies of a ~30 kb *hCD2* locus (13). Comparison of the relative positions of RNA polymerase II sites confirms the spatial preservation afforded by cryo-FISH, as transcription factories detected before FISH resist the procedure and maintain their relative positions (see Fig. 1c–e).

3.1. Preparation of Samples for Sucrose-Embedded Tokuyasu Cryosections

Cells are grown in optimal conditions, either in suspension or attached to cell culture plastic Petri dishes. Tissues are obtained either after perfusion fixation or dissected directly from organs and quickly transferred to fixative for cross-linking.

3.1.1. Fixation of Cells Grown in Suspension

1. Grow cells in appropriate culture medium, harvest cells and centrifuge at ideal speed, before removing medium supernatant

(do not wash in PBS). Ideal numbers of cells should be optimized to yield a pellet with a volume equivalent to 30 μL.

2. Resuspend cells in 1–1.2 mL of 4% PFA in 0.25 M HEPES (pH 7.6), transfer to 1.5 mL nonstick eppendorf tube, centrifuge 300×g, 5 min, remove supernatant, add 1 mL 4% PFA fixative without resuspending pellet, and incubate for 10 min at 4°C.

3. Replace fixative with 8% PFA in 0.25 M HEPES (pH 7.6) and incubate over a total period of 2 h at 4°C. During fixation, make cell pellet more compact by centrifuging at increasing speeds without resuspending pellet. After 20 min in the 8% PFA fixative, use a series of increasing speeds such as: 300×g, 5 min; 500×g, 2 min; 1,000×g, 2 min; 2,000×g, 2 min; 4,000×g, 2 min; 8,000×g, 2 min; 10,000×g, 10 min at 4°C; allow fixation to continue in the pellet for total of 2 h (see Note 3). Samples can be stored in 1% PFA in 0.25 M HEPES (pH 7.6) for a short period (overnight <2 days).

3.1.2. Fixation of Cells Grown Attached to Plastic

1. Grow cells in 90 mm tissue culture Petri dishes (~75% on the day of collection is ideal density), rinse briefly with 4% PFA in 0.25 M HEPES (pH 7.6), and incubate the fixative for 10 min at 4°C.

2. Further fix cells with 8% PFA in 0.25 M HEPES (pH 7.6) for 2 h at 4°C. During fixation, scrape Petri dish gently and unidirectional using cell scrapers previously soaked in fixative, to avoid cells sticking.

3. Collect cell suspension in fixative into a 1.5 mL nonstick hydrophobic Eppendorf tube and centrifuge to pellet cells at increasing speeds as above (see Note 3).

3.1.3. Fixation of Tissue

1. Tissues are obtained either after perfusion fixation or quickly dissected directly from organs.

2. Place immediately into a Petri dish containing 4% PFA in 0.25 M HEPES (pH 7.6) fixative.

3. Cut tissue into small cubes with sharp scalpel (under dissection microscope) to minimize the physical damage. The cubes must be ~1–2 mm square to allow the fixative to penetrate into the tissue.

4. Trimmed tissues are transferred to 1.5 mL Eppendorf tube with fresh 4% PFA in 0.25 M HEPES (pH 7.6) fixative as quickly as possible to fix for 30 min at 4°C.

5. Further fix the tissue in 8% PFA in 0.25 M HEPES (pH 7.6) for 2 h at 4°C.

3.1.4. Sucrose Embedding and Cryosectioning

1. Transfer the cell pellet or tissue pieces through a few drops of 2.1 M sucrose to be infiltrated over 2–4 h and then carefully place on a metal block (see Notes 4 and 5).

2. Remove excess sucrose with filter paper and shape the cell pellet into a cone with forceps under dissecting light microscope.

3. Freeze cell pellets or tissue pieces by immersion in liquid nitrogen with forceps and continuous shaking until the fizzing stops (see Note 6). The blocks in liquid nitrogen can be stored indefinitely.

4. Cut Tokuyasu cryosections (100–200 nm thickness) on an ultracryomicrotome with glass or Diatome 45° dry cryo-diamond knife; the thickness of cryosections is deduced from interference color (see Note 7).

5. Cryosections are collected with drops of 2.1 M sucrose on metal bacteriological loop with a small (~2 mm) diameter and transferred onto glass coverslips. It is recommended to place one section drop on to the centre of the coverslip so that sections can more easily be found under the microscope. Cryosections on coverslips can be stored at –20°C for a few weeks. We frequently use coverslips (thickness 0.17 cm, Nr1.5) with diameter 10 mm and 6 μL final probe volume, as described below.

3.2. Cryo-immunolabeling of RNA Polymerase II

This protocol can be used for sections of any cells or tissues.

1. Wash off sucrose drop from sections (3×) with PBS (over 20 min).

2. Quench free aldehyde groups with 20 mM glycine in PBS, for 15 min; then wash (3×) in PBS.

3. Permeabilize with 0.1% Triton X-100 in PBS for 10 min and then wash 3× in PBS.

4. Incubate in blocking solution (1% BSA, 0.05% casein, 0.2% gelatin, in PBS, pH 8.0) for 0.5–1 h, to reduce nonspecific background staining.

5. Incubate with the RNA polymerase II primary antibody (diluted in blocking solution) for 2 h.

6. Wash 4× times in blocking solution for 1 h.

7. Incubate with the secondary antibody (diluted in blocking solution) for 1 h.

8. Wash 4× in blocking solution for 1 h, then wash 3× in PBS, and subsequently wash in PBS containing 0.05% Tween-20.

9. Counterstain the nuclei with 2 μM TOTO-3 in PBS/0.05% Tween-20.

10. Wash 3× in PBS (labeled sections can be stored in PBS at 4°C for a few weeks).

11. Mount in VectaShield or equivalent mounting media immediately before imaging.

3.3. Immuno-cryoFISH for the Simultaneous Detection of Proteins and DNA Sequences

The following protocol exemplifies immuno-cryoFISH for labeling RNA pol II and the *CD2* locus in thymic sections.

3.3.1. Probe Preparation

1. Precipitate 1 μg of human CD2 (hCD2-cos1) construct DNA using sodium acetate and ethanol, wash in 70% ethanol, air dry briefly, then add 16 μL of double distilled water (Sigma), and vortex for 30 min to dissolve DNA.

2. Add 4 μL of DIG-Nick Translation mixture (Roche), mix, centrifuge briefly, and then incubate for 3 h at 15°C on a heat block. Incubate reaction mixture on ice after reaction, while DNA probe length is investigated.

3. Length of probe is checked by taking 1–2 μL aliquot from the reaction, add gel loading buffer, denature it at 95°C, for 3 min, then put it on ice, and run the sample on an agarose minigel along with 100 bp DNA molecular weight marker (see Note 7).

4. The nick-translation reaction is stopped by adding 1 μL of 0.5 M EDTA (pH 8.0) per 20 μL reaction volume and heating to 65°C for 10 min.

5. DIG-labeled hCD2 probe is purified by separation of unincorporated nucleotides using microBioSpin P-30 chromatography columns or ethanol precipitation.

6. Purified hCD2 probe is dissolved in the double distilled water or 1× TE buffer and can be stored at –20°C.

7. 2–4 μL of DIG-labeled hCD2 probe is coprecipitated with 5 μg human C_0t1 DNA and 1 μg salmon sperm DNA final concentration. The probe pellet is air-dried briefly and resuspended in 3 μL deionized formamide by shaking for 30 min first, then add 3 μL homemade 2× hybridization mixture for further 30 min on a shaker until mixed well. After mixing, the probe is denatured at 70°C for 10 min, put on ice and then re-annealed at 37°C for at least 30 min before hybridization.

8. For simultaneous detection of multiple DNA sequences in combination with chromosome paints, the multiple DNA probe can be labeled with different haptens of fluorophores using the biotin-, DIG-, FITC- Rhodamine- nick translation. The multiple-labeled probes are coprecipitated with C_0t1 DNA and salmon sperm DNA as described above (7), then resuspended in commercial chromosome paints. If multiple

chromosome paints are used alone, $C_0t\ 1$ DNA (1–5 µg/µL) may still be added to reduce the background signal.

3.3.2. Cryo-FISH

1. Before in situ hybridization, cryosections immunolabeled with H5 antibody to detect the active Serine2-phosphorylated form of RNA polymerase II are fixed with 8% PFA in PBS for 1 h at room temperature to preserve the antibody labeling, prior to the harsher FISH treatments.

2. Wash 3× with PBS over 1 h.

3. Wash 3× in 2× SSC.

4. Incubate with 250 µg/mL RNase A for 1–2 h at 37°C and then wash 3× in 2× SSC.

5. Permeabilize cryosections (10 min) with 0.2% Triton X-100 in 2× SSC and then wash 3× in 2× SSC.

6. Treat with 0.1 M HCl for 10 min and then wash 3× in 2× SSC.

7. Dehydrate in an ethanol series (30%, 50%, 70%, 90%, 100% series) on ice, 3 min each, then air-dry briefly at 37°C.

8. Denature DNA in samples using denaturation buffer for 8–12 min at 80°C (e.g., on an Omnislide In Situ Hybridization System); time of denaturation is optimized for each tissue and should be the minimum time that gives maximum signal.

9. Dehydrate in ethanol series (30–100% series) on ice, 3 min each, then air-dry briefly at 37°C.

10. Apply 6 µL drops of hybridization mixture containing the hCD2 DIG-labeled probe on Hybrislips, before overlaying coverslips containing sections. Seal with rubber cement and carry hybridization on the Omnislide In Situ Hybridization System or in a humid chamber at 37°C for 36–60 h.

11. Posthybridization washes are as follows : wash 3× (15 min, 2× 5 min) with 50% formamide in 2× SSC for a total of 25 min at 42°C, 3× (10 min, 2× 5 min) in 0.1× SSC for 20 min at 60°C, and 4× SSC with 0.1% Tween-20 for 10 min at 42°C.

12. DIG signal is amplified by immunodetection as described above (Subheading 3.2) using sheep Fab fragments anti-digoxigenin (1:100) and AlexaFluor488 donkey antibodies anti-sheep IgG (1:1,000).

4. Notes

1. Incubate paraformaldehyde solution in a water bath, at 60°C, for 1–2 h, with a stirrer. Temperature is tightly controlled to

keep it below 62°C. Add a few drops of 1 M NaOH until solution is clear. Let solution cool down quickly on ice and adjust pH to ~7.5, before filtering solution through 0.45 μm filters. Prepare 4 or 8% PFA in 0.25 M HEPES by dilution of the 16% PFA stock (in water) by 1/4 or 1/2 with 0.5 M HEPES and distilled water.

2. Premade PBS is added to make the solution (the correct final concentration of phosphate is lower than in 1× PBS).

3. Do not disturb the cell pellet and keep tube orientation in the centrifuge. If side of tube has a trail of cells formed on the tube, dislodge the pellet carefully with a sharpened wooden toothpick (previously soaked in the fixative) without breaking the pellet.

4. The pellet and tissue should become transparent after appreciable infiltration in sucrose, which is important to prevent ice-crystal damage of biological structures.

5. Wear safety protection.

6. The approximate thickness of cryosections is indicated by the interference color such as yellow: ~100 nm; yellow/pink: ~140 nm; blue: ~200 nm.

7. The smear of probe sizes should range between 200 and 300 nucleotides. If necessary, reincubate the reaction at 15°C and re-check the fragment size until it reaches correct probe length.

Acknowledgments

We thank Michael Hollinshead, Kate Liddiard, and Miguel R. Branco for their contribution to the development and application of cryoFISH approaches, and Miguel R. Branco and Tiago Branco for their contributions to the development of macros for quantitative analyses of cryoFISH images. We thank the Medical Research Council (UK) for funding.

References

1. Pawley, J. B. (ed.) (1995) *Handbook of biological confocal microscopy*. Plenum Press, New York.

2. Shaw, P. J. (1995) Comparison of wide-field/deconvolution and confocal microscopy for 3D imaging, in *Handbook of biological confocal microscopy* (Pawley, J. B., ed.), Plenum Press, New York, pp. 373–387.

3. Wilson, T. (1995) The role of the pinhole in confocal imaging system, in *Handbook of biological confocal microscopy* (Pawley, J. B., ed.), Plenum Press, New York, pp. 167–182.

4. Tokuyasu, K. T. (1973) A technique for ultra-cryotomy of cell suspensions and tissues. *J. Cell Biol.* **57**, 551–565.

5. Tokuyasu, K. T. (1980) Immunochemistry on ultrathin frozen sections. *Histochem. J.* **12**, 381–403.

6. Pombo, A., Hollinshead, M., and Cook, P. R. (1999) Bridging the resolution gap: imaging

the same transcription factories in cryosections by light and electron microscopy. *J. Histochem. Cytochem.* **47**, 471–480.

7. Branco, M. R. and Pombo, A. (2006) Intermingling of chromosome territories in interphase suggests role in translocations and transcription-dependent associations. *PLoS Biol.* **4**, e138.

8. Xie, S. Q., Martin, S., Guillot, P. V., Bentley, D. L., and Pombo, A. (2006) Splicing speckles are not reservoirs of RNA polymerase II, but contain an inactive form, phosphorylated on Serine2 residues of the C-terminal domain. *Mol. Biol. Cell* **17**, 1723–1733.

9. Xie, S. Q. and Pombo, A. (2006) Distribution of different phosphorylated forms of RNA polymerase II in relation to Cajal and PML bodies in human cells: an ultrastructural study. *Histochem. Cell Biol.* **125**, 21–31.

10. Branco, M. R., Branco, T., Ramirez, F., and Pombo, A. (2008) Changes in chromosome organization during PHA-activation of resting human lymphocytes measured by cryo-FISH. *Chromosome Res.* **16**, 413–426.

11. Simonis, M., Klous, P., Splinter, E., Moshkin, Y., Willemsen, R., de Wit, E., van Steensel, B., and de Laat, W. (2006) Nuclear organization of active and inactive chromatin domains uncovered by chromosome conformation capture-on-chip (4C). *Nat. Genet.* **38**, 1348–1354.

12. Palstra, R. J., Simonis, M., Klous, P., Brasset, E., Eijkelkamp, B., and de Laat, W. (2008) Maintenance of long-range DNA interactions after inhibition of ongoing RNA polymerase II transcription. *PLoS ONE* **3**, e1661.

13. Lang, G., Wotton, D., Owen, M. J., Sewell, W. A., Brown, M. H., Mason, D. Y., Crumpton, M. J., and Kioussis, D. (1988) The structure of the human CD2 gene and its expression in transgenic mice. *EMBO J.* **7**, 1675–1682.

14. Solovei, I., Cavallo, A., Schermelleh, L., Jaunin, F., Scasselati, C., Cmarko, D., Cremer, C., Fakan, S., and Cremer, T. (2002) Spatial preservation of nuclear chromatin architecture during three-dimensional fluorescence in situ hybridization (3D-FISH). *Exp. Cell Res.* **276**, 10–23.

15. Hepperger, C., Otten, S., von Hase, J., and Dietzel, S. (2007) Preservation of large-scale chromatin structure in FISH experiments. *Chromosoma* **116**, 117–133.

16. Guillot, P. V., Xie, S. Q., Hollinshead, M., and Pombo, A. (2004) Fixation-induced redistribution of hyperphosphorylated RNA polymerase II in the nucleus of human cells. *Exp. Cell Res.* **295**, 460–468.

17. Branco, M. R., Xie, S. Q., Martin, S., and Pombo, A. (2005) Correlative microscopy using Tokuyasu cryosections applications for immunolabeling and in situ hybridization, in *Cell imaging* (Stephens, D., ed.), Scion Publishing, Oxford, UK, pp. 201–217.

Chapter 16

Characterization of Chromosomal Rearrangements Using Multicolor-Banding (MCB/m-band)

Thomas Liehr, Anja Weise, Sophie Hinreiner, Hasmik Mkrtchyan, Kristin Mrasek, and Nadezda Kosyakova

Abstract

Molecular cytogenetics and especially fluorescence in situ hybridization (FISH) banding approaches are nowadays standard for the exact characterization of simple, complex, and cryptic chromosomal aberrations within the human genome. FISH-banding techniques are any kind of FISH techniques, which provide the possibility to characterize simultaneously several chromosomal subregions smaller than a chromosome arm. FISH banding methods fitting that definition may have quite different characteristics, but share the ability to produce a DNA-specific chromosomal banding. While the standard techniques such as G-bands by Trypsin using Giemsa banding lead to a protein-related black and white banding pattern, FISH-banding techniques are DNA-specific, more colorful, and thus, more informative. At present, the most frequently applied FISH banding technique is the multicolor banding (MCB/m-band) approach. MCB/m-band is based on region-specific microdissection libraries, producing changing fluorescence intensity ratios along the chromosomes. Here we describe the FISH-banding technique MCB/m-band and illustrate how to apply it for characterization of chromosomal breakpoints with a minimal number of FISH experiments.

Key words: FISH-banding, Multicolor banding, Microdissection, Marker chromosomes, Chromosomal breakpoints

1. Introduction

One main focus of human cytogenetics is to be able to characterize chromosomal rearrangements and their breakpoints by simple, rapid, and reliable approaches. Thus, the GTG banding (G-bands by Trypsin using Giemsa) technique (1) is still the gold-standard in cytogenetics as it is easy to perform and relatively inexpensive.

Joanna M. Bridger and Emanuela V. Volpi (eds.), *Fluorescence in situ Hybridization (FISH):*
Protocols and Applications, Methods in Molecular Biology, vol. 659,
DOI 10.1007/978-1-60761-789-1_16, © Springer Science+Business Media, LLC 2010

231

Nonetheless, as GTG-banding has some well-known technical limitations (2), the introduction of the fluorescence in situ hybridization (FISH) technique in the 1980s allowed advanced cytogenetics dramatically (3). The new field of "molecular cytogenetics" (4) further advanced during the last few decades with the development of approaches such as multicolor FISH using all human whole chromosome painting probes with 24 colors (5, 6). Further, FISH-based banding methods were developed in the last 10 years and were recently defined as any kind of molecular cytogenetic technique that is used to simultaneously characterize several chromosomal subregions smaller than a chromosome arm (7). Although FISH banding methods may have quite different characteristics, they share the ability to produce DNA-specific chromosomal banding (8). In contrast to standard chromosome banding techniques, which are based on an enzymatic reaction (i.e., trypsin), FISH banding methods are DNA-specific. FISH banding methods recently developed include interspersed PCR multiplex-FISH (IPM-FISH), cross-species color banding (Rx-FISH), different chromosome bar codes (CBC) such as the somatic cell hybrid-based CBC, spectral color banding (SCAN), M-FISH using chromosome region-specific probes (CRP), the microdissection-based multicolor banding (MCB/m-band), and multitude MCB (mMCB) (7). Some of the aforementioned FISH banding methods are still under development for the entire human genome (7, 8).

These FISH banding methods have been successfully applied to evolution and radiation biology, in studies of the nuclear architecture and for diagnostic purposes in prenatal, postnatal, and tumor cytogenetics (7). As summarized on the m-FISH literature homepage (http://www.med.uni-jena.de/fish/mFISH/mFISHlit.htm), these FISH banding studies led to contributions in >200 international publications. Thus, FISH banding methods are an integral part of cytogenetic research and diagnostics. Among them, the MCB/m-band approach is providing ~60% of the published studies/applications and are available as human and murine probes.

However, until recently, all FISH banding approaches, except for the YAC/BAC-based ones (2), have had no connection to the relative map position in the human genome sequence. This problem was recently solved for our array-proven MCB probe set (aMCB), by hybridizing DNA from region-specific libraries, on which the MCB approach is based, to BAC-based array comparative genomic hybridization (9).

Here we present protocols how to apply the commercially available m-band (10) and our homemade MCB/aMCB probe sets (9, 11, 12) to human chromosomes.

2. Materials

2.1. Slide Pretreatment and Denaturation

1. Coverslips 24×50 mm and 24×24 mm (Menzel).

2. Formalin-buffer (1%): 5 mL formaldehyde solution (Merck)+4.5 mL 1× PBS+500 µL 1 M MgCl$_2$-solution.

3. Slides, precleaned (Menzel).

4. 1× PBS – set up from PBS Dulbecco (Instamed) w/o Ca2+, w/o Mg2+ (Seromed-Biochrom).

5. Pepsin stock solution: 1.0 g pepsin (Sigma) in 50 mL distilled water – aliquot and store at 20°C.

6. Pepsin solution: Prewarm 95 mL of distilled water to 37°C in a Coplin jar; add 5 mL of 0.2 N HCl and 0.5 mL of pepsin stock solution – make fresh when required.

2.2. Probe Pretreatment and Denaturation

1. m-band probes can be purchased from MetaSystems, Altlussheim, Germany, as XCyte 1-22, X and Y.

2. Homemade aMCB/MCB/mMCB probe set; available on request from our laboratory on a collaborative basis.

2.3. FISH Procedure, Washing, and Detection

1. Detection solution (per slide): 0.5 µL Streptavidin-Cyanine 5 (Amersham)+50 µL 4× SSC/0.2% Tween; make fresh as required.

2. Rubber cement: trade mark Fixogum, company Marabu.

3. 4× SSC/0.2%Tween: 100 mL 20× SSC (Invitrogen, 15557-036)+400 mL ddH$_2$O+250 µL Tween 20 (Sigma, P-1379), adjust to pH 7–7.5.

4. 1× SSC-solution; set up from 20× SSC (Invitrogen, 15557-036).

2.4. Evaluation

1. MetaSystems is essential for MCB/m-band evaluation. No other image analyzing system has the features to evaluate the fluorochrome profiles and create informative pseudocolor banding along the chromosomes.

3. Methods

MCB/m-band is performed normally on metaphase spreads. However, it may also be applied in interphase FISH for research purposes (13, 14). Chromosome preparation is already reported within the core protocols of this book and in (15). Here we outline how to pretreat slides with metaphase spreads on them (See subheading 3.1), how to prepare the MCB/m-band probe set before hybridization (See subheading 3.2), how to perform the FISH-procedure itself (See subheading 3.3), and how to evaluate the results (See subheading 3.4). When using commercial m-banding probes (MetaSystems), one can either follow the manufacturers' instructions or the protocol provided here.

3.1. Slide Pretreatment and Denaturation

Similar to the conventional FISH approach, pretreatment of the slides, or better, the metaphase spreads on the slides, should be performed with pepsin followed by a postfixation with formalin-buffer to reduce background. In some protocols also the pretreatment with RNase A is suggested (16). However, according to our experience, this step does not lead to any significant improvements and can be missed out.

1. Incubate slides in 100 mL 1× PBS at room temperature (RT) for 2 min on a shaker.
2. Put slides for 5–10 min in pepsin solution at 37°C in a Coplin jar (see Note 1).
3. Repeat step 1 twice.
4. Postfix metaphase spreads on the slide surface by replacing 1× PBS with 100 mL of formalin-buffer for 10 min (RT, with gentle agitation).
5. Repeat step 1 twice.
6. Dehydrate slides in an ethanol series (70, 90, 100%, 3 min each) and air-dry.
7. Add 100 µL denaturation-buffer to each slide and cover with 24×50 mm coverslip.
8. Incubate slides on a warming plate for 2–3 min at 75°C (see Note 2).
9. Remove the coverslip immediately and place slides in a coplin jar filled with 70% ethanol (4°C) to conserve target DNA as single strands.
10. Dehydrate slide in ethanol (70%, 90%, 100%, 4°C, 3 min each) and air-dry.

3.2. Probe Pretreatment and Denaturation

1. Take an aliquot of probe (see Note 3).
2. To denature incubate the 1.5 µL microtube containing the probe-DNA at 75°C for 5 min.
3. Perform a prehybridization step at 37°C for 30 min.
4. Cool down to 4°C.

3.3. FISH Procedure, Washing, and Detection

1. Add 20 µL of probe-solution onto each denatured slide, put 24×50 mm coverslip on the drops, and seal with rubber cement (see Note 4).

2. Incubate slides for 1–3 nights at 37°C in a humid chamber (see Note 5).

3. Take the slides out of 37°C, remove rubber cement with forceps and coverslips by letting them swim off in 4× SSC/0.2%Tween (RT, 100 mL Coplin jar).

4. Postwash the slides 1–5 min in 1× SSC-solution (62–65°C) with gentle agitation (see Note 6).

5. Put the slides in 4× SSC/0.2% Tween (100 mL, RT), for a few seconds.

6. Add 50 µL of detection solution to each slide, cover with 24×50 mm coverslip and incubate at 37°C for 30 min in a humid chamber.

7. Remove the coverslip and wash 3×5 min in 4× SSC/0.2% Tween (RT, with gentle agitation).

8. Counterstain the slides with DAPI-solution (100 mL in a Coplin jar, RT) for 8 min.

9. Wash slides three times in water for a few seconds and air-dry.

10. Add 15 µL of antifade, cover with coverslip, and look at the results in a fluorescence microscope.

3.4. Evaluation

The evaluation is not only dependent on the availability of a fluorescence microscope with at least six different color channels but also on the image analyzing software of MetaSystems. For examples of how MCB/m-band results look like, we refer to http://www.metasystems.de/ and to (10–13) (see Note 7).

Three to five different fluorochromes are used to label the partial chromosome painting probes: SpectrumOrange, SpectrumGreen, TexasRed, Cyanine 5, and diethylaminocoumarine. The fluorochrome intensity ratios along a chromosome can be quantified and translated into pseudocolors by computer using the isis software (MetaSystems). The resulting MCB/m-band pattern have to be aligned with the GTG banding pattern. The evaluation can be done either by analyzing the reproducible fluorescence profiles along a chromosome or by creating pseudocolors, which are based on the fluorescence profiles. The pseudocolors can be adjusted according to the necessary resolution adequate for the individually analyzed cases. As the number of pseudocolored bands per chromosome can freely be assigned using the isis software (MetaSystems), a resolution even higher than that of GTG banding of the corresponding chromosome can be achieved. This is a unique possibility provided only by the MCB/m-band technique, to change the resolution simply by application of software, without additional sophisticated chromosome preparations and repetition of a staining technique.

Per chromosomal aberration to be characterized by MCB/ m-band we recommend analyzing 10–20 metaphases. As the practical evaluation is highly dependent on handling of the software, we omit to go into details here (see Note 8).

4. Notes

1. Pepsin treatment time has to be adapted in each lab. Success of pretreatment must be controlled by microscopic inspection. A balance between chromosome preservation and digestion must be found. If chromosomes are preserved too well, it might be that the DNA-probes cannot penetrate and no FISH-results are obtained. In the case of too much chromosome digestion, one might have still FISH-signals but is unable to correlate them to a specific chromosomal region. Also complete loss of the chromosomes during the FISH-procedure might occur. For beginners, it is recommended to start with target samples which are not limited in availability.

2. Denaturation times of 2–3 min only is suggested for the maintenance of available metaphase chromosomes. Also here a lab specific adaptation must be done and a balance between chromosome preservation and denaturation/disaggregation must be found. In case one works with tissues without metaphase spreads, this aspect is of no significance.

3. There is no difference in this step 3.2 for commercial and homemade probes.

4. Use of 24 × 24 mm coverslips is also possible to hybridize different probes on the same slide using smaller coverslips. The amount of probe/probe-solution has to be reduced according to the coverslip-size.

5. Here 1–3 days of FISH-hybridization is recommended. If incubation is already stopped after 48 or 96 h, the following may happen: while in the first case weaker signals are possible, in the second case some cross hybridization problems may arise.

6. During the FISH washing steps, it is important to prevent the slide surfaces drying out, otherwise background problems may arise.

7. When the hybridization is less than optimal quality, in most cases it is possible to achieve results using the fluorochrome profiles and real colors instead of pseudocolor bands for evaluation.

8. In summary, we want to make clear that MCB/m-band is more or less a normal multicolor FISH-procedure with specific probes. Every laboratory with experience in multicolor FISH using whole chromosome painting probes can easily adapt also FISH-banding.

Acknowledgments

Supported in parts by the DFG (436 RUS 17/109/04, 436 RUS 17/22/06, WE 3617/2-1, 436 ARM 17/5/06, LI820/11-1, LI820/17-1), Boehringer Ingelheim Fonds, Evangelische Studienwerk e.V. Villigst, IZKF Jena (Start-up S16), IZKF together with the TMWFK (TP 3.7 and B307-04004), Stiftung Leukämie, and Stefan-Morsch-Stiftung.

References

1. Seabright, M. (1971) A rapid banding technique for human chromosomes. *Lancet* **30**, 971–972.

2. Liehr, T., Weise, A., Heller, A., Starke, H., Mrasek, K., Kuechler, A., Weier, H. U., and Claussen, U. (2002) Multicolor chromosome banding (MCB) with YAC/BAC-based probes and region-specific microdissection DNA libraries. *Cytogenet Genome Res* **97**, 43–50.

3. Pinkel, D., Straume, T., and Gray, J. W. (1986) Cytogenetic analysis using quantitative, high-sensitivity, fluorescence hybridization. *Proc Natl Acad Sci U S A* **83**, 2934–2938.

4. Chang, S. S., and Mark, H. F. L. (1997) Emerging molecular cytogenetic techniques. *Cytobios* **90**, 7–22.

5. Speicher, M. R., Gwyn Ballard, S., and Ward, D. C. (1996) Karyotyping human chromosomes by combinatorial multi-fluor FISH. *Nat Genet* **12**, 368–375.

6. Schröck, E., du Manoir, S., Veldman, T., Schoell, B., Wienberg, J., Ferguson-Smith, M. A., Ning, Y., Ledbetter, D. H., Bar-Am, I., Soenksen, D., Garini, Y., and Ried, T. (1996) Multicolor spectral karyotyping of human chromosomes. *Science* **273**, 494–497.

7. Liehr, T., Starke, H., Heller, A., Kosyakova, N., Mrasek, K., Gross, M., Karst, C., Steinhaeuser, U., Hunstig, F., Fickelscher, I., Kuechler, A., Trifonov, V., Romanenko, S. A., and Weise, A. (2006) Multicolor fluorescence in situ hybridization (FISH) applied to FISH-banding. *Cytogenet Genome Res* **114**, 240–244.

8. Liehr, T., Gross, M., Karst, C., Glaser, M., Mrasek, K., Starke, H., Weise, A., Mkrtchyan, H., and Kuechler, A. (2006) FISH banding in tumor cytogenetics. *Cancer Genet Cytogenet* **164**, 88–89.

9. Weise, A., Mrasek, K., Fickelscher, I., Claussen, U., Cheung, S. W., Cai, W. W., Liehr, T., and Kosyakova, N. (2008) Molecular definition of high-resolution multicolor banding probes: first within the human DNA sequence anchored FISH banding probe set. *J Histochem Cytochem* **56**, 487–493.

10. Chudoba, I., Plesch, A., Lörch, T., Lemke, J., Claussen, U., and Senger, G. (1999) High resolution multicolor-banding: a new technique for refined FISH analysis of human chromosomes. *Cytogenet Cell Genet* **84**, 156–160.

11. Liehr, T., Heller, A., Starke, H., Rubtsov, N., Trifonov, V., Mrasek, K., Weise, A., Kuechler, A., and Claussen, U. (2002) Microdissection based high resolution multicolor banding for all 24 human chromosomes. *Int J Mol Med* **9**, 335–339.

12. Weise, A., Heller, A., Starke, H., Mrasek, K., Kuechler, A., Pool-Zobel, B. L., Claussen, U., and Liehr, T. (2003) Multitude multicolor chromosome banding (mMCB) – a comprehensive one-step multicolor FISH banding method. *Cytogenet Genome Res* **103**, 34–39.

13. Lemke, J., Claussen, J., Michel, S., Chudoba, I., Mühlig, P., Westermann, M., Sperling, K.,

Rubtsov, N., Grummt, U.W., Ullmann, P., Kromeyer-Hauschild, K., Liehr, T., and Claussen, U. (2002) The DNA-based structure of human chromosome 5 in interphase. *Am J Hum Genet* **71**, 1051–1059.

14. Iourov, I. Y., Liehr, T., Vorsanova, S. G., Kolotii, A. D., and Yurov, Y. B. (2006) Visualization of interphase chromosomes in postmitotic cells of the human brain by multicolour banding (MCB). *Chromosome Res* **14**, 223–229.

15. Liehr, T., and Claussen, U. (2002) FISH on chromosome preparations of peripheral blood, in: FISH-Technology (Rautenstrauss, B. and Liehr, T., eds.) Springer, Berlin, pp 73–81.

16. Liehr, T., Thoma, K., Kammler, K., Gehring, C., Ekici, A., Bathke, K. D., Grehl, H., and Rautenstrauss, B. (1995) Direct preparation of uncultured EDTA-treated or heparinized blood for interphase FISH analysis. *Appl Cytogenet* **21**, 185–188

Chapter 17

Visualizing Nucleic Acids in Living Cells by Fluorescence In Vivo Hybridization

Joop Wiegant, Anneke K. Brouwer, Hans J. Tanke, and Roeland W. Dirks

Abstract

The analysis of the spatial–dynamic properties of DNA and RNA molecules in living cells will greatly extend our knowledge of genome organization and gene expression regulation in the cell nucleus. The development of hybridization methods allowing detection of specific endogenous DNA and RNA sequences in living cells has therefore been a challenge for many years. However, there are many technical issues that have proven so far to be difficult, or even impossible, to overcome. As a result, in most situations, the application of in vivo hybridization methods is currently limited to the visualization of highly repetitive DNA sequences or abundant RNA species. We describe a protocol that enables the visualization and tracking of telomeres in living cells by hybridization with a fluorescent peptide nucleic acid (PNA) probe. Furthermore, we describe a method that allows the detection of abundant endogenous RNAs in living cells by microinjecting fluorescently labeled complementary $2'$-O-methyl RNA probes.

Key words: Microinjection, Glass bead loading, In vivo hybridization, Peptide nucleic acid, $2'$-O-Me RNA, Live cell imaging

1. Introduction

The cell nucleus is a multifunctional organelle in which chromatin is dynamically organized to regulate gene expression in a spatial and temporal fashion (1, 2). Fluorescence in situ hybridization (FISH) is an essential technique to study the spatial distribution of nucleic acids, either DNA or RNA, in the cell nucleus. For example, multicolor FISH has been used to study the spatial arrangement of multiple genomic sequences relative to each other in the nucleus (3). In some cases, such arrangements were shown to correlate with cell function or disease (4). In addition, FISH techniques have been used to study the spatial organization of RNA synthesis, RNA processing, and RNA transport in single

Joanna M. Bridger and Emanuela V. Volpi (eds.), *Fluorescence in situ Hybridization (FISH):*
Protocols and Applications, Methods in Molecular Biology, vol. 659,
DOI 10.1007/978-1-60761-789-1_17, © Springer Science+Business Media, LLC 2010

cells to obtain insight into the coordination and regulation of gene expression in normal and diseased cells (5, 6). An intrinsic limitation of the FISH technique is, however, that the sample material has to be morphologically maintained by fixation. As a result, information concerning the dynamics of chromatin organization or gene expression cannot be obtained.

Therefore, various methods have been developed to visualize nucleic acids in living cells. These methods include the delivery of fluorescently labeled exogenous RNA into cells but also the incorporation of DNA sequences or expression of RNA molecules that are flanked by a sequence motif that can be recognized by proteins fused to green fluorescent protein (GFP) (7). A possible limitation of these approaches is that it involves the detection of artificial molecules. Only a few methods have been described that allow the analysis of specific endogenous DNA or RNA sequences (8–10, 13). These methods rely on the introduction of fluorescently labeled probes into a cell and their subsequent hybridization to the target molecule. Here, we describe a protocol for the detection of telomeric DNA sequences in living cells, which is based on the introduction of a Cy3-labeled peptide nucleic acid (PNA) probe into cells by glass bead loading (11). In addition, we describe a protocol for the detection of abundant RNAs in the cell nucleus using $2'$-O-methyl RNA probes. These probes are resistant to nuclease degradation and are introduced into the cell by microinjection.

2 Materials

2.1 Cell Culture

1. Dulbecco's Modified Eagle's Medium (DMEM) (Invitrogen) supplemented with 10% fetal calf serum (FCS, Invitrogen), 2 mM L-glutamine, 100 U/mL penicillin, and 100 U/mL streptomycin (see Note 1).

2. DMEM cell culture medium buffered with 25 mM Hepes (Invitrogen) to pH 7.2.

3. Phosphate buffered saline (PBS): Prepare a 10× stock solution with 1.37 M NaCl, 27 mM KCl, 100 mM Na_2HPO_4, and 18 mM KH_2PO_4, pH 7.4. Store this solution at room temperature. A solution of 1× PBS is prepared by making a 1:10 dilution in distilled water, which has been autoclaved.

4. Solution of trypsin (0.25% w/v) in sterile PBS. Prepare this solution from a 2.5% trypsin stock solution (Invitrogen) just before use. The 2.5% trypsin stock solution is stored as aliquots (1 mL/tube) at –20°C.

5. Sterile glass bottom petri dishes, 3.5 cm, either with or without grid (MatTek).

2.2. Probes

1. Telomere repeat sequence-specific Cy3-labeled (C3TA2)3 PNA (8 μM stock solution in water)(DakoCytomation) (see Note 2). Store in aliquots at –20°C.

2. Linear 2'-O-methyl $(U)_{22}$ RNA probe (Eurogentec, Seraing, Belgium) that is labeled at the 5' end via a succinimidyl ester derivative with carboxy-tetramethylrhodamine (TAMRA) (see Note 3). Store the probe at –20°C.

2.3. Bead Loading

1. Glass beads (~75 mm in size, washed in HCl, Sigma-Aldrich).

2. 2× microbead buffer: 80 mM KCl, 10 mM K_2PO_4, 4 mM NaCl, pH 7.2. Filter sterilize this buffer and store at –20°C in 1 mL aliquots.

2.4. Microinjection

1. Prepulled sterile microinjection glass needles (Eppendorf).

2. Borosilicate glass capillaries (MTW 100 F-3, 1.0 mm outside diameter, 0.75 mm inside diameter, and 76 mm in length containing a filament for rapid filling, World Precision Instruments) (see Note 4).

3. Microloader (Eppendorf).

4. Microinjector and micromanipulator (type IM 300, Narishige).

5. Micropipet puller (model P-97, Sutter Instrument Co.).

6. Nitrogen gas cylinder.

7. 2× microinjection buffer containing 80 mM KCl, 10 mM Na_2PO_4, and 4 mM NaCl, pH 7.2.

2.5. Microscopy

1. Inverted fluorescence microscope (Axiovert 135 TV, Zeiss) with oil-immersion objective PlanApo 40×, 63×, and 100× phase contrast lenses, a lens heater (Bioptechs), and a heating ring that surrounds the culture chamber (Bioptechs). The microinjector and micromanipulator are firmly mounted on the microscope frame. To image TAMRA, the microscope is equipped with a filter set consisting of a 546/10 nm BP filter, a 580 nm DM, and a 590 LP barrier filter and a charge-coupled device (CCD) camera (Photometrics).

2. Inverted confocal laser scanning microscope (TCS SPE, Leica Microsystems) for obtaining high-resolution optical images of injected cells. Cy3 and TAMRA are excited at 543 nm. This system also contains a fluorescence recovery after photobleaching (FRAP) mode to analyze the kinetics of labeled RNAs (see Note 5). The microscope is placed in an incubator to control the temperature, CO_2, and moisture.

3. Methods

3.1. Preparation of Cells

1. Use adherent cells, like U2OS, Hela or dermal fibroblast cells, which grow on a glass surface.

2. Grow the cells in cell culture flasks at 37°C, 5% CO_2 until nearly confluent, aspirate the culture medium, rinse the cells with PBS, and trypsinize them.

3. Seed the cells in a 1:20 dilution in 3.5 mm glass bottom petri dishes. A petri dish may contain up to 2 ml culture medium. Allow cells to adhere and grow them for at least 24 h at 37°C, 5% CO_2 to reach ~70% confluency at the day of bead loading or microinjection.

3.2. Bead Loading PNA Probe

1. Soak the glass beads overnight in 4 M NaOH.

2. Wash the beads repeatedly in distilled water until the water reaches a pH of 7.0.

3. Aspirate the culture medium from cells grown in a glass bottom petri dish and replace it with serum-free DMEM, 37°C. Repeat this twice to rinse the cells.

4. Aspirate the medium and replace it with a mixture of 10 μL 2× loading buffer, 7 μL glass-bead solution (containing 0.15 g beads, Sigma), and 3 μL PNA solution (final concentration ~1 μM) on the glass surface (see Note 6).

5. Shake the petri dish twice with some force to rock the beads over the cells. It is through the holes that are created by the beads that the PNA probe gains access into the cytoplasmic compartment before the plasma membrane reseals (see Note 7).

6. Carefully wash the cells twice with serum-free medium.

7. Incubate the cells in DMEM culture medium and allow the cells to recover at 37°C and 5% CO_2. Optimal labeling intensity is obtained about 2 h after loading. An example of telomeres detected by a Cy3-labeled PNA probe is shown in Fig. 1.

3.3. Microinjection of 2'-O-methyl RNA Probe

1. Prepare the probe solution by mixing 24.5 μL H_2O, 25 μL 2× microinjection buffer, and 0.5 μL probe. Final probe concentration is 20–150 ng/μL.

2. Microcentrifuge this solution for 10 min at room temperature to remove aggregates and keep the solution on ice.

3. Remove the culture medium from the petri dish and add 1 mL DMEM culture medium to rinse the cells.

4. Remove the medium and add 3 mL Hepes-buffered culture medium. Hepes is important here to maintain the pH of the culture medium outside the tissue culture incubator.

Fig. 1. A confocal section showing the localization of telomere sequences in a U2OS cell. Cells were grown in a glass bottom petri dish and bead-loaded with a Cy3-labeled $(C_3TA_2)_3$ PNA probe. This image was taken 2 h after loading, using a confocal microscope. The *bright dots* indicate the position of telomeres, which vary in size. The cytoplasm and nucleoplasm are only weakly fluorescent due to unbound PNA probe.

5. Place the petri dish in the heating ring of an inverted fluorescence wide-field microscope after adding a drop of emersion oil on top of the 100× lens.

6. Backfill a prepulled microinjection needle with probe solution using a microloader. Alternatively, pull microinjection needles from glass capillaries using a puller (see Note 8). Make sure that the tip of the needle is filled completely and avoid any air bubble.

7. Set a compensation pressure at 0.2 psi to prevent flow of the culture medium into the injection needle due to the capillary force.

8. Fix the needle in the needle holder on the microscope and place it in a 45° position relative to the bottom of the petri dish such that the tip of the needle sticks in the culture medium.

9. Look in the microscope and gradually lower the needle to the bottom of the petri dish. Use the micromanipulator to carefully position the needle over the cytoplasm of a cell. This is done best by keeping the cells in focus and moving the needle down until the tip is in focus on top of a cell. Position the needle using phase–contrast illumination on top of a cell.

10. Set the pressure at 15 psi (100 hPa). The injection solution flows continuously through the needle.

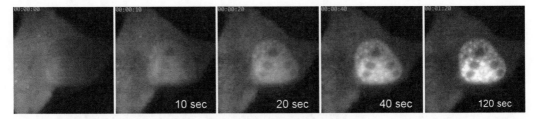

Fig. 2. Probe sequences move after microinjection into the cytoplasm rapidly into the cell nucleus. A 2′-O-methyl (U)22-Cy3 RNA probe was microinjected into the cytoplasm of a U2OS cell and images were taken at regular time intervals. The images shown were taken immediately after injection and 10 s, 20 s, 40 s, and 2 min thereafter, respectively. Immediately after injection, the probe is still present in the cytoplasm. Then, the probe rapidly accumulates in the cell nucleus and hybridizes to poly(A) RNA. The speckled distribution pattern is characteristic for poly(A) RNA localization in the cell nucleus (14).

11. Microinject the cytoplasm of a cell with the fluorescent probe solution by quickly moving the needle tip in and out of the cell. A gentle flow of probe solution in the injected cell should be visible without creating distortion of the cell (see Note 9). By switching the microscope to the fluorescence mode, the injected probe can be visualized. The best results are obtained when the injection volume is in the order of 100 fL (femtoliters) (see Note 10). In a typical experiment, approximately 100 cells can be injected in a petri dish. As shown in Fig. 2, probes injected in the cytoplasm rapidly enter the cell nucleus and hybridize to target molecules.

3.4. Imaging of Cells

For microinjection of probes into cells, we use a Zeiss Axiovert 135 TV inverted fluorescence microscope that also allows for phase–contrast imaging. TAMRA can be imaged using a filter set consisting of a 546/10 nm BP filter, a 580 nm DM and a 590 LP barrier filter, and a CCD camera (Photometrics); we image our cells in more detail using a confocal laser scanning microscope or a fluorescence wide-field microscope designed for multidimensional live cell imaging.

4. Notes

1. Many cell types grow well in this culture medium but some may require a specific medium.

2. This probe sequence is complementary to the 3′ telomere overhang that consists of repeat sequences. The complementary PNA probe results in poor hybridization signals, probably because it has limited accessibility due to the presence of proteins that specifically bind to telomere ends.

3. 2′-O-methyl RNA probes perform better than regular oligodeoxynucleotide probes because they form more stable hybrids with their target sequences and are less prone to degradation by nucleases. As a result, more specific hybridization signals are obtained using 2′-O-methyl RNA probes (12). When designing a probe, it should be taken into account that the probe sequence should be specific for the selected target RNA only. These probes can be designed as linear molecules or as molecular beacons. Molecular beacons are in principle a closed nonfluorescent conformation in the cell when they are not hybridized to a target sequence. In practice, however, these probes may open up due to their interaction with nonspecific sequences or certain proteins giving rise to less optimal signal to noise ratio after delivery into cells (12). Therefore, a careful design of molecular beacon probes should prevent such a problem.

4. Prepulled needles give a better reproducibility in microinjection. From a cost perspective, it might be convenient to pull needles from glass capillaries.

5. The type of microscope system is related to the research question. When fast imaging of several objects in different colors is required, a live cell fluorescence imaging system equipped with external filter wheels might be a good choice, while a sophisticated confocal microscope system could be the choice when detailed spatial information in 3D and time is needed. It should be taken into account that photobleaching during tracking of fluorescent probes can be a serious problem.

6. It is important to keep the volume of the probe–glass bead mixture as small as possible for efficient labeling.

7. Loading efficiency depends on cell density and cell type and can be improved by using larger glass beads (up to 450 μm) or by increasing the frequency by which the glass beads are rocked over the cells. It should be taken into account that cell viability after bead loading decreases with increasing bead size.

8. The needle tip of a pulled needle is sometimes too thin to allow a proper flow of fluid. By gently pushing the tip of the glass needle to the bottom of the petri dish, a part of the needle tip can be broken-off creating a slightly larger opening.

9. It requires some skill and practice to find the optimal settings to obtain consistently good needles. Needles should be stored in a dust- and moist-free container and discarded after several weeks when they are not used.

10. The injection volume can be determined by microinjecting a fluorescent solution in a drop of oil. By measuring the size of the injected solution, which is visible as a droplet, the volume

can be calculated. However, because there is a continuous flow of injection volume, the microinjection volume cannot be accurately determined. It is important that the injection volume does not cause a visible deformation of the injected cell.

Acknowledgments

We wish to thank Dr. K. Vang Nielsen for providing PNA probes. This work was supported by Cyttron grant no. BSIK03036.

References

1. Lanctôt, C., Cheutin, T., Cremer, M., Cavalli, G., and Cremer, T. (2007) Dynamic genome architecture in the nuclear space: regulation of gene expression in three dimensions. *Nature Rev Genet* **8**, 104–115.

2. Foster, H. A., and Bridger, J. M. (2005) The genome and the nucleus: a marriage made by evolution. Genome organisation and nuclear architecture. *Chromosoma* **114**, 212–229.

3. Bolzer, A., Kreth, G., Solovei, I., Koehler, D., Saracoglu, K., Fauth, C., et al. (2005) Three-dimensional maps of all chromosomes in human male fibroblast nuclei and prometaphase rosettes. *PloS Biol* **3**, e157.

4. Meaburn, K.J., and Misteli, T. (2008) Locus-specific and activity-independent gene repositioning during early tumorigenesis. *J Cell Biol* **180**, 39–50.

5. Dirks, R.W., Hattinger, C.M., Molenaar, C., and Snaar, S.P. (1999) Synthesis, processing, and transport of RNA within the three-dimensional context of the cell nucleus. *Crit Rev Eukaryot Gene Expr* **9**, 191–201.

6. Smith, K.P., Byron, M., Johnson, C., Xing, Y., and Lawrence, J.B. (2007) Defining early steps in mRNA transport: mutant mRNA in myotonic dystrophy type I is blocked at entry into SC-35 domains. *J Cell Biol* **178**, 951–964.

7. Trinkle-Mulcahy, L., and Lamond, A.I. (2008) Nuclear functions in space and time: Gene expression in a dynamic, constrained environment. *FEBS Lett* **582**, 1960–1970.

8. Dirks, R.W., and Tanke, H.J. (2006) Advances in fluorescent tracking of nucleic acids in living cells. *Biotechniques* **40**, 489–496.

9. Santangelo, P., Nitin, N., and Bao, G. (2006) Nanostructured probes for RNA detection in living cells. *Ann Biomed Eng* **34**, 39–50.

10. Dirks, R.W., Molenaar, C., and Tanke, H.J. (2003) Visualizing RNA molecules inside the nucleus of living cells. *Methods* **29**, 51–57.

11. McNeil, P.L., and Warder, E. (1987) Glass beads load macromolecules into living cells. *J Cell Sci* **88**, 669–678.

12. Molenaar, C., Marras, S.A., Slats, J.C.M., Truffert, J.-C., Lemaître, M., Raap, A.K., Dirks, R.W., and Tanke, H.J. (2001) Linear 2' O-Methyl RNA probes for the visualization of RNA in living cells. *Nucleic Acids Res* **29**, e89.

13. Molenaar, C., Wiesmeijer, K., Verwoerd, N.P., Khazen, S., Eils, R., Tanke, H.J., and Dirks, R.W. (2003) Visualizing telomere dynamics in living mammalian cells using PNA probes. *EMBO J* **22**, 6631–6641.

14. Molenaar, C., Abdulle, A., Gena, A., Tanke, H.J., and Dirks, R.W. (2004) Poly(A)⁺ RNAs roam the cell nucleus and pas through speckle domains in transcriptionally active and inactive cells. *J Cell Biol* **165**, 191–202.

Part III

Translational FISH: Applications for Human Genetics and Medicine

Chapter 18

Quality Control in FISH as Part of a Laboratory's Quality Management System

Ros Hastings

Abstract

Quality control in the laboratory setting requires the establishment of a quality management system (QMS) that covers training, standard operating procedures, internal quality control, validation of tests, and external quality assessment (EQA). Laboratory accreditation through inspection by an external body is also desirable as this provides an effective ... uring quality and also reassures the patient that ... dards. The implementation of fluorescence in si... ory requires rigorous quality control with atten... systematic approach to the validation of prob... dity of all FISH tests performed, technical proc... results. Knowledge of the limitations of any ... e being examined, since errors of analysis and ... A structured QMS with internal quality con...

..., Quality management system, Specificity

1. In

Quality control includes internal quality control and external quality assessment (EQA). Both are essential for ensuring the reliability and accuracy of a laboratory undertaking diagnostic or research procedures, while the competency of a laboratory and its adherence to accepted guidelines and standards is overseen by accreditation (1). All aspects of quality control for fluorescence in situ hybridization (FISH) tests require particular consideration, since the results may have relevance to important lifetime decisions both for the individuals being tested and for their family.

Joanna M. Bridger and Emanuela V. Volpi (eds.), *Fluorescence in situ Hybridization (FISH): Protocols and Applications*, Methods in Molecular Biology, vol. 659,
DOI 10.1007/978-1-60761-789-1_18, © Springer Science+Business Media, LLC 2010

Not all individuals tested will have clinical phenotype or symptoms; hence, the accuracy of the tests and the results are of paramount importance. This is equally essential whether FISH is used as a stand-alone diagnostic test or as an adjunct to chromosome analysis, microarray, or other form of analysis. Many clinical tests involving genetics are only done once and therefore processes must be in place to minimise diagnostic errors.

FISH can be applied to a wide variety of specimen and sample preparation types, e.g. paraffin-embedded tissue sections, disaggregated cells from paraffin blocks, bone marrow, blood smears, touch-preparations of cells from lymph nodes or solid tumour and metaphase preparations from cultured cells. Routine cytogenetic analyses can yield normal results because of an inherent low mitotic index and/or lack of response to stimulation of certain cell types, for example, mature B cells; therefore, FISH is a useful tool for detecting chromosomal aberrations in these "non-dividing" interphase cells. Assessment of interphase nuclei from uncultured preparations allows for a rapid screening for specific chromosome rearrangements or numerical abnormalities associated with haematologic malignancies or in prenatal follow-up tissue.

1.1. Internal Quality Control

Internal quality control covers the documentation of the internal processes of a laboratory such as reagent batch and expiry dates, specificity and sensitivity of FISH probes, probe validation, standard operating procedures (SOPs) for technical and analytical aspects of the laboratory service, and audit. The internal quality control parameters differ depending on the sample type or FISH probe type. However, some general principles always apply, such as the need for comprehensive documentation of all reagents, probes, and processes used on the sample, to permit a comprehensive vertical audit trail from sample receipt to the final report.

1.2. External Quality Assessment

EQA is undertaken by an organisation or body independent from the laboratory and measures technical, analytical, and interpretative performance. FISH EQA usually involves the distribution of samples or images, where the results are known to the EQA scheme, to laboratories for analysis and reporting (2, 3). Using assessors who are experts in the field, laboratory performance is marked against pre-determined marking criteria, which include a definition of the level at which satisfactory performance is achieved. EQA is recognised by international standards (ISO) and accreditation bodies as a tangible measure of the quality of a laboratory's performance (1, 4). EQA also gives assurance, both to patients and referring clinicians, that the diagnostic laboratory is competent to produce results that are reliable and accurate. Participation in EQA should be continuous for all aspects of the diagnostic service and is required if a laboratory is to become accredited.

1.3. Accreditation

Accreditation involves inspection by a body independent of the laboratory. The accreditation process examines the laboratory according to international standards with respect to the premises, facilities, information system, training, health and safety, and quality management systems (QMSs), including internal quality control and participation in EQA. Audits of all stages of the examination process are undertaken during the inspection (4, 5).

2. Materials

2.1. Internal Quality Control

2.1.1. Records

A record of all probe details including probe name, manufacturer, catalogue number, batch number, expiry date, total volume, number of tests taken from a batch, and date on which each aliquot was used should be maintained. In addition, a reference document should be kept, containing photographs or drawings of normal and abnormal signal patterns including simple and more complex variant forms, for all probes and combination sets in use in the laboratory.

2.1.2. Analysis Forms

An analysis form should record the case details, specific probe(s) used including the manufacturer, microscope used, signal patterns observed, and signatures of the first analyst and the checking analyst. For each case, records of FISH analysis should be filed with any other test results.

2.1.3. Standing Operating Procedures

SOPs should be written for each laboratory process and must be available to staff at the point of use. SOPs include:

- Location at which the procedure takes place
- Staff trained to undertake the procedure
- Personal protective equipment required
- Reagents
- Material and equipment including their locations
- Risk assessments
- Procedure(s) explained in simple unambiguous steps

2.2. External Quality Control

If no national EQA programme exists, then participation in a European or an international EQA scheme is appropriate. CAP, CEQA, UK NEQAS are examples of such international schemes offering EQA in various aspects of the diagnostic service (6–8).

2.3. Accreditation

National accreditation bodies affiliated to the European Co-operation for Accreditation Bodies (EA) should have examples of content required for some of the documentations such as the Quality Manual, Quality Policy, and Health and

Safety Records. Information on national accreditation bodies can be found on the European Co-operation for accreditation website (9).

3. Methods

3.1. Internal Quality Control

Whether a FISH probe is purchased from a commercial supplier, manufactured within the laboratory, or acquired from elsewhere, validation is needed when it is introduced into the laboratory, including specific validation of the probe itself (*probe validation*) and validation of the procedures using the probe (*analytical evaluation*).

3.1.1. Probe Validation (Specificity and Sensitivity)

Sensitivity is the proportion of targets demonstrating a signal, while specificity is the proportion of signals at the target site compared with other chromosome regions. Probe specificity validation includes determination of the signal intensity and signal pattern as well as excluding any significant cross-hybridisation or contamination, which might confound the results. Sensitivity and specificity must be high (>95%) to avoid misdiagnosis. For most commercially available probes, the supplier has usually established these parameters. For home grown probes, these checks must be done internally before the probe can be used diagnostically. All internal validation data should be fully documented for later audit.

3.1.2. Analytical Evaluation of FISH Probes

3.1.2.1. Interphase Nuclei

The first time a new batch of probes is used in a diagnostic department, it should be tested to check that it hybridises to the correct location on the chromosome(s) using metaphase preparations. This validation can be achieved by assessing the probe characteristics on known positive and/or negative samples using reverse DAPI or sequential G-banding (3, 10–12). While the availability of normal chromosome samples does not pose a problem, where possible it is preferable to include a sample with the specific aberration as a control.

To determine the test parameters used for the interpretation of the interphase signals, the analytical validation also requires an understanding of the type of probe, the topography of the DNA in interphase, and the disease characteristics. The topography of the chromosomes at interphase is important because the condensation of some chromosome regions is variable; for example, the alpha-satellite DNA of chromosome 18 is relatively under-condensed compared to the chromosome X and Y probes used in conjunction with it in the rapid aneuploidy screening kit, resulting in dispersed signals that must not be confused with an additional signal. Single locus-specific probes, fusion probes, or break-apart probes, all have different signal patterns, and depending on the type of sample, tumour load or cell involvement in the different diseases, different analysis regimes are likely to be required.

For interphase FISH analysis, the extent of signal variation in normal samples needs to be determined so that distinction can be made between normal and abnormal results. To determine the limits for designating whether a result is normal or abnormal (in some cases mosaic), the background level of false positive and negative signals need to be established. One approach to validate a dual-colour, dual-fusion probe, for example PML/RARA, would be to analyse 500 interphase nuclei from bone marrow in a minimum of ten normal (karyotype) individuals plus a minimum of ten AML individuals who are PML-RARA fusion positive (13). However, sometimes the probes, e.g. BCR/ABL, are fusion positive in several diseases (AML, CML, ALL) and then the validation is done across the different diseases, requiring the evaluation of more fusion positive cases. Care must be taken that the cut off for one disease type is in the same range as other disease types. Wiktor and collaborators suggested that statistical analysis of the normal and abnormal cut-offs could be calculated using the Microsoft Excel β inverse function, =BETAINV (95% confidence level) (14). Once the limits have been established, a few normal and abnormal cases should be assessed to ensure they comply. Continuous quality monitoring of the probes and annual EQA auditing should be performed to ensure the limits set are still applicable.

3.1.2.2. Metaphase Chromosomes

Provided there is an internal control probe, there is no need to validate a new probe as the normal chromosome and the control probe act as reliable internal controls.

3.1.3. Probe Lots

New batches of probe or reagents used in the preparation of slides must be tested prior to use. Validation can be achieved by comparing new and old batches side by side and ensuring the results are equivalent. To monitor FISH testing over time and assess any adverse technical trends, a regular audit of the FISH assay performance is recommended to ensure that the sensitivity and specificity do not change.

3.1.4. Analysis

It is essential to know the probe locus and fluorochrome(s), the expected signal pattern, and the criteria for definition of signal distance to define split signals (for break-apart probes) and fused signals (for dual-colour fusion), prior to use by reference to the relevant SOP or manufacturer's instructions. The signal distance required to define split or fused signals is different for different probes, as is the signal pattern. It is important to be aware that, for the same probe name, different manufacturers may use different loci that may result in different signal patterns or results for the same abnormal sample e.g. IGH/CCND1 probe.

For interphase nuclei, particular emphasis should be placed on distinguishing between a duplex signal (two signals in very close proximity, representing two chromatids from a single chromosome) and two real separate signals. For the dual-colour fusion probes, the emphasis is upon distinguishing between two signals overlapping (co-localised) by chance, or genuinely fused by translocation (or other structural rearrangement), as well as recognising other clinically significant variant patterns. For the break-apart probes, where an overlapping red/green or fused yellow signal represents the normal pattern, and separate red and green signals indicate the presence of a rearrangement, the emphasis should be on judging the distance between signals necessary to call them split. Manufacturers' guidance is usually that one signal distance apart is sufficient, but there are exceptions where the distance has to be greater e.g. those probes involving the IGH locus, because signals tend to be more diffuse. It is important to stress that the expected normal signal patterns are totally different for fusion and break-apart probes, and the wrong interpretation could be made if the analyst is not aware of the type of probe and conversant with the expected normal signal pattern.

In prenatal diagnosis, the presence of maternal contamination (MCC) should be noted and recorded accordingly as with excessive MCC the interphase FISH result cannot be interpreted. Amniotic fluid samples have a non-homogeneous cell population with different cell types and, in later gestation, samples may contain debris making analysis of interphase nuclei more difficult.

Visualising the probes on metaphases may be used to confirm an abnormal result detected on the interphase nuclei, particularly if an atypical abnormal FISH pattern is observed. Understandably, some samples, such as haematologic blood and direct preparations of chorionic villus samples, may not have metaphases available, and use of an internal control (for another locus on the same chromosome, for example) may be necessary for atypical FISH patterns. It is important to remember that interphase FISH analysis can only detect a subset of chromosome abnormalities and may not provide a complete result, or may be misleading in the absence of conventional banded cytogenetic analysis on metaphases. When interpreting interphase FISH results, an understanding of the range of different aberrations that could result in abnormal, variant, and normal results is essential. Hence, it is also appropriate to ensure that staff are aware of any previous cytogenetic or molecular genetic results for the patient. This particularly applies to analysis of oncology samples.

To obtain the requisite number of analysed cells (interphase or metaphase), a methodical scan of non-overlapping rows should be undertaken, so the entire target area is covered. Avoid nuclei in contact with others, nuclei in which signals are obscured by debris or excessive cytoplasm, and nuclei with an interrupted nuclear membrane (interphase analysis) or "broken" metaphases (metaphase analysis). Be aware that the abnormal metaphases may be of poorer quality than the normal metaphases in haemato-oncology samples. Also, some interphase nuclei, particularly mature neutrophils in blood preparations, can look like small clusters of overlapping nuclei. Care should be taken to avoid unnecessarily excluding such nuclei, where selective scoring of mature neutrophils is justified (15). Where possible, it is helpful to confirm an abnormal atypical signal pattern on interphase nuclei analysis with metaphases.

The signal patterns observed must be recorded on the analysis sheet, and the analysis should comply with National or International recommendations. The primary analysis for interphase nuclei (i.e. excluding the checkers' analysis) is usually 100 nuclei for oncology samples, 20 nuclei for prenatal samples, and 10 nuclei for microduplication/deletion in postnatal samples. For metaphase analysis, only five metaphases are required for all tissue types. However, in oncology, or other clonal or mosaic situations, the level of analysis may need to be increased in keeping with the known or likely proportions of abnormal cells. The American College of Medical Genetics, Association of Clinical Cytogeneticists, and European Cytogenetic Guidelines have provided guidance for the number of cells (interphase nuclei/metaphases) that should be included in FISH studies for microdeletions, enumeration of chromosomes in interphase cells from prenatal samples, and interphase screening of haemato-oncology samples for numerical and structural abnormalities (10–12, 16).

The accuracy of any FISH test is related to the number of cells (nuclei/metaphases) scored and the analytic sensitivity of the probes used. Caution should be taken in interpreting breakpoint positions on metaphases from whole chromosome FISH (paint) results and should only be done in conjunction with banding studies. It should be noted that the resolution of chromosome painting may vary between different paints. Small rearrangements may not be detected since whole chromosome paints may not be uniformly distributed along the full length of the target chromosome.

If microduplication is suspected, results should preferably be confirmed by alternative methodologies (e.g. molecular analysis, densitometry). Duplication of, for example, the Prader–Willi syndrome region may only be reliably detected in interphase nuclei or occasionally in the extended chromosomes of prophase or prometaphase (3).

All FISH analyses should be checked by a second trained member of staff. The checker should have no knowledge of the first person's analysis. The checker should examine 30–70% of the total of cells used for the analysis (11). For metaphase FISH, the checker needs to analyse two metaphases. If the results from both individuals are discrepant, a third analyst should be brought in.

If hybridisation is not optimal, the test needs to be repeated. When a deletion or another rearrangement is suspected, if no internal control probe is present, e.g. SRY, then the results must be confirmed with at least one other probe. There may be variation in probe signals both between slides (depending on age, quality, etc. of metaphase spreads) and within a slide. Where a deletion or a rearrangement is suspected, the signal on the normal chromosome constitutes the best control of hybridisation efficiency, and the control probe also provides an internal control for the efficiency of the FISH procedure.

The completed analysis form, signed by the analyst and checker, should make note of any case where the analysis is suboptimal or does not conform to best practise guidelines. For the benefit of the referring clinician, the limitations of any FISH analysis should be documented on the final report issued from the laboratory, in particular as further tests may be required.

When there are discrepancies between the expected laboratory findings, and the clinical referral, results should preferably be followed up by karyotype analysis.

3.1.5. Result Reporting

A system for FISH nomenclature, the International System for Human Cytogenetic Nomenclature (ISCN) (2009), including both metaphase and FISH analysis, has been developed to convey the precise nature of a result (17). Although the system may seem confusing to those not working directly with chromosomes, correct nomenclature designations are important.

The laboratory report should contain ISCN, a written description and an interpretation of the findings. It is important to include a statement as to whether the FISH result is normal or abnormal as the ISCN nomenclature may be misinterpreted by the recipient of the report. The report should also clearly indicate the number of abnormal and normal cells and whether FISH results are from metaphases or interphase nuclei (or from both). The probe sets used to obtain results (see ISCN, 2009 for preferred options) must also be included on the written report, including the name of the manufacturer. The report must indicate any specific limitations of the assay, some of which may be described in the probe manufacturer's information. The report should clearly indicate both the diagnostic and/or prognostic significance of the FISH findings, including their relevance to the reason for referral, the patient's age, or other clinical factors, if relevant. Longitudinal studies of patients should be compared with their previous genetic test findings, and the report should make clear recommendations concerning future testing of such patients, e.g. metaphase chromosome analysis, interphase FISH, and molecular reverse transcription-PCR testing (18). The report requires a summary of the FISH signals, a written description of the signal pattern, and an interpretation of the findings in relation to the clinical referral. The conventions included in the standard nomenclature, ISCN 2009, allow a succinct and precise portrayal of the karyotype and any anomalies it might contain (17).

3.2. External Quality Assessment

Annual participation in EQA is required to confirm ongoing ascertainment of a satisfactory performance. If poor performance is identified then an internal audit is required to identify what corrective action(s) is needed to improve the quality of the diagnostic service.

Common errors identified through EQA include failure to follow routine procedures, which results in incorrect interpretation of the results. These include unawareness of breakpoint heterogeneity; underestimating the level of residual disease in chronic myeloid leukaemia by not analysing neutrophils in follow-up FISH BCR/ABL studies (15); not taking into account the tumour load of a sample e.g. a low-level clone in a sample with a small tumour load is significant but may not be if the tumour load is high (depending on clinical indication); and misinterpreting colocalised signals as fused, or diffuse fusions as split signals or misunderstanding of the characteristics of the probe set used in the analysis.

3.3. Accreditation

The QMS involves documentation of all the processes within a laboratory, including the procedures and policies that are required for the planning and execution of a quality service and to satisfy the accreditation body. There should also be a policy and procedure in place that can be implemented when the laboratory detects that any aspect of its examination process does not conform with its own procedure. Procedures for corrective action should include an investigation process to determine the underlying cause(s) of the problem. If preventive action is required, then action plans should be developed.

A good QMS leads to a better organisation of laboratory workflow, more traceability, and more accurate results. A QMS may include the following areas:

- Personnel training and qualification
- Document control
- Purchasing control mechanisms
- Product identification and traceability at all stages of the procedure
- Controlling and defining the FISH procedure/process
- Defining and controlling inspection, measuring and test equipment
- Validating processes
- Product acceptance criteria
- Instituting corrective and preventive action when errors occur

- Labelling and packaging controls
- Handling, storage, distribution, and installation
- Records
- Servicing

The QMS should comply with OECD guidelines and international standards such as ISO 15189 and ISO 9001. Most accreditation bodies accredit laboratories under the standard ISO 15189.

4. Notes

Laboratories should regularly audit sample success rates and overall preparation quality. Where standards fall below the agreed criteria, it should be possible to investigate the underlying reasons and then investigate measures to rectify any deficiency. Any procedural, analytical, or reporting errors should be checked regularly and how these are rectified should be documented.

Acknowledgments

The author would like to thank Rod Howell and Mike Griffiths (UK NEQAS Steering Committee), for reviewing and giving helpful suggestions on this manuscript.

References

1. OECD (2005) Quality assurance and proficiency testing for molecular genetic testing: survey of 18 OECD member countries. www.oecd.org/sti/biotechnology/ (also applicable to Cytogenetic laboratories).

2. Hastings, R.J., Maher, E.J., Quellhorst-Pawley, B., Howell. R. (2008) An internet-based External Quality Assessment (EQA) in cytogenetics that audits a laboratory's analytical and interpretative performance. *Eur. J. Hum. Genet.* 16, 1217–1224.

3. Mascarello, J.T., Brothman, A. R., Davison, K., Dewald, G.W., Herrman, M., McCandless, D., Park , P., Persons, D. L., Rao, K.W., Schneider, N. R., Vance, G. H., Cooley, L. D. (2002) Proficiency testing for laboratories performing fluorescence in situ hybridization with chromosome-specific DNA probes. *Arch. Pathol. Lab. Med.* 126, 1458–1624.

4. ISO 15189. (2003) Medical laboratories – particular requirements for quality and competence. http://www.iso.org/iso/standards_development/technical_committees/list_of_iso_technical_committees.htm

5. Howell, R., Hastings, R.J. (2006) The current scope of cytogenetics external quality assessment schemes and key recommendations for harmonization of external quality assessment in Europe. www.eurogentest.org

6. College of American Pathologists website. www.cap.org

7. European Cytogenetic Quality Assurance website. www.cyto-ceqa.eu

8. UK NEQAS for Clinical Cytogenetics website. www.ccneqas.org.uk

9. EA website.www.european-accreditation.org/content/database/Links.mpi

10. Bethesda, M.D. ed. (1999) ACMG Standards and Guidelines for Clinical Genetics Laboratories. 2nd: American College of Medical Genetics.

11. Hastings, R.J., Cavani, S., Dagna Bricarelli, F., Patsalis, P.C., Kristoffersson, U. (2006) Cytogenetic guidelines and quality assurance: A common European framework for quality assessment for constitutional and acquired cytogenetic investigations. *ECA newsl.* **17**, 13–32. www.biologia.uniba.it/eca and www.eurogentest.org

12. Hastings, R.J., Cavani, S., Dagna Bricarelli, F., Patsalis, P.C., Kristoffersson, U. (2007) Cytogenetic guidelines and quality assurance: A common European framework for quality assessment for constitutional and acquired cytogenetic investigations. A summary. *Eur. J. Hum. Genet.* **15**, 525–527.

13. Wolff, D.J., Bagg, A., Cooley, L.D., Dewald, G.W., Hirsch, B.A., Jacky, P.B., Rao, K.W., Rao, P.N. (2007) Guidance for fluorescence in situ hybridization testing in hematologic disorders. *J. Med. Diagn.* **9**, 134–143.

14. Wiktor, A.E., Van Dyke, D.L., Stupca, P.J., Ketterling, R.P., Thorland, E.C., Shearer, B.M., Fink, S.R., Stockero, K.J., Majorowicz, J.R., Dewald, G.W. (2006) Preclinical validation of FISH assays for clinical practice. *Genet. Med.* **8**, 16–23.

15. Reinhold, U., Hening, E., Leiblein, S., Niederweiser, D., Deininger, N.W.N. (2003) FISH for BCR-ABL interphases of peripheral blood neutrophils but not of unselected white cells correlates with bone marrow cytogeneitcs in CML patients treated with imatinib. *Leukaemia* **17**, 1925–1929.

16. Association for Clinical Cytogenetics. Professional guidelines for clinical cytogenetics.www.cytogenetics.org

17. Shaffer, L.G., Slovak, M.L., Campbell, L.J. eds. (2009) ISCN (2009): An International System for Human Cytogenetic Nomenclature. S. Karger, Basel

18. Tefferi, A., Dewald, G.W., Litzow, M.L., Cortes, J., Mauro, M.J., Talpaz, M., Kantarjian, H.M. (2005) Chronic myeloid leukemia: Current application of cytogenetics and molecular testing for diagnosis and treatment. *Mayo Clin. Proc.* **80**, 390–402.

Chapter 19

Flash FISH: "Same Day" Prenatal Diagnosis of Common Chromosomal Aneuploidies

Sherry S.Y. Ho and Mahesh A. Choolani

Abstract

Fluorescence in situ hybridization (FISH) and quantitative fluorescence (QF)-PCR are rapid molecular methods that test for common chromosomal aneuploidies in prenatal diagnosis. While cytogenetic analysis requires approximately 7–14 days before fetal karyotypes are available, these molecular methods release results of sex chromosome aneuploidies, Down syndrome, Edward's syndrome, and Patau's syndrome within 24–48 h of fetal sampling, alleviating parental anxiety. However, specific diagnosis or exclusion of aneuploidy should be available within the same day of amniocentesis. We developed "*Flash*FISH," a low cost FISH method that allows accurate results to be reported within 2 h of fetal sampling. Here, we report our experience of using *Flash*FISH in prenatal diagnosis, and we illustrate in detail the protocols used for the purpose in our laboratory.

Key words: Cytogenetics, Down syndrome, FISH, Molecular diagnosis, Prenatal diagnosis

1. Introduction

Prenatal diagnosis is offered to pregnant women at increased risk of having a fetus with chromosomal abnormalities such as Down syndrome. Since chromosome 13, 18, 21, X, and Y aneuploidies account for 60–80% of fetal abnormalities detected during prenatal diagnosis (1, 2), rapid aneuploidy detection (RAD) methods were developed to report results of these common aneuploidies within 24–48 h of fetal sampling. These RAD methods such as fluorescence in situ hybridization (FISH) (3, 4), quantitative fluorescence PCR (QF-PCR) (5, 6), and multiplex ligation-dependent probe amplification (MLPA) (7, 8) obviate the need for cell culture to prepare metaphase spreads for karyotype analysis. Although RAD methods are already an improvement to cytogenetic

Joanna M. Bridger and Emanuela V. Volpi (eds.), *Fluorescence in situ Hybridization (FISH): Protocols and Applications*, Methods in Molecular Biology, vol. 659,
DOI 10.1007/978-1-60761-789-1_19, © Springer Science+Business Media, LLC 2010

analysis, which typically takes 7–14 days before release of results, prenatal diagnostic results that are available within the same day – as anomalies are suspected after abnormal ultrasound findings and/or abnormal maternal serum screening results – can alleviate significant parental anxieties (9). FISH relies on visual counting of fluorescent signals within target fetal cells rather than comparing ratios of informative genetic markers for quantitative assessment in QF-PCR and MLPA. Results from FISH are informative in all cases where hybridization is successful, and the technique can also be applied for chromosomal rearrangements such as Di George syndrome (DGS) (10). However, commercially available probes that target common chromosomal aneuploidies are costly when compared to fluorescence-labeled primers used in QF-PCR (11). We developed a cost-effective FISH method – *Flash*FISH – that diagnoses common chromosomal aneuploidies within the same day as fetal sampling. In addition to reducing the amount of probe used, without compromising signal quality, we perform routine testing for trisomy 21 and sex chromosome abnormalities, but targeted testing for trisomy 13 and 18 when ultrasound evidence of abnormality is apparent. In a separate modeling study of 268 cases of trisomy 13, 18, 21 and sex chromosome abnormalities, we found that 80% trisomy 18 and 90% trisomy 13 fetuses can be identified by ultrasound while 42.4% trisomy 21 and 50.9% fetuses with sex chromosomal abnormalities had no detectable ultrasonographic markers. Routine testing for sex chromosomes and chromosome 21 with targeted testing for chromosomes 13 and 18 allows 94% of common aneuploidies to be identified. Such strategies will lower cost per sample for FISH. Therefore, in addition to using the Food and Drug Administration-approved AneuVysion® Prenatal Test for targeted testing of trisomy 13 and 18 (Table 1), we used CEP® X/Y Direct Labeled

Table 1
Details of probes in the AneuVysion® Prenatal Test

Probe name	Probe location	Fluorophore
Vysis CEP 18	D18Z1 alpha-satellite DNA probe corresponding to 18p11.1–q11.1	SpectrumAqua™
Vysis CEP X	DXZ1 alpha-satellite DNA probe corresponding to Xp11.1–q11.1	SpectrumGreen™
Vysis CEP Y	DYZ3 alpha-satellite DNA probe corresponding to Yp11.1–q11.1	SpectrumOrange™
Vysis LSI 13	Locus-specific DNA probe corresponding to retinoblastoma gene at 13q14	SpectrumGreen™
Vysis LSI 21	Locus-specific DNA probe corresponding to 21q22.13–q22.2	SpectrumOrange™

Table 2
Details of probes in the CEP® X/Y Direct Labeled Fluorescent DNA Probe Kit

Probe name	Probe location	Fluorophore
CEP X	DXZ1 alpha-satellite DNA probe corresponding to Xp11.1–q11.1	SpectrumOrange™
CEP Y	DYZ1 satellite III DNA probe corresponding to Yq12	SpectrumGreen™

Fluorescent DNA Probe Kit (Table 2) together with LSI13/21 probe mixture from the AneuVysion Prenatal Test kit for routine testing of trisomy 21 and sex chromosomal abnormalities. The time taken from sample collection to the release of results had been shown to be less than 2 h (12). *Flash*FISH represents the necessary development for a same-day prenatal diagnostic service that is cost-effective.

2. Materials

2.1. Sample Pretreatment

2.1.1. Amniotic Fluid Cells (Amniocytes)

1. Prewarmed (37°C) KCl (0.075 M).
2. Carnoy's fixative, methanol:glacial acetic acid (3:1) (v/v) (see Note 1).

2.1.2. Chorionic Villi

1. Collagenase (2,400 U/mL).
2. Phosphate buffered saline (PBS).
3. 1% Sodium citrate:KCl (1:1) (v/v).
4. Carnoy's fixative: methanol:glacial acetic acid (3:1) (v/v) (see Note 1).

2.1.3. Blood (Fetal, Umbilical, Neonatal, Adult)

1. Phosphate buffered saline (PBS).
2. Density gradient Ficoll 1,077 g/mL (Histopaque-1077; Sigma–Aldrich).
3. Carnoy's fixative: methanol:glacial acetic acid (3:1) (v/v) (see Note 1).

2.2. FlashFISH

2.2.1. Slide-Making

1. Frosted end glass slides.
2. Glass cover slips, 22 × 22 mm.
3. Hot plate set at 60°C.

2.2.2. Hybridization

1. AneuVysion® Prenatal Test consisting of CEP® 18/X/Y and LSI®13/21 probe mixtures (Abbott Laboratories, Vysis) (Table 1).

2. CEP® X/Y Direct Labeled Fluorescent DNA Probe Kit (Table 2).

3. Hybridization buffer (50% formamide and 10% dextran sulphate in 2× SSC, pH 7.0).

4. Thermal block (MJ Research PTC-200).

2.2.3. Posthybridization

1. 50 mL 0.4× SSC/0.3% NP-40 (pH 7.0 ± 0.2) in Coplin jar at room temperature.

2. 50 mL 2× SSC/0.1% NP-40 (pH 7.0 ± 0.2) in Coplin jar at 72°C.

2.2.4. Visualization

1. Automated microscope (Olympus BX61) with single Bandpass (SpectrumOrange, SpectrumGreen, SpectrumAqua) FISH filters (Vysis), DAPI filter cube (ASI) fitted with a mercury lamp.

2. Images are captured and analyzed using the CytolabView, version 5.0.

3. Methods

3.1. Sample Pretreatment

3.1.1. Amniotic Fluid Cells (Amniocytes)

1. Centrifuge 2 mL of amniotic fluid at $500 \times g$ for 5 min in a 15 mL tube to pellet the amniocytes.

2. Remove supernatant and add 3 mL KCl.

3. Incubate in a water bath at 37°C for 30 min.

4. Add 2 mL Carnoy's fixative dropwise and continuously tap tube to mix well.

5. Centrifuge at $500 \times g$ for 5 min.

6. Remove supernatant and add 3 mL Carnoy's fixative.

7. Repeat steps 5 and 6 and proceed to slide-making.

3.1.2. Chorionic Villi

1. Remove maternal or anomalous material by careful dissection in a Petri dish with PBS using fine needles (see Note 2).

2. Dissect sorted chorionic villi into fragments with fine needles (see Note 3).

3. Collect fragments into 15 mL tubes containing 480 units collagenase in 1 mL PBS.

4. Incubate in a water bath at 37°C for 45 min (tap tube once to mechanically dissociate fragments during incubation).

5. Centrifuge cell suspension at $500 \times g$ for 5 min and remove supernatant.

6. Resuspend cell pellet with 2 mL Sodium citrate:KCl and incubate for 37°C for 20 min.

7. Add 2 mL Carnoy's fixative dropwise and continuously tap tube to mix well.

8. Centrifuge at $500 \times g$ for 5 min.

9. Remove supernatant and add 3 mL Carnoy's fixative.

10. Repeat steps 5 and 6 and proceed to slide-making.

3.1.3. Blood (Fetal, Umbilical, Neonatal, Adult)

1. Dilute 1 mL blood with 1 mL 1× PBS (1:1).

2. Carefully layer diluted blood onto equal volume of density gradient (2 mL).

3. Centrifuge at $300 \times g$ for 30 min at 15°C.

4. Collect the white interface layer using a 3 mL syringe attached with a 23G needle into 15 mL tube.

5. Add 5 mL PBS to lymphocytes (collected white layer) and invert tube gently to wash cells.

6. Centrifuge at $500 \times g$ for 5 min.

7. Discard the supernatant and resuspend the cell pellet with 2 mL PBS.

8. Centrifuge at $500 \times g$ for 5 min and discard supernatant.

9. Tap gently to dissociate cell pellet and add 1 mL KCl.

10. Incubate in water bath at 37°C for 30 min.

11. Add 2 mL Carnoy's fixative dropwise and continuously tap tube to mix well.

12. Centrifuge at $500 \times g$ for 5 min.

13. Remove supernatant and add 3 mL Carnoy's fixative.

14. Repeat steps 5 and 6 and proceed to slide-making.

3.2. FlashFISH

3.2.1. Slide-Making

1. Drop cells directly onto two cold glass slides and place immediately onto a warmed 60°C hot plate until dry (approximately 1 min) (see Note 4).

2. Mark the area of cells with a diamond pen.

3.2.2. Hybridization

1. Add 5 µL probe mixture onto the area of marked cells and place the coverslips on gently, making sure that there are no air bubbles (see Note 5).

2. Place slides on a thermal block for denaturation at 80°C for 90 s followed by 15-min hybridization at 42°C.

3.2.3. Posthybridization

1. Gently agitate a Coplin jar containing hybridized slides in 0.4× SSC/0.3% NP-40 at 72°C for 2 min.

2. Immediately remove slides and place in a Coplin jar containing 2× SSC/0.1% NP-40. Gently agitate Coplin jar at room temperature for 2 min (see Note 6).

3. Air-dry slides and mount in fluorescence antifade medium containing DAPI.

4. Seal with a coverslip and clear nail varnish.

3.2.4. Visualization

1. Score 50 nuclei directly from the microscope without image enhancement for each probe (see Note 7).

2. Results are considered informative if ≥80% nuclei displayed the same normal/abnormal hybridization pattern for any specific probe (see Note 8).

4. Notes

1. Ensure that the glacial acetic acid and methanol used in preparing Carnoy's fixative are anhydrous (AnalaR grade) and not of HPLC grade. The high water content of HPLC grade chemicals (0.1%) is ten times higher than the anhydrous chemicals (0.01%) and will affect the morphology of the cells during fixation.

2. Chorionic villi are to be collected in a medium that does not contain EDTA, a chelating agent that will remove calcium ions required for collagenase activity.

3. During the dissection of chorionic villi, wet needles or scalpels with collection medium or 1× PBS to prevent adhesion of tissue resulting in cell loss.

4. Slides have to be cold during slide-making so that condensation will form a thin water film to spread the cells.

5. For AneuVysion® Prenatal Test probe mixtures, use 5 μL instead of the recommended 10 μL for each of the probe mixtures LSI 13/21 and CEP 18/X/Y for each cell spot. For CEP X/Y probe mixture, use 5 μL probe mixture consisting of 2 μL CEP probes mixed with 3 μL hybridization buffer.

6. Slides with LSI probe gave better signals when washes were only 1 min each with less agitation.

7. The signal intensity of SpectrumOrange™ LSI probes is stronger when mercury burner lamp is used instead of the xenon burner lamp. This is because unlike mercury lamps, xenon lamps do not have very high spectral intensity peaks. Much of the intensity of the mercury lamp is expended in the near ultraviolet, with peaks of intensity at 313, 334, 406, 435, 546, and 578 nm. The excitation and emission peaks of SpectrumOrange™ are at 559 and 588 nm, respectively (Table 3).

Table 3
Spectral technical data of Vysis probes (adapted from abbottmolecular.com)

Probes	Excitation		Emission	
	Peak	FWHM	Peak	FWHM
SpectrumOrange	559	38	588	48
SpectrumGreen	497	30	524	56
SpectrumAqua	433	53	480	55
DAPI	367	61	452	92

FWHM full width of the spectral band at half of the peak intensity, *DAPI* 4′,6-diamidino-2-phenylindole

All units are nanometers (nm)

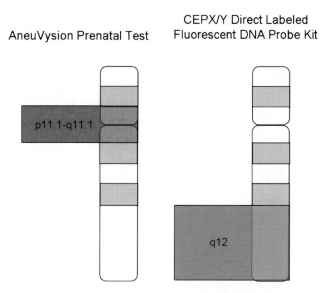

Fig. 1. Different probe target loci for chromosome Y in AneuVysion® Prenatal Test and CEPX/Y Direct Labeled Fluorescent DNA Probe Kit.

8. The corresponding locations of CEP Y probes in the probe mixture CEP18/X/Y (in the AneuVysion Prenatal Test) and the CEPX/Y probe mixture (of CEPX/Y Direct Labeling Fluorescent DNA Probe Kit) are different and may lead to different results in the presence of mutations that affect Yq12 (Fig. 1).

References

1. Evans, M. I., Henry, G. P., Miller, W. A., Bui, T. H., Snidjers, R. J., Wapner, R. J., Miny, P., Johnson, M. P., Peakman, D., Johnson, A., Nicolaides, K., Holzgreve, W., Ebrahim, S. A., Babu, R., Jackson, L. (1999) International, collaborative assessment of 146,000 prenatal karyotypes: Expected limitations if only chromosome-specific probes and fluorescent in-situ hybridization are used. *Hum. Reprod.* **14**, 1213–1216.

2. Caine, A., Maltby, A. E., Parkin, C. A., Waters, J. J., Crolla, J. A. (2005) Prenatal detection of Down's syndrome by rapid aneuploidy testing for chromosomes 13, 18, and 21 by FISH or PCR without a full karyotype: A cytogenetic risk assessment. *Lancet* **366**, 123–128.

3. Klinger, K., Landes, G., Shook, D., Harvey, R., Lopez, L., Locke, P., Lerner, T., Osathanondh, R., Leverone, B., Houseal, T., et al. (1992) Rapid detection of chromosome aneuploidies in uncultured amniocytes by using fluorescence in situ hybridization (FISH). *Am. J. Hum. Genet.* **51**, 55–65.

4. Weise, A., Liehr, T. (2008) Rapid prenatal aneuploidy screening by fluorescence in situ hybridization (FISH). *Methods Mol. Biol.* **444**, 39–47.

5. Pertl, B., Yau, S. C., Sherlock, J., Davies, A. F., Mathew, C. G., Adinolfi, M. (1994) Rapid molecular method for prenatal detection of Down's syndrome. *Lancet* **343**, 1197–1198.

6. Mann, K., Petek, E., Pertl, B. (2008) Prenatal detection of chromosome aneuploidy by quantitative fluorescence PCR. *Methods Mol. Biol.* **444**, 71–94.

7. Schouten, J. P., McElgunn, C. J., Waaijer, R., Zwijnendburg, D., Diepvens, F., Pals, G. (2002) Relative quantification of 40 nucleic acid sequences by multiplex ligation-dependent probe amplification. *Nucleic Acids Res.* **30**, e57.

8. Kooper, A. J., Faas, B. H., Kater-Baats, E., Feuth, T., Janssen, J. C., van der Burgt, I., Lotgering, F. K., Geurts van Kessel, A., Smits, A. P. (2008) Multiplex Ligation-dependent Probe Amplification (MLPA) as a standalone test for rapid aneuploidy detection in amniotic fluid cells. *Prenat. Diagn.* **28**, 1004–1010.

9. Tercyak, K. P., Johnson, S. B., Roberts, S. F., Cruz, A. C. (2001) Psychological response to prenatal genetic counseling and amniocentesis. *Patient Educ. Couns.* **43**, 73–84.

10. Jouannic, J. M., Martinovic, J., Bessières, B., Romana, S., Bonnet, D. (2003) Fluorescence in situ hybridization (FISH) rather than ultrasound for the evaluation of fetuses at risk for 22q11.1 deletion. *Prenat. Diagn.* **23**, 607–608.

11. Shaffer, L. G., Bui, T-H. (2007) Molecular cytogenetic and rapid aneuploidy detection methods in prenatal diagnosis. *Am. J. Med. Genet. C. Semin. Med. Genet.* **145**C, 87–98.

12. Choolani, M., Ho, S. S., Razvi, K., Ponnusamy, S., Baig, S., Fisk, N. M., Biswas, A., Rapid Molecular Testing in Prenatal Diagnosis Group. (2007) FastFISH: Technique for ultrarapid fluorescence in situ hybridization on uncultured amniocytes yielding results within 2 h of amniocentesis. *Mol. Hum. Reprod.* **13**, 355–359.

Chapter 20

FISH for Pre-implantation Genetic Diagnosis

Paul N. Scriven and Caroline Mackie Ogilvie

Abstract

Pre-implantation genetic diagnosis (PGD) is an established alternative to pre-natal diagnosis, and involves selecting pre-implantation embryos from a cohort generated by assisted reproduction technology (ART). This selection may be required because of familial monogenic disease (e.g. cystic fibrosis), or because one partner carries a chromosome rearrangement (e.g. a two-way reciprocal translocation). PGD is available for couples who have had previous affected children, and/or in the case of chromosome rearrangements, recurrent miscarriages, or infertility. Oocytes aspirated following ovarian stimulation are fertilized by in vitro immersion in semen (IVF) or by intracytoplasmic injection of individual spermatocytes (ICSI). Pre-implantation cleavage-stage embryos are biopsied, usually by the removal of a single cell on day 3 post-fertilization, and the biopsied cell is tested to establish the genetic status of the embryo.

Fluorescence in situ hybridization (FISH) on the fixed nuclei of biopsied cells with target-specific DNA probes is the technique of choice to detect chromosome imbalance associated with chromosome rearrangements, and to select female embryos in families with X-linked disease for which there is no mutation-specific test. FISH has also been used to screen embryos for sporadic chromosome aneuploidy (also known as PGS or PGD-AS) in order to try and improve the efficiency of assisted reproduction; however, due to the unacceptably low predictive accuracy of this test using FISH, it is not recommended for routine clinical use.

This chapter describes the selection of suitable probes for single-cell FISH, assessment of the analytical performance of the test, spreading techniques for blastomere nuclei, and in situ hybridization and signal scoring, applied to PGD in a clinical setting.

Key words: Fluorescence in situ hybridization, Pre-implantation genetic diagnosis, PGD, Sex determination, Translocations, Chromosome aneuploidy

1. Introduction

The application of fluorescence in situ hybridization (FISH) to a single embryo cell (blastomere) presents special challenges both in practicalities and in interpretation of the signal pattern. The biopsied cell needs to be spread within a pre-defined area on the slide in

Joanna M. Bridger and Emanuela V. Volpi (eds.), *Fluorescence in situ Hybridization (FISH): Protocols and Applications*, Methods in Molecular Biology, vol. 659,
DOI 10.1007/978-1-60761-789-1_20, © Springer Science+Business Media, LLC 2010

order to facilitate its localization following FISH; extreme care needs to be taken in ensuring that the cell is lysed, that the cytoplasm has been dispersed, and that the nucleus is visible and intact; and, as the diagnosis depends on the results from this single cell, stringent scoring and interpretation guidelines should be applied. However, in experienced hands, FISH is a robust technique for pre-implantation genetic diagnosis (PGD) in clinical practice.

The principle of PGD by FISH is that target-specific DNA probes labelled with different fluorochromes or haptens can be used to detect the copy number of specific loci, and thereby to detect chromosome imbalance associated with Robertsonian translocations, reciprocal translocations, inversions, and other chromosome rearrangements. FISH can also be used to select female embryos in families with X-linked disease, for which there is no mutation-specific test. More controversially, FISH has also been used to screen for sporadic chromosome aneuploidy. This chapter concentrates on the technical aspects and the limitations of FISH applied to clinical single-cell diagnosis.

2. Materials

1. Amine-coated slides (Genetix).
2. Diamond-tipped pen.
3. Cell lysis buffer (for spreading cells): 0.1% Tween-20 in 0.01 M HCl, pH 2.0, at 20°C.
4. Phosphate-buffered saline (PBS), pH 7.0: 0.14 M NaCl, 3 mM KCl, 10 mM Na_2HPO_4, 2 mM KH_2PO_4.
5. Sterile distilled water.
6. Ethanol series: 70, 90, and 100%.
7. Low-power phase contrast microscope (×100).
8. Fluorochrome- or hapten-labelled in situ hybridization (ISH) DNA probes (Vysis, Oncor, CytoCell). Available fluorochromes/haptens and strategies for discriminating probes include the following: TexasRed (TR), fluorescein isothiocyanate (FITC), SpectrumGreen (Vysis), SpectrumOrange (Vysis), SpectrumAqua, biotinylated probes detected with TR-avidin, FITC-avidin, or Cy-5 streptavidin (visualized using a FarRed filter), a mix of red and green probes to produce a yellow signal, a second round of hybridization and a third colour created by sequential capturing of SpectrumOrange using a TexasRed and a SpectrumGold filter Note 1).
9. No. 0 cover slips (22 × 22 mm and 11 × 11 mm).
10. Denaturation hot block at 75°C (Hybrite or other brand calibrated to the exact temperature).

11. Rubber cement (Cow Gum; Cow Proofing).

12. Incubator and hybridization chamber at 37°C.

13. 4× standard saline citrate (SSC)/Tween-20 wash solution: 0.05% Tween in 4× SSC, pH 7.0 (0.6 M NaCl, 60 mM $C_6H_5Na_3O$).

14. 0.4× SSC stringent wash solution: 0.4× SSC, pH 7.0, at 71°C.

15. Parafilm®.

16. 4N,6-Diamidino-2-phenylindole (DAPI)/Vectashield: 160 ng of DAPI in 1 mL of Vectashield mounting medium (Vector Laboratories).

17. Clear nail varnish.

18. Fluorescence microscope fitted with appropriate filters.

19. Imaging software (e.g. Isis, MetaSystems, Altlussheim, Germany; CytoVision, Genetix).

20. Where indirectly labelled probes are used, fluorescent avidin and fluorescent anti-digoxygenin can be employed.

3. Methods

3.1. Selection of Probes

Probe mixes can combine directly labelled and indirectly labelled probes, and probes from different manufacturers. Probes for known polymorphic chromosome regions (1, 2), or those known to cross-hybridize significantly with other chromosomes (3) should be avoided, although can be used if shown to be specific and suitable for PGD by prior testing on both reproductive partners.

3.1.1. Sex Determination

A probe set containing three probes, specific for the centromere regions of the X and Y chromosomes and one autosome, is recommended for sex determination (4); the autosomal probe is used to establish ploidy and thereby to differentiate between trisomy X (2n, 47,XXX) and triploidy (3n, 69,XXX), and between tetrasomy X (48,XXXX) and tetraploidy (4n, 92,XXXX). A typical probe set applied in this setting is the Abbott AneuVysion mix containing alpha-satellite X, Y, and 18; this probe set has been demonstrated to have a very low polymorphism rate, and therefore pre-treatment work-up of the reproductive partners is not required (4).

3.1.2. Chromosome Rearrangements

The probe mix for any specific rearrangement should:

- Ideally contain probes at least sufficient to detect all the expected products of the rearrangement with chromosome imbalance. If this is not possible, probe mixes where the undetected unbalanced products have been assessed to be non-viable in a recognizable pregnancy and are likely to have very low prevalence may be acceptable (4).

- Be tested on cultured lymphocyte metaphases from both reproductive partners. At least ten metaphase spreads should be examined for probe specificity, polymorphisms and cross-hybridization, and, for the chromosome rearrangement carrier, to ensure that the probes hybridize as expected to the different segments of the rearrangement. In addition, at least 100 interphase nuclei from these preparations should be scored to assess signal specificity, brightness, and discreteness (4).

3.1.3. Sporadic Aneuploidy Testing

Commercial probe sets are available (e.g. Abbott MultiVysion PB or PGT), targeting the chromosomes most frequently found to be aneuploid in products of conception, and comprising a single probe per chromosome targeted. Typically the nucleus may have a second hybridization with probes for additional chromosomes providing an assay for chromosomes 13, 15, 16, 18, 21, 22, and XY (4).

3.2. Assessing the Analytical Performance of a Single-Cell Test

For the purpose of the assessment of performance it is assumed that the cell sampled is representative of the embryo, although in practice this is often not the case; whole chromosome mosaicism is particularly prevalent in day 3 human embryos (5–7), probably due to post-zygotic events, typically anaphase lag or non-disjunction of chromosomes. However, mosaicism for a genetic defect on a single chromosome such as a single gene mutation or arising from a chromosome rearrangement involving only two chromosomes is less likely.

The analytical performance of the test depends on

- The accuracy of scoring normal and abnormal copy number for each probe.
- The prevalence of the condition tested (e.g. sex chromosome complement or chromosome aneuploidy).

Sensitivity, specificity, and positive and negative predictive accuracy are recognized to be the most valuable indicators of analytical performance.

- Sensitivity and specificity give a measure of the quality of the test.
- Positive and negative predictive accuracy give measures of the effect that different practical situations have on the test and give the post-test probability of being affected or unaffected.

The use of these statistics to assess different single-cell applications of FISH are given below and solutions are presented in Table 1, where K = the proportion of embryos with a male or aneuploid chromosome complement, L = the proportion of

Table 1

The analytical performance of different PGD FISH applications

FISH application	Sex determination	Two-way reciprocal translocation	Sporadic chromosome aneuploidy	Sporadic chromosome aneuploidy	Sporadic chromosome aneuploidy
Chromosome pairs tested	2	2	7	7	7
Aneuploidy per chromosome (%)	1	–	3	18.7	3
Disomy accuracy per probe (%)	94	94	94	94	99.62
Prevalence	0.509	0.529	0.192	0.743	0.192
Sensitivity	0.999	0.997	0.945	0.970	0.997
Specificity	0.884	0.857	0.648	0.648	0.974
Positive predictive accuracy	0.899	0.887	0.390	0.900	0.900
Negative predictive accuracy	0.998	0.997	0.980	0.870	0.999

normal female or normal/balanced embryos with a male or aneuploid (positive) test result (false abnormal) and M = the proportion of male or aneuploid embryos that have a normal female or normal/balanced (negative) test result (false normal).

The proportion of unaffected embryos that have a normal (negative) test result for the chromosomes tested is:

$$\text{Specificity} = 1 - L$$

The proportion of affected embryos that have a male or aneuploid (positive) test result for the chromosomes tested is:

$$\text{Sensitivity} = 1 - M$$

The proportion of abnormal (positive) test results for the chromosomes tested that have an abnormal (or male) chromosome complement is:

$$\text{Positive predictive accuracy (PPA)}$$
$$= [K - (K \times M)] / [(K - (K \times M)) + L(1 - K)]$$

The proportion of normal (negative) test results for the chromosomes tested that have a normal chromosome complement is:

$$\text{Negative predictive accuracy (NPA)}$$
$$= [(1 - K) - L(1 - K)] / [(1 - K) - (L(1 - K) + K \times M)]$$

3.2.1. Sex Determination

In general, PGD for sex determination is carried out for couples where the female partner carries a recessive single gene defect on one of her X chromosomes (e.g. Duchenne muscular dystrophy). Any male embryos are therefore at 50% risk of inheriting the defect and manifesting the genetic condition. It is therefore important to estimate the likelihood of misdiagnosing male embryos as normal females, and recommending them for transfer, resulting in an affected pregnancy. Two errors (failure to detect the Y chromosome signal in addition to detection of an extra X chromosome signal) in addition to accurately scoring a normal copy number for the autosome are required to make such a misdiagnosis.

For example, assuming that:

- The FISH gives 94% disomy accuracy per chromosome tested.
- The prevalence of male or aneuploid embryos is 50.9% (based on a 1:1 sex ratio and only 1% aneuploidy per chromosome pair).

Then, using the calculations given above:

- Sensitivity = 99.9%
- Specificity = 88.4%
- NPA = 99.8%
- PPA = 89.9%

However, if the disomy accuracy is only 90%, then the performance parameters change to:

- Sensitivity = 99.7%
- Specificity = 81.0%
- NPA = 99.6%
- PPA = 84.5%

3.2.2. Reciprocal Translocations

There are 32 different products of a two-way reciprocal translocation (8); a normal and a heterozygote (balanced) complement and 30 unbalanced products (Table 2). Studies have shown that ~53% of reciprocal translocation products are abnormal in pre-implantation embryos (9, 10). The analytical performance of a test that employs probes for both translocated segments and one of the two centric segments (Note 2) will therefore depend on the disomy accuracy of each probe in the probe mix.

For example, assuming that :

- The FISH gives 94% disomy accuracy per chromosome tested.
- The prevalence of embryos with unbalanced products is 53.0%.

Then, using the calculations given above:

- Sensitivity = 99.7%
- Specificity = 83.1%

Table 2

Two-way reciprocal translocation segregation products: segment counts for the translocated and centric segments (TS and CS) for chromosome A and chromosome B (Note 3)

Segregation mode	Zygote 2n complement	Segment count TSA	TSB	CSA	CSB
4:0	A,B	1	1	1	1
3:1 Tertiary monosomy	A,der(A),B	1	2	2	1
3:1 Interchange monosomy	A,B,B	1	2	1	2
2:2 Adjacent-2 cross-over in A	A,der(A),der(A),B	1	3	3	1
2:2 Adjacent-1	A,der(A),B,B	1	3	2	2
2:2 Adjacent-2 cross-over in B	A,B,B,B	1	3	1	3
3:1 Cross-over in A	A,der(A),der(A),B,B	1	4	3	2
3:1 Cross-over in B	A,der(A),B,B,B	1	4	2	3
3:1 Interchange monosomy	A,A,B	2	1	2	1
3:1 Tertiary monosomy	A,B,der(B)	2	1	1	2
2:2 Adjacent-2	A,A,der(A),B	2	2	3	1
2:2 Alternate (normal)	A,A,B,B	2	2	2	2
2:2 Alternate (balanced)	A,der(A),B,der(B)	2	2	2	2
2:2 Adjacent-2	A,B,B,der(B)	2	2	1	3
Anaphase II non-disjunction	A,A,der(A),der(A),B	2	3	4	1
3:1 Tertiary trisomy	A,A,der(A),B,B	2	3	3	2
3:1 Cross-over in A	A,der(A),der(A),B,der(B)	2	3	3	2
3:1 Interchange trisomy	A,der(A),B,B,der(B)	2	3	2	3
3:1 Cross-over in B	A,A,B,B,B	2	3	2	3
Anaphase II non-disjunction	A,B,B,B,der(B)	2	3	1	4
2:2 Adjacent-2 cross-over in A	A,A,A,B	3	1	3	1
2:2 Adjacent-1	A,A,B,der(B)	3	1	2	2
2:2 Adjacent-2 cross-over in B	A,B,der(B),der(B)	3	1	1	3
Anaphase II non-disjunction	A,A,A,der(A),B	3	2	4	1
3:1 Interchange trisomy	A,A,der(A),B,der(B)	3	2	3	2
3:1 Cross-over in A	A,A,A,B,B	3	2	3	2
3:1 Tertiary trisomy	A,A,B,B,der(B)	3	2	2	3
3:1 Cross-over in B	A,der(A),B,der(B),der(B)	3	2	2	3
Anaphase II non-disjunction	A,B,B,der(B),der(B)	3	2	1	4
4:0	A,A,der(A),B,B,der(B)	3	3	3	3
3:1 Cross-over in A	A,A,A,B,der(B)	4	1	3	2
3:1 Cross-over in B	A,A,B,der(B),der(B)	4	1	2	3

- NPA = 99.6%
- PPA = 86.9%

However, if the disomy accuracy is only 90%, then the performance parameters change to:

- Sensitivity = 99.6%
- Specificity = 72.9%
- NPA = 99.3%
- PPA = 80.6%

3.2.3. Sporadic Chromosome Aneuploidy

Using FISH to establish the prevalence of sporadic aneuploidy in pre-implantation embryos is complicated by the errors of the technique and mosaicism and multinucleation in cleavage-stage embryos. Therefore, the true prevalence of clinically significant aneuploidy in this material is likely to be overestimated.

However, the prevalence can be estimated using combined female and male gamete frequencies (11, 12), and is assumed for the purposes of this illustration to be 3% per chromosome pair tested. Typically, 7 chromosome pairs are tested; the overall prevalence is therefore 19.2%

For example, assuming that:

- The FISH gives 94% disomy accuracy per chromosome tested.
- The prevalence of embryos with unbalanced products is 19.2%.

Then, using the calculations given above:

- Sensitivity = 94.5%
- Specificity = 64.8%
- NPA = 98.0%
- PPA = 39.0%

However, if the disomy accuracy is only 90%, then the performance parameters change to:

- Sensitivity = 90.9%
- Specificity = 47.8%
- NPA = 95.7%
- PPA = 29.3%

3.2.4. Comparison of the Different Applications of the FISH Technique

Table 1 shows a comparison of the different applications above assuming 94% disomy accuracy per probe, i.e. 94/100 diploid nuclei with a normal or balanced chromosome complement have a normal signal score. This is a level of performance for the FISH technique that should be achievable in most hands using commercial probes.

With these conditions, the performance parameters for both sex determination and chromosome rearrangements are acceptable. In particular, the PPA for sex determination is 89.9%, which means that 12% of normal female embryos will be rejected (6 embryos per 100 tested) due to FISH error. Similarly, for chromosome rearrangements, the PPA is 86.9, leading to rejection of 17% of normal/balanced embryos (8 embryos per 100 tested). However, in the case of sporadic aneuploidy testing, the PPA is 39%, meaning that 35% of normal embryos will be rejected due to FISH error (28 normal embryos for every 100 embryos tested), substantially reducing the couple's chance of establishing a pregnancy.

In order to increase the PPA for sporadic aneuploidy testing to a level similar to sex determination and translocations, either

- The degree of aneuploidy would have to be 18.7% per chromosome pair tested.

- The disomy accuracy would have to be 99.6% per chromosome.

It is considered extremely unlikely that any patient group would have this degree of aneuploidy; in addition, 99.6% FISH accuracy per chromosome is very unlikely to be achievable in most peoples' hands.

Whilst a reduction of disomy accuracy to 90.0% has only a marginal effect on the performance of sex determination and chromosome rearrangement tests, the effect is greater for sporadic aneuploidy testing (see above), resulting in an even less acceptable performance. Even with 94% disomy accuracy, the probability that, for sporadic aneuploidy testing, an abnormal test result is correct, is less likely than flipping a coin. For these reasons, FISH for pre-implantation detection of sporadic aneuploidy is not recommended for routine clinical practice.

3.3. Preparation of Blastomeres

Methods of spreading and fixing single blastomeres include methanol/acetic acid (13, 14), Tween/HCl (15), and a combination of Tween/HCl and methanol/acetic acid (16). Variations include hypotonic treatment of cells prior to spreading and/or pepsin and paraformaldehyde treatment after fixation. The method should be appropriately validated for the laboratory (4). The Tween/HCl method is described in greater detail next.

3.3.1. Tween/HCl

The Tween/HCl method is technically simple and highly reproducible in different laboratories. This method can be used to prepare single nuclei for the FISH diagnosis of sex determination and chromosome rearrangements with acceptable diagnostic performance:

1. Score a small circle (approximately 5 mm) on the underside of an amine-coated slide and pre-label the slide with the case number, unique slide number, and biopsy date. Use a separate slide for each blastomere in numerical order, and label with the embryo number.

2. Place a small volume of lysis buffer within the circle.

3. Transfer the blastomere into the lysis buffer; remove and add lysis buffer as necessary to lyse the cell.

4. Observe the nucleus to ensure that it remains within the circle and is not lost; if the cell is anucleate or multinucleated, biopsy another cell.

5. Leave the slide to air-dry at room temperature.

3.4. In Situ Hybridization and Signal Scoring

Co-denaturation of the probe mix with the target DNA is the technique of choice. The ISH method and signal-scoring criteria should be appropriately validated for the laboratory (4). The variation of the co-denaturation protocol described next is technically simple and can be used for sex determination and chromosome rearrangements and has acceptable analytical performance from single cells:

1. Defrost probes, vortex, and centrifuge. Pipet volumes as required to make up the probe mixture.

2. Pre-wash fixed nuclei on slides using PBS for 5 min at room temperature.

3. Rinse twice in sterile distilled water.

4. Dehydrate with ethanol series (70, 90, and 100%) for 2 min each at room temperature and air-dry.

5. Record the position of the nucleus within the circle by visualizing with a phase contrast microscope.

6. Dehydrate with 100% ethanol for 2 min at room temperature and air-dry.

7. Apply 2 μL of probe mixture, and cover with a 9 × 9 mm no. 0 cover slip.

8. Seal the edges of the coverslip with rubber solution.

9. Codenature the slides on a hot block (e.g. Hybaid Omnislide or Vysis Hybrite) at 75°C for 5 min, and then hybridize the slides overnight (16–20 h) in a humidified chamber at 37°C. Probe mixes that consist entirely of centromere probes (i.e. for sex-linked cases) will give a satisfactory result after 60 min of hybridization.

10. Carefully remove the rubber solution from each slide and rinse off the coverslip using 4× SSC/0.05% Tween-20 at room temperature.

11. Wash the slides in a 0.4× SSC stringent wash at 71°C for 5 min.

12. Wash the slides in 4× SSC/0.05% Tween-20 at room temperature for 2 min.

13. If the probe mix contains indirectly labelled probe(s), drain the slides of excess liquid and apply 20 μL of fluorescently

conjugated antibody under a 20×20 mm square of Parafilm. Incubate in a humidified chamber at 37°C for 15 min. Remove the Parafilm and wash once in 4× SSC/0.05% Tween-20 at room temperature for 2 min.

14. Wash twice for 2 min in PBS at room temperature and drain the slides.

15. Apply 6 μL of DAPI to a 22×22 mm no. 0 cover slip and invert the slide over the cover slip.

16. Blot and seal the edges of the cover slip with clear nail varnish.

17. Analyze using a fluorescence microscope suitably equipped with the appropriate filter sets for the probes being used.

3.4.1. Signal Scoring

Score signals by direct visualization using a fluorescence microscope and single band-pass filters for each fluorochrome in the assay. A general guideline should lead to scoring of a single signal where two closely spaced signals are less than one domain (signal) apart; however, judgment based on experience needs to be exercised to interpret signals of varying size, intensity, and separation. Use imaging software to capture an image of the nucleus for confirmation of the visual diagnosis, and for image archiving as part of the laboratory quality assurance plan (4).

4. Notes

1. An elegant method to generate a third colour exploits the observation that Abbott SpectrumOrange probes can be visualized using a red filter and a yellow filter and Cytocell TexasRed-labelled probes can only be seen using the red filter (Personal communication, Julie Oliver, Genetic Technologist, Genetics & IVF Institute, Fairfax, VA).

2. There must be a balance of priorities between establishing a successful pregnancy and minimising the risk of an affected outcome; an overcautious strategy may result in marginal reduction in the risk of misdiagnosis at the expense of live births. One-cell biopsy and informed selection of three FISH probes (one centric segment and both translocated segments, see Table 2) is recommended as the optimum strategy for most translocations. However, incorporation of a fourth probe and/or transfer based on a concordant two-cell result will be indicated in some cases. It is prudent to consider transfer based on a concordant result from two biopsied cells where imbalance for small derivative chromosomes is known, or is likely, to be viable. A small derivative chromosome may be lost from a balanced chromosome complement or 3:1 segregation resulting in tertiary trisomy for the small derivative chromosome.

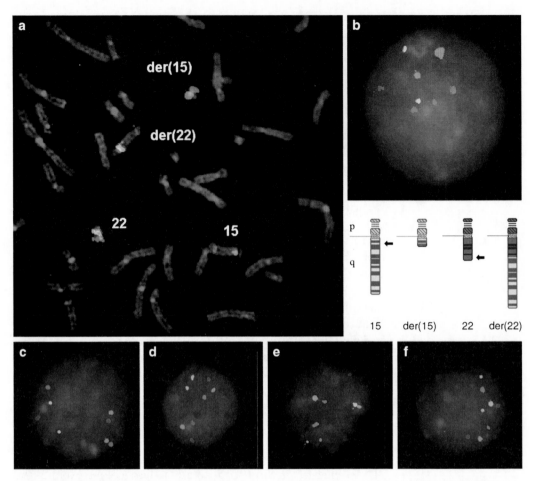

Fig. 1. Pre-implantation genetic diagnosis (PGD) for a two-way reciprocal translocation: 46,XY,t(15;22)(q13;q13.3). The position of the translocation breakpoints is indicated by *arrows* on the normal homologues in the partial ideogram. The fluorescence in situ hybridization (FISH) probes used were: Abbott CEP 15 (15p11.2, D15Z1 satellite III, SpectrumAqua), Cytocell subtel 15q (15q26.3, D15S936, TexasRed), Abbott LSI TUPLE1 (22q11.2, SpectrumOrange), and Abbott LSI ARSA (22q13.3, SpectrumGreen) (Note 4). (**a**) Cultured peripheral blood metaphase spread from the translocation carrier hybridized with probes specific for each centric (*blue CS15* and *orange CS22*) and translocated (*red TS15* and *green TS22*) segment. (**b**) Cultured peripheral blood interphase nucleus hybridized with the test probe mix; scoring 100 interphase nuclei from each partner was used to assess the performance of the probe mix. (**c–f**) Blastomere nuclei spread from single blastomeres biopsied from the couple's in vitro embryos 3 days after fertilization (the 8-cell stage). (**c**) Has two signals for each probe, which indicates two copies of each of the four segments and is consistent with a normal or balanced (carrier) complement for the translocation chromosomes. (**d**) Has one *green signal*, two *blue* and *orange signals*, and three *red signals*; this is consistent with adjacent-1 segregation of the translocation resulting in trisomy for the translocated segment of chromosome 15 (15q13→15qter) and monosomy for the translocated segment of chromosome 22 (22q13.3→22qter). (**e**) Has one *red signal*, two *blue* and *orange signals*, and three *green signals*; this is consistent with adjacent-1 segregation resulting in monosomy for the translocated segment of chromosome 15 and trisomy for the translocated segment of chromosome 22. (**f**) Has one *blue signal*, two *red* and *green signals*, and three *orange signals*; this is consistent with adjacent-2 segregation resulting in monosomy for the centric segment of chromosome 15 (15pter→15q13) and trisomy for the centric segment of chromosome 22 (22pter→22q13.3).

3. Two-way reciprocal translocations between chromosomes A and B have two centric segments (CSA and CSB) and two terminal translocated segments (TSA and TSB). All the meiotic segregation products of the translocation can be expressed in terms of the copy number of these segments. A diploid (2n) normal or balanced translocation chromosome complement has two copies of every segment; all diploid segregation products with chromosome imbalance have between one and four copies of at least two of the segments (Table 2).

4. Including probes for both centric segments and both translocated segments for a two-way reciprocal translocation ensures that two scoring errors are required to misdiagnose any of the unbalanced products with chromosome imbalance. Figure 1 shows an example for a reciprocal translocation between the long arms of chromosomes 15 and 22 with breakpoints at 15q13 and 22q13.3. FISH studies of cultured peripheral blood lymphocytes from the translocation heterozygote (Fig. 1a) confirm that the probes selected are informative with respect to the breakpoints (i.e. one probe hybridizes to each of the four segments) and scoring interphase nuclei (Fig. 1b) allows the assessment of the performance of the probe mix.

References

1. Hsu, L. Y., Benn, P. A., Tannenbaum, H. L., Perlis, T. E., Carlson, A. D. (1987) Chromosomal polymorphisms of 1, 9, 16, and Y in 4 major ethnic groups: a large prenatal study. *Am. J. Med. Genet.* **26,** 95–101.

2. Shim, S. H., Pan, A., Huang, X. L., Tonk, V. S., Varma, S. K., Milunsky, J. M., Wyandt, H. E. (2003) FISH variants with D15Z1. *J. Assoc. Genet. Technol.* **29,** 146–151.

3. Knight, S. J. and Flint, J. (2000) Perfect endings: a review of subtelomeric probes and their use in clinical diagnosis. *J. Med. Genet.* **37,** 401–409.

4. Thornhill, A. R., deDie-Smulders, C. E., Geraedts, J. P., Harper, J. C., Harton, G. L., Lavery, S.A., Moutou, C., Robinson, M.D., Schmutzler, A.G., Scriven, P.N., Sermon, K.D., Wilton, L.; ESHRE PGD Consortium (PGS) (2005) ESHRE PGD Consortium 'Best practice guidelines for clinical preimplantation genetic diagnosis (PGD) and preimplantation genetic screening (PGS)'. *Hum. Reprod.* **20,** 35–48.

5. Delhanty, J. D., Harper, J. C., Ao, A., Handyside, A. H., Winston, R. M. (1997) Multicolour FISH detects frequent chromosomal mosaicism and chaotic division in normal preimplantation embryos from fertile patients. *Hum. Genet.* **99,** 755–760.

6. Bahce, M., Cohen, J., Munné, S. (1999) Preimplantation genetic diagnosis of aneuploidy: were we looking at the wrong chromosomes? *J. Assist. Reprod. Genet.* **16,** 176–181.

7. Trussler, J. L., Pickering, S. J., Ogilvie, C. M. (2004) Investigation of chromosomal imbalance in human embryos using comparative genomic hybridization. *Reprod. Biomed. Online* **8,** 701–711.

8. Scriven, P. N., Handyside, A. H., and Ogilvie, C. M. (1998) Chromosome translocations: segregation modes and strategies for preimplantation genetic diagnosis. *Prenat. Diagn.* **18,** 1437–1449.

9. Estop, A. M., Van Kirk, V., and Cieply, K. (2005) Segregation analysis of four translocations, t(2;18), t(3;15), t(5;7), and t(10;12), by sperm chromosome studies and a review of the literature. *Cytogenet. Cell Genet.* **70,** 80–87.

10. Mackie Ogilvie, C., and Scriven, P. N. (2002) Meiotic outcomes in reciprocal translocation carriers ascertained in 3-day human embryos. *Eur. J. Hum. Genet.* **10,** 801–806.

11. Pellestor, F., Andréo, B., Arnal, F., Humeau, C., and Demaille, J. (2003) Maternal aging and chromosomal abnormalities: new data drawn from in vitro unfertilized human oocytes. *Hum Genet.* **112**, 195–203.

12. Shi, Q., and Martin, R. H. (2000) Aneuploidy in human sperm: a review of the frequency and distribution of aneuploidy, effects of donor age and lifestyle factors. *Cytogenet. Cell Genet.* **90**, 219–226.

13. Tarkowski, A. K. (1966) An air drying method for chromosome preparations from mouse eggs. *Cytogenetics* **5**, 394–400.

14. Munné, S., Lee, A., Rosenwaks, Z., Grifo, J., and Cohen, J. (1993) Diagnosis of major chromosome aneuploidies in human preimplantation embryos. *Hum. Reprod.* **8**, 2185–2192.

15. Harper, J. C., Coonen, E., Ramaekers, F. C. S., Delhanty, J. D. A., Handyside, A. H., Winston, R. M. L., and Hopman, A. H. N. (1994) Identification of the sex of human preimplantation embryos in two hours using an improved spreading method and fluorescent in situ hybridisation using directly labelled probes. *Hum. Reprod.* **9**, 721–724.

16. Dozortsev D. I., and McGinnis K. T. (2001) An improved fixation technique for fluorescence in situ hybridization for preimplantation genetic diagnosis. *Fertil. Steril.* **76**, 186–188.

Chapter 21

PNA–FISH on Human Sperm

Franck Pellestor, Cécile Monzo, and Samir Hamamah

Abstract

Peptide nucleic acids (PNAs) constitute a remarkable new class of synthetic nucleic acids analogs, based on peptide-like backbone. This structure gives PNAs the capacity to hybridize with high affinity and specificity to complementary RNA and DNA sequences. Over the last few years, the use of PNAs has proven its efficacy in cytogenetics for the rapid in situ identification of human chromosomes. Multicolour PNA–FISH protocols have been described and their adaptation to human spermatozoa has allowed the development of a new and fast procedure, which can advantageously be used for the assessment of aneuploidy in male gametes.

Key words: PNA, FISH, In situ labeling, Spermatozoa, Chromosomes, Aneuploidy

1. Introduction

Peptide nucleic acids (PNAs) constitute a remarkable new class of synthetic nucleic acids analogs, in which the phosphodiester backbone is replaced by a noncharged peptide-like backbone (1). This unique structure confers PNAs the ability to hybridize to complementary RNA and DNA sequences with high affinity and specificity, and an improved resistance to nucleases and proteinases (2). The remarkable physico-chemical properties of PNAs have led to the development of a large variety of research and diagnostic assays in the field of genetics, including genome mapping and mutation detection (3). The PNA labeling reaction presents several advantages (specificity, speed, and discrimination) that make it very attractive for cytogenetic purposes. Recent studies have reported the development of fast, simple, and robust PNA–Fluorescence in situ hybridization (PNA–FISH) assays for in situ chromosomal investigations, and the successful use

Joanna M. Bridger and Emanuela V. Volpi (eds.), *Fluorescence in situ Hybridization (FISH): Protocols and Applications*, Methods in Molecular Biology, vol. 659,
DOI 10.1007/978-1-60761-789-1_21, © Springer Science+Business Media, LLC 2010

of chromosome-specific PNA probes on human lymphocytes, amniocytes, as well as on isolated blastomeres and human gametes, opening new and promising perspectives for PNAs in the field of genetic diagnosis (4–7).

The direct chromosomal analysis of spermatozoa is an essential approach for the investigation of the occurrence and etiology of chromosomal abnormalities in humans in a wide variety of clinical conditions. To date, numerous chromosomal analyses on human sperm have been performed using FISH (8, 9). These reports demonstrated the efficiency of the in situ labeling procedure on male gametes, but also pointed out the limitations of conventional FISH on this biological material, which are essentially linked to the peculiarity of the human sperm nucleus in terms of genomic compaction and accessibility of DNA sequences. The primed in situ (PRINS) reaction has offered an alternative approach for the direct chromosome analysis of human spermatozoa, but its current use requires sequential reactions (10). PNA technology provides an alternative for chromosomal screening on sperm nuclei. The adaptation of this new class of probes to fluorescence in situ hybridization on human spermatozoa has constituted a new step in the development of PNA methodology (11). This chapter describes this innovative procedure for in situ detection of several chromosomes on human sperm nuclei.

2. Materials

2.1. Human Sperm Preparation

1. Phosphate-buffered saline (PBS) (Gibco BRL).
2. Methanol, 99%.
3. Ethanol series: 70, 90, and 100%.
4. Glacial acetic acid.
5. 0.5 M NaOH (Merck).
6. 20× SSC: 3 M NaCl, 0.3 M trisodium citrate, pH 7.5 (can be stored for several months at room temperature).
7. Washing buffer: 2× SSC diluted from 20× SSC.
8. Twin-frost glass microscope slides (CML). The slides must be cleaned by soaking in absolute ethanol to which concentrated HCl has been added at a concentration of 1 mL/100 mL. The slides are chilled in a refrigerator. They are removed from the acid/alcohol and polished with clean piece of muslin just before dropping the sperm suspension.
9. Light microscope Leica DMLB with ×10, ×40 objectives.

2.2. PNA Reaction

1. The PNA probes are supplied ready to use in hybridization buffer (Applied Biosystems). Each PNA probe consists of a

mixture of several short synthetic sequences (15–22 base units) specific for the centromeric repeated DNA sequence of the targeted chromosome (see Note 1). PNA probes specific for chromosomes 1, X, and Y are used in the present protocol. The chromosome 1-specific probe is labeled in blue with diethlaminocoumarine. The chromosome X-specific probe is labeled in red with rhodamine, and the chromosome Y-specific probe is labeled in green with fluorescein.

2. Phosphate-buffered saline.

3. Tween 20 (Roche Diagnostics).

4. Washing buffers: 1× PBS, 0.1% Tween 20 and 2× SSC, 0.1% Tween 20.

5. 1.5 mL sterile microcentrifuge tubes (Eppendorf AG).

6. Rubber cement (Artos).

7. Coverslips (22 × 32 mm) (CML).

8. Waterbath at 73°C.

9. Humidified hybridization chamber.

10. Coplin jar (50 mL).

11. Incubator at 37°C.

2.3. Detection and Microscopy

1. 4′, 6-Diamidino-2-phenylindole dihydrochloride (DAPI) (Sigma.)

2. Propidium iodide (PI) (Sigma).

3. Antifade solution Vectashield (Vector Labs).

4. Coverslips (20 × 40 mm).

5. Rubber cement (Artos).

6. Epifluorescence Microscope Leica DMRB equipped with ×40 and ×100 Plan FluoTar objectives, and with a DAPI single band-pass filter (Leitz filter A, N° 513804), a FITC single band-pass filter (filter I3, N° 513808), a TRITC single band-pass filter (filter N2.1, N° 513812), a FITC/TRITC double band-pass filter (filter G/R, N° 513803), and a triple filter (filter B/G/R, N° 513836) for simultaneous observation of DAPI/Cascade-Blue, FITC, and TRITC signals.

7. For image capture, we use the software "Metasystem Isis Version 5.0" (Metasystem).

3. Methods

3.1. Preparation of Sperm Slide

1. Freshly ejaculated sperm sample is allowed to liquefy at room temperature for 30 min.

2. Dilute 1 mL of sperm in 1× PBS (1:10) and centrifuge for 5 min at $500 \times g$.

3. Remove the supernatant and resuspend the pellet in 10 mL of 1× PBS for a new wash by centrifugation (5 min at 500×g).

4. Resuspend the pellet in 1 mL of fresh fixative (3:1 methanol: glacial acetic acid).

5. Fix 1 h at –20°C.

6. With a Pasteur pipette, drop from a height of about 10 cm one droplet of the sperm suspension on a clean microscope slide.

7. The slide is air-dried and checked under the light microscope (×10 and ×40) to ensure that both the cell concentration and the spreading are optimum (see Note 2). Using a diamond marker, draw a circle on the underside of the slide to mark where the spermatozoa are.

8. Store the slide for 2 days at room temperature; preferably in a hermetic box in order to avoid dust deposits (see Note 3).

9. Immediately before the PNA–FISH reaction, the slide is denatured in 0.5 M NaOH at room temperature for 4 min (see Note 4).

10. Wash the slide twice for 2 min at room temperature in 2× SSC.

11. Pass the slide through an ethanol series (70, 90, and 100%), 2 min each step, and air-dry.

3.2. PNA–FISH Reaction

1. Prepare the PNA reaction mixture: aliquots of 5 μL of each PNA probe (chromosome 1-, chromosome X-, and chromosome Y-specific probes) are mixed into a microcentrifuge tube.

2. Denature the PNA probe mixture at 73°C for 6 min.

3. Apply the PNA reaction mixture to the denatured sperm slide, and cover with a 22×32 mm coverslip.

4. Seal the slide with rubber cement (see Note 5).

5. Put the slide in a humidified hybridization chamber for 60 min at 37°C.

6. At the end of the hybridization, carefully remove the coverslip from the slide using a scalpel blade.

7. Transfer the slide in a Coplin jar containing 1× PBS, 0.1% Tween 20, and wash the slide for 2 min at room temperature with gentle agitation.

8. Transfer the slide to 58°C prewarmed 1× PBS, 0.1% Tween 20 for 10 min with gentle agitation.

9. Rinse the slide in 2× SSC, 0.1% Tween 20 for 1 min.

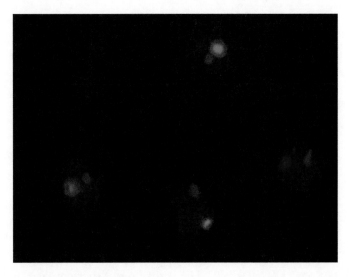

Fig. 1. Examples of PNA–FISH on human sperm nuclei, using three different chromosome-specific centromeric PNA probes. The chromosome centromere 1 is labeled in blue, the chromosome X centromere is labeled in red, and the chromosome Y centromere in green.

3.3. Detection and Microscopy

1. Drain the excess washing solution off the slide.

2. Mount the slide in Vectashield antifade mounting solution containing a mix of propidium iodide (0.3 µL/mL) and DAPI (0.3 µL/mL). Use 15–20 µL/slide.

3. Cover with a 22×40 mm coverslip and seal with rubber cement.

4. Examine the slide under the epifluorescence microscope equipped with suitable filters. First use a triple band-pass filter, and then confirm the coloring of the fluorescent spot with single band-pass filters. Figure 1 shows typical results obtained on a human sperm preparation.

4. Notes

1. PNA probes can be prepared following standard solid-phase synthesis protocols for peptides. However, this would require laboratories to have both the experience and the resources to support manual or automated peptide synthesis, and consequently it is not ideal for standard cytogenetic laboratories. The commercial availability of PNA probes for cytogenetic purposes is still limited to consensus telomeric and a few

human-specific satellite DNA probes. Until 2001, Boston Probes (Bedford, MA) was the leader in the development of PNA technology. In November 2001, the company was acquired by Applied Biosystems, which pursues the development and the commercialization of PNA probes. A custom PNA probe service, PNA design guidelines and a PNA probe order service are available on the Applied Biosystems web site (http://www.appliedbiosystems.com). DAKO A/S (Glostrup, Denmark), which was the majority owner of Boston Probe, can still provide a consensus telomeric PNA probe kit (http://www.dakocytomation.com). The PNA probes are compatible with a wide range of reporter molecules and fluorochromes including fluoresceine, rhodamine as well as cyanine and Alexa dyes available on a large variety of colors.

2. The spreading of spermatozoa on the slide must be homogeneous in order to facilitate the screening and to limit errors during the scoring procedure when performing chromosomal analysis on sperm nuclei. It is particularly important to avoid both aggregation and excessive dilution of cells on the slide. A density of 70–100 spermatozoa per field under a ×40 objective is optimal. Consequently, the washing procedure is a critical step. It must be performed taking into account the quality of the initial sperm sample, i.e. the viscosity after liquefaction, the sperm concentration, the agglutination of spermatozoa and the presence of cellular debris or various cell types (leucocytes, epithelial cells, immature germinal cells, bacteria). In the case of low quality sperm samples, one or two additional washes by centrifugation may be required for obtaining adequate smeared sperm slides.

3. The age of slides is an important parameter. Slides should be used within a week of preparation. Best results are obtained with 2-day-old spreads as these give the best signals. Using slides more than 1–2-weeks old can be successful, but may lead to reduced sensitivity.

4. The use of NaOH solution allows the simultaneous denaturation and decondensation of sperm nuclei, with the possibility of a rapid control of the degree of nucleus decondensation under the microscope. The success of the labeling technique depends to a high degree on the quality and efficiency of the decondensing protocol. Initially, we used 3 M NaOH, but numerous experiments and a comparison of the results in terms of quality of the preparation obtained led us to adopt a 0.5 M NaOH solution. This method provides homogeneous sperm decondensation, and subsequently a high level of sperm in situ labeling. The duration of 0.5 M NaOH treatment depends on the age of the sperm preparation slides. The longer

the slides are aged, the longer they need 0.5 M NaOH treatment: 2-day old, 4 min; 3-day old, 5 min; 4-day old, 6 min. The NaOH treatment induces uniform swelling of the sperm nucleus till 1.5–2 times its normal size, and maintains the characteristics and shape of the sperm nucleus, including the tail. This allows differentiating between spermatozoa and other cells such as leukocytes or immature germ cells present in the ejaculate.

5. For PNA–FISH hybridization, rubber cement provides an adequate seal, easy to remove at the end of the reaction. Nail polish gives a very secure seal, but is more difficult to remove at the end of the procedure. It is also possible to use adhesive plastic frames and coverslips (e.g. Hybaid Sure Seal) to frame the area of the slide on which to spread the hybridization mixture and ensure a good seal during the reaction. In all cases, take care not to trap any air bubbles.

References

1. Nielsen, P.E., Egholm, M., Berg, R.H. and Buchardt, O. (1991) Sequence-selective recognition of DNA by strand displacement with a thymine-substituted polyamide. *Science* **254**, 1497–1500.

2. Nielsen, P.E. and Egholm, M. (1999) An introduction to peptide nucleic acid. *Curr. Issues Mol. Biol.* **1**, 89–104.

3. Nielsen, P.E. (2001) Peptide nucleic acid: a versatil tool in genetic diagnostics and molecular biology. *Curr. Opin. Biotechnol.* **12**, 16–20.

4. Lansdorp, P.M., Verwoerd, N.P., van de Rijke, F.M., Dragowska, V., Little, M.T., Dirks, R.W., Raap, A.K. and Tanke, H.J. (1996) Heterogeneity in telomere length of human chromosomes. *Hum. Mol. Genet.* **5**, 685–691.

5. Taneja, K.L., Chavez, E.A., Coull, J., and Lansdorp, P.M. (2001) Multicolor fluorescence in situ hybridization with peptide nucleic acid probes for enumeration of specific chromosomes in human cells. *Gen. Chrom. Cancer.* **30**, 57–63.

6. Paulasova, P., Andréo, B., Diblik, J., Macek, M. and Pellestor, F. (2004) The peptide Nucleic Acids as probes for chromosomal analysis: application to human oocytes, polar bodies and preimplantation embryos. *Mol. Hum. Reprod.* **10**, 467–472.

7. Agerholm, I.E., Ziebe, S., Williams, B., Berg, C., Cruger, D.G., Petersen, G.B. and Kolvraa, S. (2005) Sequential FISH analysis using competitive displacement of labelled petide nucleic acid probes for eight chromosomes in human blastomeres. *Hum. Reprod.* **20**, 1072–1077.

8. Egozcue, J., Blanco, J. and Vidal, F. (1997) Chromosome studies in human sperm nuclei using fluorescence in situ hybridization (FISH). *Hum. Reprod. Update* **3**, 441–452.

9. Shi, Q. and Martin, R.H. (2001) Aneuploidy in human spermatozoa: FISH analysis in men with constitutional chromosomal abnormalities, and in infertile men. *Reproduction* **121**, 655–666.

10. Pellestor, F., Imbert, I. and Andréo, B. (2002) Rapid chromosome detection by PRINS in human sperm. *Am. J. Med. Genet.* **107**, 109–114.

11. Pellestor, F., Andréo, B., Taneja, K. and Williams, B. (2003) PNA on human sperm: a new approach for in situ aneuploidy estimation. *Eu. J. Hum. Genet.* **11**, 337–341.

Chapter 22

POD-FISH: A New Technique for Parental Origin Determination Based on Copy Number Variation Polymorphism

Anja Weise, Madeleine Gross, Sophie Hinreiner, Vera Witthuhn, Hasmik Mkrtchyan, and Thomas Liehr

Abstract

With the progress of array technologies and the enabled screening of individual human genomes, a new kind of polymorphism has been described – the so-called copy number variation (CNV) polymorphism. Copy number variants can be found in around 12% of the human genome sequence and have a size of up to several hundred kilobase pairs. These variants can not only differ between individuals, but also between corresponding alleles on homologous chromosomes. We recently developed a cytological assay for parental origin determination that relies on the design of CNV-based sets of probes for fluorescence in situ hybridization (POD-FISH). Here we describe an improved POD-FISH protocol that exploits "high frequency" variants for better discrimination of homologous chromosomes.

Key words: Parental origin determination fluorescence in situ hybridization, Copy number variation, Homologues, Chromosomes, Bacterial artificial chromosome

1. Introduction

Differences between individuals concerning their DNA sequence are well known since the molecular investigation of the human genome started. These differences make up 0.1% of the genome sequence. For example, single nucleotide polymorphisms (SNPs) appear every thousand base pair and are used in genotyping, pharmaco- and population genetics (1). Beside SNPs that affect only one base pair, there are mini- and micro-satellites, and small insertion or deletion polymorphisms (INDELS, <1 kb), with variable changes of up to 20 kb in size (2). Microscopically visible chromosome polymorphisms can be found in several regions of

Joanna M. Bridger and Emanuela V. Volpi (eds.), *Fluorescence in situ Hybridization (FISH):*
Protocols and Applications, Methods in Molecular Biology, vol. 659,
DOI 10.1007/978-1-60761-789-1_22, © Springer Science+Business Media, LLC 2010

the genome, including the differently sized heterochromatic regions of chromosomes 1, 9, 16, and Y, the variable satellite structure and size of the short arms of the acrocentric chromosomes 13, 14, 15, 21, and 22, the size polymorphisms in centromeres (3) and inversion variants mostly of chromosome 2 and 9 (4–6). In the absence of such microscopic polymorphisms, homologous chromosomes cannot be distinguished, as DNA sequence polymorphisms are widespread over the genome, but cannot be resolved at the chromosomal level and on a per cell basis. In fact, these sequence polymorphisms can only be detected by molecular genetic methods like microsatellite analysis (7), using isolated DNA from a cell mixture.

The discovery of a new kind of variation in the human genome that affects the copy number of euchromatic regions up to several hundred kilobases in size (8, 9) has the potential to bridge this resolution gap. This new class of polymorphism, the so-called copy number variants (CNVs, >1 kb), was initially identified by different array-based approaches (10). CNVs reflect sites of deletion or gain of euchromatic material in normal individuals and make up >60% of the reported structural variations so far collected in the database of genomic variants (http://projects.tcag.ca/variation/) (11–13). The physical size of these polymorphic variants allows for the first time to visualize at the microscopical level DNA sequence polymorphisms on chromosomes and discriminate between homologous chromosomes in single cells by a technique called "parental origin determination" fluorescence in situ hybridization (POD-FISH) (14). With the rising number of newly reported polymorphic loci, it is now possible to select CNVs that are present at a high frequency (>20%) in the investigated population and on this basis design increasingly informative POD-FISH probe sets.

2. Materials

2.1. BAC Selection and Ordering

Chromosomal regions for POD-FISH studies can be selected from the database of genomic variants (http://projects.tcag.ca/variation/). To identify suitable BAC clones from the corresponding CNV regions we recommend selecting regions of no less than 150 kb in size. Moreover, the CNV should have been reported in over 20% of the investigated individuals. This will lead to more informative hybridization patterns along the chromosome.

An exemplary list for high frequency CNV-BAC clones is given for chromosome 6 in Table 1. Once selected, there are several sources to order BAC clones, e.g. BAC-PAC Chori Oakland (http://bacpac.chori.org/home.htm) and Rosewell Park Cancer Institute (http://www.roswellpark.org/). BAC clones are normally purchased as plasmids in *Escherichia coli* with a specific antibiotic

Table1

Example of high frequency chromosome 6-specific POD-FISH probe set

Chromosome	CNV loci	BAC clone	Fluorochrome	Population frequency (%)
6p25.3	0374	RP11-328C17	Texas Red	30.00
6p21.33–32	4492	RP11-427G15	Spectrum Green	37.89
6q12	4002	RP11-80L16	DEAC	78.89
6q14.1	2644	RP11-483P24	Texas Red	24.07
6q25.1	4512	RP11-655H19	Spectrum Green	21.05
6q27	3655	RP11-37D8	Texas Red	26.67

resistance. This information will be supplied by the BAC clone provider. Usually an overnight culture, followed by plasmid isolation (e.g. with Quiagen "QIAprep Spin Miniprep Kit"), amplification (e.g. by DOP-PCR) (15) and labelling (e.g. by Nick Translation, Roche) will lead to a useful locus-specific FISH probe.

2.2. Combining CNV-BACs into Chromosome-Specific Sets

Depending on the available fluorescence microscope filter sets, BACs for chromosome-specific polymorphic CNV regions can be labelled in different colours and applied in a single hybridization step. An example of four-colour POD-FISH approach for the high frequency set of probes for chromosome 6 is shown in Fig. 1. When working with more BAC clones at the same time it is more convenient to split the probe sets into two or more successive hybridization (14).

This is recommended especially for longer chromosomes where chromosome arm-specific POD-FISH sets can be applied. After labelling of different BAC DNA in different fluorescence colours or haptens, the probes are precipitated together and dissolved in hybridization buffer (e.g. dextran sulphate (Sigma-Aldrich).

2.3. Slide Pre-treatment and Denaturation

1. 24×50 mm and 24×24 mm coverslips (Menzel or other suppliers).

2. Formalin-buffer (1%): 5 mL formaldehyde solution (Merck) +4.5 mL 1XPBS + 500μL 1 M MgCl$_2$-solution.

3. Microscope Slides, pre-cleaned (Menzel).

4. PBS (1X) – prepare from PBS Dulbecco (Instamed) w/o Ca, w/o Mg (Seromed).

5. Pepsin stock solution: 1.0 g pepsin (Sigma) in 50 mL distilled water. Aliquot and store at –20°C.

6. Pepsin-solution: Pre-warm 95 mL of distilled water to 37°C in a Coplin jar; add 5 mL of 0.2 N HCl and 0.5 mL pepsin stock solution – make fresh when required.

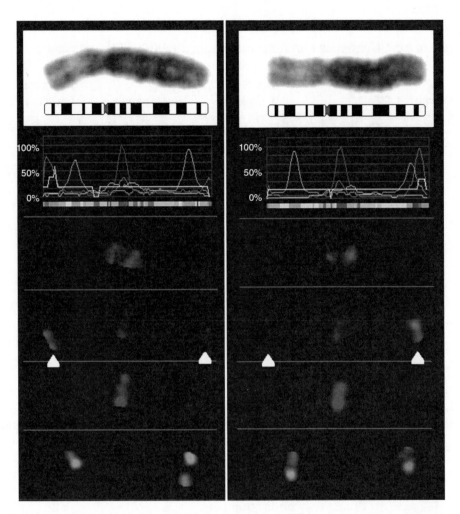

Fig. 1. High frequency POD-FISH set for chromosome 6 (see Table 1). Differences on homologous chromosomes are marked with *arrows*. From *left to right*: SpectrumGreen, SpectrumOrange, TexasRed, diethylaminocoumarine (DEAC), fluorescence profiles along the chromosome, ideogram of chromosome 6 and inverted Diamindinophenylindol (DAPI) figure of both homologous chromosomes 6.

2.4. Probe Pre-treatment and Denaturation	1. High frequency chromosome-specific POD-FISH probe sets are created as described in Subheadings 2.1 and 2.2. 2. C_0t-1 DNA (Roche).
2.5. FISH-Procedure and Washing	1. Rubber cement: Fixogum, Marabu. 2. 4× SSC/0.2%Tween: 100 mL 20× SSC + 400 mL ddH$_2$O + 250 μL Tween 20 (Sigma), adjust to pH 7–7.5. 3. 1× SSC-solution; prepare from 20× SSC.
2.6. Evaluation	1. MetaSystems software or SCION software (http://www.scioncorp.com) is recommended for evaluation.

2. Fluorescence microscope (e.g. Zeiss) with appropriate number of fluorescence filters according to the labelled POD-FISH probes equipped with a CCD camera (e.g. F-view, Olympus).

3. Methods

POD-FISH is performed on metaphase spreads even though signal differences can also be obtained and evaluated on interphase nuclei. However, the latter should be done with respect to corresponding cut-off values obtained for non-polymorphic probes in interphase, as interphase FISH is very sensitive to methodical variations and also the signals are sometimes at different microscope focus levels caused by the flattened spherical character of interphase nuclei. Routine chromosome preparation for FISH analysis is reported within the core protocols of this book and in (16). Here we outline how to pre-treat slides with metaphase spreads on them, how to prepare the POD-FISH probe set before hybridization, how to perform the FISH-procedure itself, and how to evaluate the results.

3.1. Slide Pre-treatment and Denaturation

Slide pre-treatment is performed like in a conventional FISH approach with a pepsin treatment followed by post-fixation with formalin-buffer to reduce the background.

1. Incubate slides in 100 mL 1× PBS at RT for 2 min on a shaking platform.
2. Put slides for 5–10 min in pepsin-solution at 37°C in a Coplin jar (see Note 1).
3. Repeat step 1 twice.
4. Post-fix metaphase spreads on the slide surface by replacing 1× PBS with 100 mL of formalin-buffer for 10 min (RT, with gentle agitation).
5. Repeat step 1 twice.
6. Dehydrate slides in an ethanol series (70, 90, 100, 3 min each) and air-dry.
7. Add 100 µL of denaturation buffer to each slide and cover with a 24 × 50 mm coverslip.
8. Incubate slides on a hot plate for 2–3 min at 75°C (see Note 2).
9. Remove the coverslip immediately and place slides in a Coplin jar filled with 70% ethanol (4°C) to conserve target DNA as single strands.
10. Dehydrate slide in ethanol (70, 90, 100%, 4°C, 3 min each) and air-dry.

3.2. Probe Pre-treatment and Denaturation

According to the size of the hybridization area on the slide (e.g. 5 µL probe mix plus 7 µL dextran sulphate for 24×24 mm) (see Note 3) the probe set is pre-annelaed with C_0t-1 DNA (e.g. 10 µg lyophilized) (see Note 4) in a thermo cycler at 75°C for 5 min, at 37°C for 30 min and cool down to 4°C.

3.3. FISH-Procedure, Washing, and Detection

1. Add the probe-solution onto each denatured slide, cover with an appropriate coverslip over the probe and seal with rubber cement (see Notes 3–5).

2. Incubate slides overnight at 37°C in a humid chamber.

3. Take the slides out of 37°C humid chamber, remove rubber cement with forceps, and coverslips by letting them swim off in 4× SSC/0.2% Tween (RT, 100 mL Coplin jar).

4. Wash the slides 1×5 min in 1× SSC-solution (62–65°C) with gentle agitation (see Note 6) followed by 3×5 min in 4× SSC/0.2% Tween (RT, with gentle agitation).

5. Dehydrate slide in ethanol (70, 90, 100%, 4°C, 3 min each) and air-dry.

6. Counter-stain the slides with DAPI-solution (100 mL in a Coplin jar, RT) for 8 min.

7. Wash slides three times in water for a few seconds and air-dry.

8. Add 15 µL of antifade, cover with coverslip and look at the results with a fluorescence microscope.

3.4. Evaluation

For the evaluation of the POD-FISH hybridization patterns a fluorescence microscope with an appropriate filter set is needed. The high frequency set for chromosome 6 (Table 1 and Fig. 1) requires a five-filter fluorescence microscope with filters for Spectrum Orange, SpectrumGreen, TexasRed, DEAC, and 4′,6-Diamidin-2′-phenylindoldihydrochlorid (DAPI). Image capture is performed with a digital camera device and suitable software. The analysis can be done (1) directly by eyes using a fluorescence microscope, (2) by analyzing fluorescence profiles with appropriate software (Fig. 1), and (3) by measuring signal intensity and area with a software that has been shown to be suitable for measuring FISH intensity signal before, like the SCION (14). We recommend analyzing 10–20 metaphase spreads (see Note 6).

4. Notes

1. Pepsin treatment time has to be adapted in each lab. Success of pre-treatment must be controlled by microscopic inspection. A balance between chromosome preservation and

digestion must be found. For beginners it is recommended to start with target samples which are not limited in availability.

2. Denaturation time of 2–3 min only is suggested for the maintenance of available metaphase chromosomes. Also in this case, a balance between chromosome preservation and denaturation must be found. In case, one works with tissues without metaphase spreads this aspect is of no significance.

3. Use of 24×24 mm coverslips is also possible; to hybridize different probes on the same slide using smaller coverslips. The amount of probe/probe-solution has to be reduced according to the coverslip size.

4. When working with directly labelled fluorescence probes avoid direct light exposure, which will lead to fluorochrome fading and weak FISH signals.

5. During the FISH washing steps it is important to prevent the slide surfaces drying out; otherwise; background problems may arise.

6. Not all applied BAC clones from CNV regions will show a distinctive pattern in one individual, depends on the frequency of the polymorphism in the population. In our high frequency POD-FISH approach the probability of finding differences is increased by employing only CNV-BACs with over 20% population frequency. Here the POD-FISH approach is similar to the expected results of a microsatellite analysis (14). Even if a signal difference is not directly visible by eye it could be measured by software approaches like SCION. Once one finds a distinctive signal pattern in one individual the parents can be tested with the same POD-FISH set in order to get a hint on the chromosome segregation. Or this can be used to differentiate between different cell lines, e.g. in cases of maternal contamination in prenatal diagnostics, or in follow up studies after bone marrow transplantations (14).

Acknowledgments

Supported in parts by the DFG (436 RUS 17/109/04, 436 RUS 17/22/06, WE 3617/2-1, 436 ARM 17/5/06, LI820/11-1, LI820/17-1), Boehringer Ingelheim Fonds, Evangelische Studienwerk e.V. Villigst, IZKF Jena (Start-up S16), IZKF together with the TMWFK (TP 3.7 and B307-04004), University Jena, Stiftung Leukämie, and Stefan-Morsch-Stiftung.

References

1. Lee, C. (2005) Vive la différence! *Nat Genet* **37**, 660–661.

2. Schlötterer, C. (2004) The evolution of molecular markers-just a matter of fashion? *Nat Rev Genet* **5**, 63–69.

3. Liehr, T., Ziegler, M., Starke, H., Heller, A., Kuechler, A., Kittner, G., Beensen, V., Seidel, J., Hassler, H., Musebeck, J., Claussen, U. (2003) Conspicuous GTG-banding results of the centromere-near region can be caused by alphoid DNA heteromorphism. *Clin Genet* **64**, 166–167.

4. Müller, H., Klinger, H.P., Glasser, M. (1975) Chromosome polymorphism in a human newborn population II Potentials of polymorphic chromosome variants for characterizing the idiogram of an individual. *Cytogenet Cell Genet* **15**, 239–255.

5. Gardner, R.G.M., Sutherland, G.R. (2004) Oxford Monographs on Medical Genetics No 46: Chromosome abnormalities and genetic counselling. 3rd edition *Oxford University Press*, Oxford, New York.

6. Shaffer, L.G., Tommerup, N. (eds) (2005) ISCN: An international System for Human Cytogenetic Nomenclature. *S. Karger*, Basel, Switzerland.

7. Liehr, T., Nietzel, A., Starke, H., Heller, A., Weise, A., Kuechler, A., Senger, G., Ebner, S., Martin, T., Stumm, M., Wegner, R., Tonnies, H., Hoppe, C., Claussen, U., von Eggeling, F. (2003) Characterization of small marker chromosomes (smc) by recently developed molecular cytogenetic approaches. *J Assoc Genet Technol* **29**, 5–10.

8. Iafrate, A.J., Feuk, L., Rivera, M.N., Listewnik, M.L., Donahoe, P.K., Qi, Y., Scherer, S.W., Lee, C. (2004) Detection of large-scale variation in the human genome. *Nat Genet* **36**, 949–951.

9. Sebat, J., Lakshmi, B., Troge, J., Alexander, J., Young, J., Lundin, P., Månér, S., Massa, H., Walker, M., Chi, M., Navin, N., Lucito, R., Healy, J., Hicks, J., Ye, K., Reiner, A., Gilliam, T.C., Trask, B., Patterson, N., Zetterberg, A., Wigler, M. (2004) Large-scale copy number polymorphism in the human genome. *Science* **23**, 525–528.

10. Feuk, L., Marshall, C.R., Wintle, R.F., Scherer, S.W. (2006) Structural variants: changing the landscape of chromosomes and design of disease studies. *Hum Mol Genet* **15**, 57–66.

11. Redon, R., Ishikawa, S., Fitch, K.R., Feuk, L., Perry, G.H., Andrews, T.D., Fiegler, H., Shapero, M.H., Carson, A.R., Chen, W., Cho, E.K., Dallaire, S., Freeman, J.L., González, J.R., Gratacòs, M., Huang, J., Kalaitzopoulos, D., Komura, D., MacDonald, J.R., Marshall, C.R., Mei, R., Montgomery, L., Nishimura, K., Okamura, K., Shen, F., Somerville, M.J., Tchinda, J., Valsesia, A., Woodwark, C., Yang, F., Zhang, J., Zerjal, T., Zhang, J., Armengol, L., Conrad, D.F., Estivill, X., Tyler-Smith, C., Carter, N.P., Aburatani, H., Lee, C., Jones, K.W., Scherer, S.W., Hurles, M.E. (2006) Global variation in copy number in the human genome. *Nature* **444**, 444–454.

12. Wang, K., Li, M., Hadley, D., Liu, R., Glessner, J., Grant, S.F., Hakonarson, H., Bucan, M. (2007) PennCNV: an integrated hidden Markov model designed for high-resolution copy number variation detection in whole-genome SNP genotyping data. *Genome Res* **17**, 1665–1674.

13. Perry, G.H., Ben-Dor, A., Tsalenko, A., Sampas, N., Rodriguez-Revenga, L., Tran, C.W., Scheffer, A., Steinfeld, I., Tsang, P., Yamada, N.A., Park, H.S., Kim, J.I., Seo, J.S., Yakhini, Z., Laderman, S., Bruhn, L., Lee, C. (2008) The fine-scale and complex architecture of human copy-number variation. *Am J Hum Genet* **82**, 685–695.

14. Weise, A., Gross, M., Mrasek, K., Mkrtchyan, H., Horsthemke, B., Jonsrud, C., Von Eggeling, F., Hinreiner, S., Witthuhn, V., Claussen, U., Liehr, T. (2008) Parental-origin-determination fluorescence in situ hybridization distinguishes homologous human chromosomes on a single-cell level. *Int J Mol Med* **21**, 189–200.

15. Telenius, H., Carter, N.P., Bebb, C.E., Nordenskjöld, M., Ponder, B.A., Tunnacliffe, A. (1992) Degenerate oligonucleotide-primed PCR: general amplification of target DNA by a single degenerate primer. *Genomics* **13**, 718–725.

16. Liehr, T., Claussen, U. (2002) FISH on chromosome preparations of peripheral blood, in: FISH-Technology (Rautenstrauss, B., Liehr, T., eds) Springer, Berlin, 73–81.

Chapter 23

Sequence-Based High Resolution Chromosomal Comparative Genomic Hybridization (CGH)

Agata Kowalska, Eva Bozsaky, and Peter F. Ambros

Abstract

We aimed to devise an appropriate method to directly link the fluorescence profile of chromosomal copy number alterations detected by chromosomal comparative genomic hybridization (cCGH) or any other hybridization or staining information with the genome sequence data. Our goal was to establish an internal anchoring system that could facilitate profile alignment and thus increase the resolution of cCGH. We were able to achieve the alignment of chromosomes with gene mapping data by superimposition of (a) the fluorescence intensity pattern of a sequence-specific fluorochrome (GGCC binding specificity), (b) the cCGH fluorescence intensity profile of individual chromosomes, and (c) the GGCC motif density profile extracted from a genome sequence database. The adjustment of these three pieces of information allowed us to precisely localize, in cytobands and mega base pairs (Mb), regions of genomic alterations such as gene amplifications, gains, or losses. The combined visualization of sequence information and cCGH data together with application of the Warp tool, presented here, considerably improves the cCGH accuracy by increasing its resolution from 10 to 20 Mb to less than 2 Mb.

Key words: Chromosome, CGH, Chromomycin A3, Sequence banding, Warp tool

1. Introduction

Comparative Genomic Hybridization (CGH) is one of the most widely used methods for detecting amplifications and large-scale gains or losses of genetic material, as it enables genome-wide screening for imbalances in a single experiment (1–6). Chromosomal CGH (cCGH) involves competitive hybridization of differentially labeled test and reference whole genomic DNAs in equal amounts to normal metaphases. Regions of lost or gained DNA sequences, i.e., deletions, duplications, or amplifications appear as relative changes in the fluorescence intensities of the two samples along the target chromosomes.

Joanna M. Bridger and Emanuela V. Volpi (eds.), *Fluorescence in situ Hybridization (FISH): Protocols and Applications*, Methods in Molecular Biology, vol. 659,
DOI 10.1007/978-1-60761-789-1_23, © Springer Science+Business Media, LLC 2010

Although cCGH provides a fluorescence profile of the entire genome, the method has so far been restricted by the relatively low resolution mainly due to the fact that the results are averages from profiles of a number of inconsistently contracted chromosomes. cCGH experiments have shown that single copy deletions can be reliably detected if they are in the range of 10–20 Mb (mega base pairs) (7–10). Moreover, the classical cCGH technique lacks any direct linkage of the cCGH fluorescence pattern with gene map data. Thus, due to the absence of any adequate anchoring system, it has so far not been feasible to precisely ascertain which genomic loci are represented by the gained, amplified, or lost chromosomal regions.

The sequence-specific fluorochrome Chromomycin A3 (CMA3), by highlighting GGCC-specific chromosomal regions (11–13), yields the chromosome banding information that can be directly interpreted in terms of the genome sequence information (14). This so-called "sequence banding," i.e., CMA3 fluorescence intensity pattern revealing the GGCC density alterations along the chromosome, is superimposed on the GGCC motif density profile extracted from the Ensembl genome sequence browser (http://www.ensembl.org). Such rapid and direct assignment of chromosomal bands with the gene mapping data makes it possible to define every chromosomal locus in not only cytobands but also Mb. To achieve this goal, software termed ISIS "External Profiles Function" and the "Warp tool" were developed (15).

The simultaneous application of cCGH with the sequence banding and the subsequent analysis with the Warp tool considerably elevates the cCGH resolution to less than 2 Mb and provides a direct and more precise assignment of gains, amplifications, and losses to genomic regions (16).

2. Materials

2.1. Chromosome Comparative Genomic Hybridization

2.1.1. DNA Isolation

1. QIAamp DNA blood mini kit (Qiagen).

2.1.2. DNA Labeling (Nick-Translation Procedure)

1. Nick-translation labeling mixture (50 µL) contains the following:
   - 5 µL 500 µM DTT (AppliChem).
   - 5 µL 10× NT buffer containing 0.5 M Tris–HCl (pH 8.0), 50 mM $MgCl_2$, and 0.5 mg/mL bovine serum albumin (pH 7.8).

- 5 µL 10× dNTP mixture containing 0.5 mM dGTP, 0.5 mM dATP, 0.5 mM CTP and 0.15 mM dTTP (Sigma-Aldrich).
- Labeled nucleotides: 1 µL 1 mM Alexa Fluor-568-dUTP (Molecular Probes) to reference sample, 1 µL 1 mM DEA Coumarin 5-dUTP to test sample (New England Nuclear) (see Note 1).
- 3.5 µL DNase solution (3 ng/µL) (Sigma-Aldrich).
- 1.5 µL (=13.5 U) DNA Polymerase (9 U/µL) (Promega).
- 1 µg DNA template.
- Distilled water to a final volume of 50 µL.

2. Agarose and 1× TBE buffer in order to prepare 1.2% agarose gel.

3. 0.5 M EDTA pH 8.0 as stop buffer.

2.1.3. Preparation of the Hybridization Mixture

1. Human C_0t-1 DNA (Roche).

2. 3 M sodium acetate pH 5.5 (Merck).

3. Ice-cold 100 and 70% ethanol.

4. Hybridization master mix: 50% formamide, 20% dextran sulphate, and 5% 20× SSC in distilled water.

2.1.4. Hybridization to Metaphase Spreads

1. 4% paraformaldehyde (Roth).

2. Phosphate buffered saline (PBS).

3. Pepsin working solution: 0.05% pepsin (Sigma-Aldrich) in 0.01 N HCl.

4. 70% formamide (Merck) in 2× SSC.

5. A series of 70, 95, and 100% ethanol.

2.1.5. Washing the Slides

1. 2× concentrate saline–sodium citrate (SSC) buffer.

2. 50% formamide (Merck) in 2× SSC.

3. 0.1% Tween (Merck) in 4× SSC.

4. PBS

5. Series of 70, 95, and 100% ethanol.

2.2. Chromomycin A3 Staining

1. McIlvane's citric acid–Na_2HPO_4 buffer (pH 7): mix 82 mL of 0.1 M citric acid with 18 mL of 0.2 M disodium hydrogen-phosphate to adjust for pH 7.

2. CMA3 stock solution: prepare 0.5 mg/mL solution in McIlvane's buffer (pH 7) diluted 1:1 with distilled water containing 1 M $MgCl_2$ (to give 5 mM final concentration). Add the sterile buffer slowly to the CMA3 powder (Calbiochem) without stirring it and dissolve slowly by leaving at 4°C (see Note 2). Store at 4°C. CMA3 is highly toxic and should be handled with caution. Wear gloves to avoid exposure.

3. CMA3 staining solution: add paraformaldehyde to the CMA3 stock solution to achieve a final concentration of 4% (see Note 3).

4. Distamycin A (DA) solution (Serva): 0.1 mg/mL in McIlvane's buffer (pH 7). Store at –20°C in 1 mL aliquots. DA should be handled using gloves.

5. DAPI (4′, 6-diamidino-2-phenylindole) (Sigma) stock solution: 2 mg/mL in McIlvane's buffer (pH 7). Store it frozen at –20°C.

6. DAPI staining solution: stock solution diluted 1:1,000 with PBS. Store at 4°C.

7. Antifade solution: prepare a mixture of 87% glycerol plus McIlvane's buffer (pH 7) 1:1 (v/v), the latter should contain 5 mM $MgCl_2$ to give a 2.5 mM final concentration. This solution is mixed with a Vectashield mounting medium for fluorescence (Vector Laboratories) in a 1:1 ratio.

3. Methods

3.1. Chromosome Comparative Genomic Hybridization

3.1.1. DNA Isolation

Genomic DNA is extracted using a QIAamp DNA blood mini kit according to the manufacturer's protocol.

3.1.2. DNA Labeling

Labeling of DNA probes is performed by Nick-translation procedure.

1. Prepare the nick-translation mixture according to materials subheading 2.1.2 and incubate the reaction for 1.5 h at 15°C. Stop the reaction by placing the samples on ice after incubation.

2. The fragment length of the labeled DNA samples is verified by agarose gel electrophoresis.

3. Adjust the length of the labeled probes to a size between 300 and 600 bp by prolonging/reducing the 15°C incubation time and/or increasing/decreasing the concentration of DNase in the labeling mix.

4. Once the appropriate fragment length of the labeled DNA template is achieved, add 0.1 volume of stop buffer (0.5 M EDTA, pH 8.0) to stop the reaction.

3.1.3. Preparation of the Hybridization Mixture

1. Mix equal amounts of the labeled reference and test samples and add human C_0t-1 DNA to suppress cross-hybridization to human repetitive sequences. 50-fold excess of human

C_0t-1 DNA compared to the amount of probe DNA is recommended.

2. Add 0.1 volume of 3 M sodium acetate (pH 5.5), vortex, and add 2.5 volume of ice-cold 100% ethanol and vortex again.

3. Precipitate the labeled samples either overnight at −20°C or for 1 h at −80°C.

4. Centrifuge at $15,700 \times g$ for 30 min at 4°C to pellet the DNA.

5. Remove the supernatant and wash the pellet by adding 1 mL 70% ethanol.

6. Centrifuge again ($15,700 \times g$ for 15–20 minutes at 4°C), remove the supernatant, and air-dry the pellet for 10–15 min at 37°C.

7. Resuspend the dry pellet in adequate volume (5–7 μL) of hybridization master mix; mix well and allow it to dissolve for 30 min or more.

3.1.4. Hybridization to Metaphase Spreads

1. Fix metaphase spreads on slides in 4% paraformaldehyde at 4°C for 10 min.

2. Wash the slides in PBS. Make sure that the slides do not dry out until the ethanol step.

3. Warm the pepsin working solution to 37°C and incubate the slides in it for an appropriate time (depending on the quality of metaphase spreads). Wash the slides in PBS for 5 min.

4. Denature the metaphase spreads on the slide in 70% formamide/2× SSC at 75°C for 3 min. (The length of denaturation can be different for each series of metaphase spread preparation.)

5. Dehydrate the slides through a series of 70, 95, and 100% ice-cold ethanol, 5 min each, and then air-dry.

6. In parallel, denature the labeled DNA samples at 76°C for 6 min in a heater block followed by immediate incubation at 37°C for 20–30 min to allow preannealing of repetitive sequence elements.

7. Apply the hybridization mixture onto the slides with the denatured chromosomes. Put a cover slip of appropriate size (e.g. 15×15 mm) on top of the hybridization droplet and seal the edges of the cover slip using rubber cement.

8. Put the slides in a humid chamber and incubate at 37°C for 8–48 h.

3.1.5 Washing

Since fluorochrome-labeled DNAs were used, all steps were performed under light protection.

1. Peel off the rubber cement, gently remove the cover slip, and transfer the slides to 2× SSC.

2. Wash the slides twice for 5 min in 2× SSC at room temperature.

3. Wash the slides in 50% formamide/2× SSC, 2× SSC, and 4× SSC/Tween 20, each for 5 min at 42°C.

4. Rinse the slides in PBS for 5 min at room temperature.

5. Dehydrate the slides in a series of 70, 95, and 100% ethanol, 3 min each, and then air-dry (see Note 4).

3.2. Chromomycin A3 Staining

1. In order to simultaneously visualize cCGH pattern and sequence banding on the chromosomes, chromomycin A3 (CMA3), distamycin A (DA), and DAPI are used as chromosome stains (CDD triple staining) (see Note 5). The CDD chromosome staining is performed according to standard protocols (11,12).

2. Apply CMA3 staining solution (containing 4% paraformaldehyde) on the slide. Cover it with a plastic coverslip and incubate overnight at room temperature in the dark in a humid chamber.

3. Remove the coverslip by rinsing the slide in deionized water, shake off excess water, and blow the slide dry.

4. Apply DA solution on the slide, cover with a plastic coverslip for 10 min (incubate in the dark).

5. Float off the coverslip, rinse the slide in deionized water and blow-dry.

6. Apply DAPI solution on the slide, cover with a plastic coverslip, and incubate for 15 min in the dark.

7. Remove the coverslip, rinse the slide in deionized water, and blow-dry.

8. Mount the slides in an antifade solution prepared as described in subheading 2.2, remove the excess of mountant by pressing the coverslip gently with filter paper and incubate for at least three days at 37°C (see Note 6).

3.3. ISIS External Profiles Function and Warp Tool

1. The following instructions comprise the use of an External Profile Function, which is available as one of the MetaSystems' ISIS software applications (MetaSystems).

2. The cCGH slides stained with CMA3 are imaged with the fluorescence microscope, which, in our studies, is equipped with a motorized reflector turret. Digital images are captured using an IMAC S30 integrating CCD camera and the ISIS software.

3. For each fluorochrome a specific fluorescence filter cube is used: TRITC filter to capture a pattern, given by the reference sample, reflected by Alexa Fluor-568-dUTP signals and DEAC

filter for DEA Coumarin 5-dUTP signals of the test sample (see Note 7). Switching between the filter cubes is under software control. For each color channel, the optimal integration time is automatically determined by the ISIS software.

4. Define, with the cursor, a longitudinal axis of an individual chromosome (see Note 8), which displays both the cCGH fluorescence pattern and the GGCC distribution highlighted by the CMA3 staining (see Fig. 1a). Set the track of the axis along one of the chromatids (see Note 9). Consequent measurement and normalization of the fluorescence intensities along the axis is performed automatically. Correlation of the chromosomal fluorescence pattern with the GGCC binding motif density from the DNA sequence is then performed using the ISIS External Profiles Function and the Warp tool (Fig. 1b, c).

5. After defining the longitudinal axis, open Profiles display by right mouse clicking. Again click right mouse button to activate the External Profiles Function. Two profiles will get graphically displayed in the lower part of the External Profile Function display (curves a and b in Fig. 1b). The blue curve (a) illustrates CMA3 fluorescence intensity reflecting GGCC motif frequency along the chromosome axis. The white curve (b) delineates the ratio between fluorescence intensities of DEAC-labeled "test" DNA and red-labeled "reference" DNA (the ratio of the test-to-the reference fluorescence intensity). The setup button offers a possibility to display either CGH ratio profile or individual (single) test and reference fluorescence intensity profiles.

6. Select the chromosome which is going to be analyzed (circle c in Fig. 1b). The GGCC sequence density information is read from the intermediate file (see Note 10), normalized to the length of the fluorescence profiles, and shown in the upper part of the display (curve d in Fig. 1b). By changing the smoothing power (circle e in Fig. 1b), the level of detail in the displayed sequence density profile can be adapted to that observed in the fluorescence intensity profiles making the matching process easier.

7. Start the Warp mode (circle f in Fig. 1b). First define the telomeres and the centromere (primary landmarks) by clicking corresponding points in the upper (the white curve of GGCC sequence density created on the basis of Ensembl database) and the lower part (the fluorescence intensity curve of the CMA3-stained chromosome) of the External Profiles Function display (see Note 11).

8. If additional characteristic positive and negative peaks, in the respective profiles, can be matched (i.e., secondary landmarks),

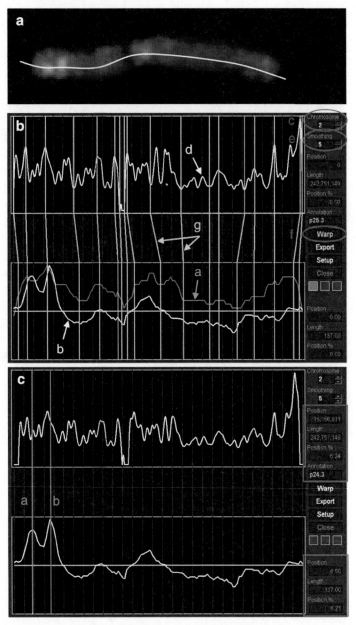

Fig. 1. Assignment of cCGH profile to the genome sequence information using External Profiles Function and Warp tool. (a) Definition of the longitudinal axis (*white line*) of normal human chromosome 2 after hybridization of differentially labeled test (neuroblastoma cell line) and reference (leukocytes) DNAs and simultaneous staining with CMA3. The *white line* indicates the track of the measured fluorescence intensity. (b)The cCGH fluorescence intensity ratio profile and the CMA3 fluorescence profile are visualized as intensity curves, the *blue curve (a)* in the *lower part* of the picture illustrates GGCC motif frequency based on the CMA3 fluorescence intensity; the *white curve (b)* corresponds to the ratio profile of test to the reference fluorescence; the *white curve (d)* in the *upper part* of the picture represents the GGCC density according to Ensembl databank; *yellow lines (g)* denote linkage of landmarks. Pter is defined as the first landmark at 0 bp (on the left side), qter as the last position (on the right side) and the centromere as GC-poor domain (low fluorescence intensity in submedial region). Most prominent peaks and valleys of GGCC density are set as secondary landmarks (see Notes 12, 13). (c) "Warped" chromosome 2 after combined cCGH and CMA3 staining and superimposing of the chromosomal sequence information to the database sequence information. The cell line used for cCGH analysis displays *two peaks* on 15.16 Mb (*a*) and 29.11 Mb (*b*) (ref. 16 with permission of S. Karger AG, Basel).

mark them in the same way as the primary landmarks. It increases the accuracy of the position measurement. All landmarks are displayed as yellow/white lines connecting corresponding positions in the profiles (*g* in Fig. 1b).

9. Terminate the Warp mode by clicking the right mouse button. Yellow/white lines are no longer displayed, and the sequence density profile is linearly transformed (stretched/squeezed) in such a way that all landmark positions in both profile areas coincide (Fig. 1c).

10. To perform a position measurement in the CGH ratio profile, move the cursor represented by a green/white vertical line to a specific location by a mouse click (*a* in Fig. 1c). The system displays the current cursor position in the warped sequence density profile in base pairs, the corresponding band number annotation as well as the total length of the sequence of the chromosome in base pairs, and the relative position of the cursor in percentage (upper red frame in Fig. 1c). In addition, the system also displays the cursor position in the fluorescence intensity ratio profiles, i.e., the absolute position in pixels, the total length of the profile in pixels, and the relative position in percentage (lower red frame in Fig.1c).

11. By using the Export button, the fluorescence intensity profiles, the GGCC density profiles, the position information, and the band number annotations can be exported to a text file for further processing and analysis.

4. Notes

1. In order to analyze the fluorescence intensities of CMA3, test, and reference DNAs, the choice of adequate combination of fluorescent dyes is crucial. We use Alexa Fluor-568-dUTP (generates red fluorescence) and DEA Coumarin 5-dUTP (light blue) for sample labeling to be able to combine them with CMA3 staining (green). Even though the excitation and emission spectra of DAPI (used together with CMA3 and distamycin A in CDD staining) and DEA Coumarin are similar, unlike expected, it seems that DAPI does not interfere with DEAC pattern. This might be explained by the quenching/displacing phenomenon of chromosomal DAPI fluorescence due to distamycin A counterstaining. However, in a situation where DAPI-DEAC interference is suspected, it is recommended to omit DA/DAPI step and stain cCGH slides only with CMA3/DA.

2. Since "aged" CMA3 solutions appeared to give a better staining, it is recommended to stabilize CMA3 solution for at least 3 days up to 1 month at +4°C.

3. Paraformaldehyde fixation is necessary to avoid chromosome morphology disruption, the so-called "swelling". Compared to the fixed ones, "swollen" chromosomes display far worse banding patterns.

4. We have introduced a modification to the standard cCGH protocol as target chromosomes to which differentially labeled DNAs are hybridized are subsequently stained with CMA3 and acquisition of all three fluorescence intensity profiles (test, reference, and CMA) is done simultaneously.

5. CMA3 is used instead of DAPI banding, which has frequently been employed for chromosome identification. We consider CMA3 peaks superior to reverse DAPI peaks which appear less well defined especially in the telomeric regions. By using distamycin/DAPI as counterstains, the contrast of the CMA3 bands is increased (17).

6. The inclusion of Mg^{2+}, as well as the ageing of slides for at least three days, is crucial to obtain a stable CMA3 fluorescence.

7. In order to fully take advantage of the mapping potential of the Warp tool, the proper alignment of all fluorescence channels of the captured metaphase is essential. Images composed of diverse fluorescence channels have to be merged with great precision since even a slight discordance may cause an error in mapping results.

8. The External Profiles Function and the Warp tool utilize individual chromosomes. Therefore, for an overall genome wide screening for copy number changes in the test samples, it is advised to perform classical chromosomal CGH, which provides an average result from a number of metaphases. However, to obtain high resolution information, individual chromosomes in combination with the Warp tool should be analyzed, i.e., for assigning the fluorescent peaks of interest to the Mb positions and thus to the gene map data.

9. With slightly separated chromatids, the track of the chromosome axis should be set along one of the chromatids and not in the space in between (see Fig. 1a). It is a requisite for achieving an appropriate and simultaneous measurement of the CMA3 sequence banding and the cCGH fluorescence intensities.

10. Regarding the Warp tool development for sequence density analyses, the 24 assembled DNA chromosome files in FASTA format were downloaded from the freely accessible Ensembl

web server (ftp://ftp.ensembl.org/pub/). The sequence of each chromosome was divided into intervals of 20,000 base pairs, and the absolute GGCC sequence motif count in each interval was saved to a single binary external profile set (XPS) file using an ISIS utility program for later use with the Warp tool. The ISCN band numbers corresponding to sequence positions were directly extracted from the Ensembl web site (http://apr2006.archive.ensembl.org/Multi/newsview?rel=38#cat2), using java methods and ENSJ, one of Ensembl's Application Programming Interfaces (API). The connection to the database was established through an Ensembl driver. The driver provides adaptors that allow the retrieval of the required annotation information that would be stored in one ANSI text file per chromosome. The ISIS utility read these 24 files and saved them as a single ISIS external profile annotation (XPA) file.

11. Despite the fact that telomeric and centromeric regions are polymorphic in length between individuals, standard sizes were allocated to these regions in different databases (Ensembl, NCBI) (18–20). Such intra- and inter-individual discordances in telomere and centromere length relative to the arbitrarily assumed sequence data may elevate heterogeneity and imprecision in defining the position of a chromosomal locus. Furthermore, the GGCC densities given in the genome sequence database are characterized by a drastic increase on pter and decrease on qter of the base frequency (from 0 to 1,000 and ~360 to 0, respectively, in the case of chromosome 1; upper part of Fig. 2). These values represent relative sequence density normalized to a maximum of 1,000. In contrast, the CMA3 fluorescence intensity reflecting the GGCC density of the chromosome does not abruptly increase and decrease, but a gradual rise and fall can be noticed at the chromosomal ends (lower part of Fig. 2). To define the exact position at which the rise/fall of the fluorescence intensity corresponds to the actual start or end of the chromosome as given in the sequence database, we analyzed the CMA3 fluorescence pattern of 180 individual chromosomes to which we hybridized six different BAC clones (30 chromosomes per clone) located close to the telomeres. Results of these BAC FISH experiments were then correlated with the sequence data. After such a learning phase, we were able to define the position of the telomere on the fluorescence curve. The most accurate results were obtained when the pter and qter were set at a 70% rise or fall of the fluorescence intensity whereby the top of the adjacent peak of the terminal region was set as 100% (lower part of Fig. 2). This reflects the situation in which the chromosomal ends were GGCC-rich. In the situation of

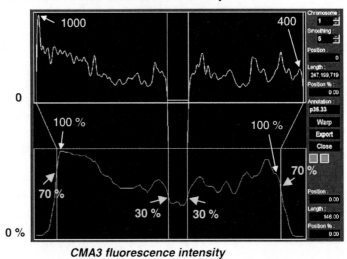

Fig. 2. Definition of telomeric and centromeric landmarks. Chromosome 1 serves as an example. *Yellow vertical lines* indicate defined pter, qter, starting, and ending position of the centromeric region (for further details see Note 11).

GGCC-poor distal regions of the chromosome, the pter or qter was determined at a 50 % rise or fall of the fluorescence intensity. Centromeric regions are clearly visible in the GGCC density profile (sequence data) as GGCC motif count at zero level (upper part of Fig. 2). For the CMA3 fluorescence intensity profile, a strategy to take a 30 % rise or fall of the fluorescence intensity as starting or ending position of the centromere was the most appropriate (the top of the last peak on p arm and the first peak on q arm of the chromosome was then defined as 100%) (lower part of Fig. 2).

12. As both the paracentromeric AT-rich heterochromatin and interchromosomal positive and negative bands can easily be identified using CMA3 staining, these regions do not cause any mapping problems and virtually all interchromosomal positions can be precisely mapped with our technique (15,16).

13. However, the fact that (a) the chromosomal ends are not well defined and (b) the subtelomeric sequences are variable in length accounts for a less accurate mapping of subtelomeric or telomeric markers. The exact definition of telomeric landmark positions is particularly problematic in the case of chromosomes, which undergo procedures employing structural disruption, even if the disruption is mild. The cCGH procedure includes several steps that may indeed easily disrupt the chromosome structure, especially the telomeric ends. It was reported previously that the DNA of chromosome ends is likely to be less tightly coiled/folded in comparison to the proximal regions (21). Therefore, telomeric ends may preferentially emanate,

"loop out" or "swell" due to, e.g., high temperature or severe enzymatic treatment which are employed during the hybridization. Optimization of digestion and denaturation of metaphase slides undergoing cCGH procedure improves the quality of chromosomes used for the analysis, thus making it possible to overcome the problem of exact telomeric landmark definition and enabling more precise mapping of the data derived from this procedure. However, since, for example, the age of metaphase slides has a particular impact, it is recommended to optimize digestion and denaturation steps for the individual experimental systems.

Acknowledgements

We thank Bettina Brunner for excellent technical assistance, Cornelia Stock for valuable suggestions and overall enormous help, and Marion Zavadil for proofreading. We also gratefully acknowledge Thomas Lörch (MetaSystems, Germany) as well as Zlatko Trajanoski, Dietmar Rieder, Gabriela Bindea, and Thomas Ramsauer (Institute for Genomics and Bioinformatics, Graz University of Technology) for the very helpful collaboration. This work was supported by St. Anna Kinderkrebsforschung.

References

1. Kallioniemi, A., Kallioniemi, O. P., Piper, J., Tanner, M., Stokke, T., Chen, L., Smith, H. S., Pinkel, D., Gray, J. W., and Waldman, F. M. (1994) Detection and mapping of amplified DNA sequences in breast cancer by comparative genomic hybridization, *Proc Natl Acad Sci U S A* **91**, 2156–2160.

2. Lichter, P., Bentz, M., and Joos, S. (1995) Detection of chromosomal aberrations by means of molecular cytogenetics: painting of chromosomes and chromosomal subregions and comparative genomic hybridization, *Methods Enzymol* **254**, 334–359.

3. Lichter, P., Joos, S., Bentz, M., and Lampel, S. (2000) Comparative genomic hybridization: uses and limitations, *Semin Hematol* **37**, 348–357.

4. Stock, C., Kager, L., Fink, F. M., Gadner, H., and Ambros, P. F. (2000) Chromosomal regions involved in the pathogenesis of osteosarcomas, *Genes Chromosomes Cancer* **28**, 329–336.

5. Wang, N. (2002) Methodologies in cancer cytogenetics and molecular cytogenetics, *Am J Med Genet* **115**, 118–124.

6. Garnis, C., Buys, T. P., and Lam, W. L. (2004) Genetic alteration and gene expression modulation during cancer progression, *Mol Cancer* **3**, 9.

7. Kallioniemi, A., Kallioniemi, O. P., Sudar, D., Rutovitz, D., Gray, J. W., Waldman, F., and Pinkel, D. (1992) Comparative genomic hybridization for molecular cytogenetic analysis of solid tumors, *Science* **258**, 818–821.

8. Kallioniemi, O. P., Kallioniemi, A., Piper, J., Isola, J., Waldman, F. M., Gray, J. W., and Pinkel, D. (1994) Optimizing comparative genomic hybridization for analysis of DNA sequence copy number changes in solid tumors, *Genes Chromosomes Cancer* **10**, 231–243.

9. Bentz, M., Plesch, A., Stilgenbauer, S., Dohner, H., and Lichter, P. (1998) Minimal sizes of deletions detected by comparative genomic hybridization, *Genes Chromosomes Cancer* **21**, 172–175.

10. Pinkel, D., Segraves, R., Sudar, D., Clark, S., Poole, I., Kowbel, D., Collins, C., Kuo, W. L., Chen, C., Zhai, Y., Dairkee, S. H., Ljung,

B. M., Gray, J. W., and Albertson, D. G. (1998) High resolution analysis of DNA copy number variation using comparative genomic hybridization to microarrays, *Nat Genet* **20**, 207–211.

11. Schweizer, D. (1976) Reverse fluorescent chromosome banding with chromomycin and DAPI, *Chromosoma* **58**, 307–324.

12. Schweizer, D., and Ambros, P. F. (1994) Chromosome banding. Stain combinations for specific regions, *Methods Mol Biol* **29**, 97–112.

13. Hou, M. H., Robinson, H., Gao, Y. G., and Wang, A. H. (2004) Crystal structure of the [Mg2+-(chromomycin A3)2]-d(TTGGCCAA)2 complex reveals GGCC binding specificity of the drug dimer chelated by a metal ion, *Nucleic Acids Res* **32**, 2214–2222.

14. Ambros, P. F., and Sumner, A. T. (1987) Correlation of pachytene chromomeres and metaphase bands of human chromosomes, and distinctive properties of telomeric regions, *Cytogenet Cell Genet* **44**, 223–228.

15. Kowalska, A., Bozsaky, E., Ramsauer, T., Rieder, D., Bindea, G., Lorch, T., Trajanoski, Z., and Ambros, P. F. (2007) A new platform linking chromosomal and sequence information, *Chromosome Res* **15**, 327–339.

16. Kowalska, A., Brunner, B., Bozsaky, E., Chen, Q. R., Stock, C., Lorch, T., Khan, J., and Ambros, P. F. (2008) Sequence based high resolution chromosomal CGH, *Cytogenet Genome Res* **121**, 1–6.

17. Schweizer, D. (1981) Counterstain-enhanced chromosome banding, *Hum Genet* **57**, 1–14.

18. Aviv, A., Levy, D., and Mangel, M. (2003) Growth, telomere dynamics and successful and unsuccessful human aging, *Mech Ageing Dev* **124**, 829–837.

19. Riethman, H., Ambrosini, A., and Paul, S. (2005) Human subtelomere structure and variation, *Chromosome Res* **13**, 505–515.

20. Ambrosini, A., Paul, S., Hu, S., and Riethman, H. (2007) Human subtelomeric duplicon structure and organization, *Genome Biol* **8**, R151.

21. Saitoh, Y., and Laemmli, U. K. (1994) Metaphase chromosome structure: bands arise from a differential folding path of the highly AT-rich scaffold, *Cell* **76**, 609–622.

Chapter 24

ImmunoFISH on Isolated Nuclei from Paraffin-Embedded Biopsy Material

Soo-Yong Tan and Goran Mattsson

Abstract

The detection of genetic abnormalities in paraffin sections by fluorescence in situ hybridization (FISH) is widely used in clinical practice to detect amplification of the *ERB2* gene in breast carcinoma and various chromosomal translocations in lymphomas and soft tissue tumors. However, interpretation of FISH signals in tissue sections may be difficult due to overlapping nuclei and nuclear truncation artifacts. Some of these shortcomings may be avoided by the use of isolated nuclear preparations. However, identification of cell populations may be difficult in detached cells removed from their histological context. We have described an optimized immunoFISH technique on isolated nuclear suspension, which combines the benefits of studying isolated cells derived from paraffin embedded tissues by FISH analysis with the ability to detect cell lineage and other markers by immunofluorescence.

Key words: ImmunoFISH, Isolated cells, Paraffin block, Chromosome paint, FICTION, Fluorescence in situ hybridization

1. Introduction

The fact that the fluorescence in situ hybridization technique (FISH) can be performed on conventional paraffin sections means that it lends itself well to be used for routine work in pathology laboratories. However, the use of tissue sections for diagnostic purposes also introduces problems. In particular, a substantial number of cells lose part of their nuclear material (nuclear truncation) during tissue sectioning, resulting in incomplete FISH labelling patterns (i.e., loss of one or more of the target sequences for hybridization). In addition, it is necessary to distinguish one nucleus from another (which is difficult when closely packed nuclei overlie each other) and to assess signals seen in different

Joanna M. Bridger and Emanuela V. Volpi (eds.), *Fluorescence in situ Hybridization (FISH):
Protocols and Applications*, Methods in Molecular Biology, vol. 659,
DOI 10.1007/978-1-60761-789-1_24, © Springer Science+Business Media, LLC 2010

focal planes. Difficulties in interpreting FISH labelling of routine tissue sections have prompted several laboratories to perform FISH analysis on nuclei isolated from tissue blocks (1,2). An additional problem with ordinary FISH analysis is that it may be difficult to distinguish between normal and neoplastic cells. ImmunoFISH procedures (such as the FICTION technique) (3,4), which combine immunohistological staining and FISH analysis, may overcome this problem by identifying the nature of individual cells, but they are technically demanding, and few laboratories use them in routine clinical practice.

We have recently described a method to perform immuno-FISH analysis on isolated nuclei derived from paraffin embedded material (5). In brief, paraffin-embedded tissues are mechanically homogenized to isolate the nuclei, which we have shown to still retain a fair amount of cytoplasm. This allows us to perform immunofluorescence (with both nuclear and cytoplasmic markers) to delineate cell populations of interest before applying a variety of split-apart and dual-fusion FISH probes to detect translocations commonly encountered in hematolymphoid neoplasms (Fig. 1). It may be particularly useful for studying tumors in which neoplastic cells are present in relatively low numbers (e.g., Hodgkin's lymphoma, T-cell-rich B-cell lymphoma). In addition, this technique simplifies chromosomal enumeration using centromeric probes and whole chromosomal paints by reducing the number of overlapping nuclei.

2. Materials

2.1. Isolation of Cells

1. 1 mm diameter tissue microarray needle (Beecher Instruments Inc.).
2. Xylene or Citroclear (HD Supplies).
3. Ethanol.
4. Plastic universal spatula (Alpha Laboratories Ltd).
5. Proteinase K (Dako).
6. Phosphate buffered saline (PBS).
7. 50 µM Filicon nylon mesh (BD Biosciences).
8. Methanol.
9. Acetic acid.

2.2. Transfer of Cell Suspensions to Microscope Slides

1. X-tra™ adhesive microscopic slides (Surgipath).
2. Wax pen (PAP-BioGenex).
3. Poly-L-lysine. Dilute to 10% with distilled water.

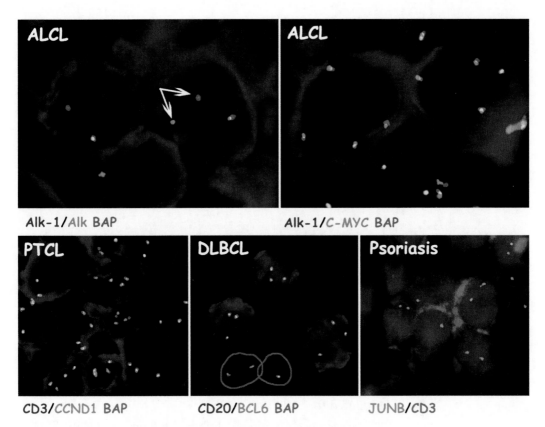

Fig.1. ImmunoFISH on isolated nuclei derived from paraffin-embedded material. *Top*: Immunofluorescence for alk1 shows cytoplasmic staining in this example of anaplastic large cell lymphoma showing *ALK* translocation and *C-MYC* amplification. *Bottom*: ImmunoFISH analysis with "split-apart" probes showing amplifications of *CCND1*/cyclin D1 in a case of peripheral T-cell lymphoma (PTCL) and that of *BCL6* in a diffuse large B-cell lymphoma (DLBCL). The last example using a probe for *JUNB* shows a normal FISH pattern in T-cells in a case of psoriasis.

2.3. Immuno-fluorescence

1. Tris-EDTA buffer, pH 9.0. This is prepared by dissolving 48 g of Tris, 4 g of EDTA in 2L of deionized water and adjusting the pH to 9.0 using 0.2 N HCl.

2. Catalyzed Signal Amplification 1 (CSA1) System (Dako A/S).

3. Biotinylated goat anti-mouse/rabbit antibody (Biotinylated Link, Dako A/S). Serum-free Protein Block (X0909, Dako A/S).

4. Alexa Fluor 350 Streptavidin (Molecular Probes, Invitrogen), diluted to 1:100 using PBS.

5. Alexa Fluor 350 goat anti-mouse IgG1, Alexa Fluor 350 goat anti-mouse IgG2a, Alexa Fluor 350 goat anti-mouse IgG3, or Alexa Fluor 350 goat anti-rabbit IgG (Molecular Probes), depending on the immunoglobulin species and subtype of the primary antibody (see Table 1). Use a working concentration of 1:100 by diluting with PBS.

Table 1
Sequence of antibody incubation for different primary antibody isotypes

Primary antibody			Third antibody (Alexa Fluor 350 conjugated)	Fourth antibody (Alexa Fluor 350 conjugated)
Examples	Isotype	Second antibody		
CD79a/JCB117	Mouse IgG1	Biotinylated goat anti-mouse/ rabbit IgG	Goat anti-mouse IgG1	Donkey anti-goat IgG
CD20/L26	Mouse IgG2a		Goat anti-mouse IgG2a	
CD25/4C9	Mouse IgG2b		Goat anti-mouse IgG2b	
CD68/PGM1	Mouse IgG3		Goat anti-mouse IgG	
CD15/LeuM1	Mouse IgM		Goat anti-mouse IgM	
CD3	Rabbit polyclonal		Goat anti-rabbit IgG	

6. Alexa Fluor 350 donkey anti-goat IgG (Molecular Probes), diluted to 1:100 using PBS.

7. Fluorescence microscope fitted with filters for DAPI, FITC/ Spectrum Green and Texas Red (Dako FISH probes) or Spectrum Orange (Vysis FISH probes).

2.4. FISH

1. The following FISH probes have been tested with this protocol:

   • "Split-apart" probes for the following genes: *BCL2*, *BCL6*, *IGH*, *C-MYC*, *CCND1*, *MALT1*, and *PAX5* (Dako A/S). Some of these probes are also available from Vysis (Abbott Molecular Inc.).

   • "Dual color and dual fusion" probes for the *IGH-CCND1* and *BCR-ABL* fusion genes (Abbott Molecular Inc.).

   • Centromeric probes against chromosomes 3, 8, and 11 (Abbott Molecular Inc.).

   • FISH probes from Dako come in ready-to-use formulations, whereas those from Vysis are concentrated and should be diluted in the ratio 1 part probe: 2 parts HPLC water (Sigma): 7 parts hybridization buffer (Abbott Molecular Inc.).

2. Whole chromosomal paint (Cambio) probes for chromosomes 3 and 8 have also been tested (see Note 8). These are available as ready-to-use preparations and may be treated like ordinary FISH probes.

3. Dako FISH Ancillary Kit (Dako A/S).

4. Vectorshield fluorescent mounting medium without DAPI (Vector Laboratories).

3. Methods

Suspensions prepared from paraffin-embedded samples for FISH analysis are commonly referred to as "isolated nuclei." In fact, such "nuclear" suspensions contain many intact or semi-intact cells, as evident from electron microscopic analysis, immunofluorescence labelling, and flow cytometry. The ability to identify cell populations by immunofluorescence labelling underlies the success of this specific application as it allows to define simultaneously at the single cell level both the cellular lineage/proliferation status and genetic anomalies identified by FISH analysis.

The immunoFISH technique described here combines immunofluorescence staining with a blue fluorophore and interphase FISH using probes that are conjugated to green and red fluorophores. Common to both immunostaining and in situ hybridization is the need for heat pre-treatment, which is performed using Tris-EDTA buffer, pH 9.0. The major difference from ordinary FISH analysis on paraffin-embedded tissues is that the protease digestion step is omitted. Although protein digestion will remove obscuring cell proteins and improve FISH signals, it is deleterious to the immunofluorescence signal. Similarly, hybridization over a longer period of time also increases the intensity of FISH signals at the expense of immunofluorescence staining. As most commercially available FISH probes are labelled with green and red fluorophores, immunostaining has to be performed using a blue fluorophore, which also tends to give weaker signal intensity. For these reasons, the immunoFISH technique works best when the primary antibody is robust and produces a strong immunostaining. Occasionally, it may be necessary to increase signal intensity using tyramide signal amplification, as described below.

3.1. Preparation of Isolated Cells

1. Mark the area of interest on a microscope slide stained with haematoxylin and eosin.

2. Using a 1 mm tissue microarray needle, remove one or two cores of tissue from the paraffin section corresponding to the area of interest (see Note 1). Alternatively, if preselection is not essential, 10–30 μm thick sections can simply be cut from paraffin blocks.

3. Place the tissue cores or sections in 1.5 mL micro-centrifuge tubes.

4. Add 1–1.5 mL of xylene (or similar solvent such as Citroclear) and incubate at room temperature for 10 min. Centrifuge to form pellet and decant supernatant. Add xylene and repeat the process another two times.

5. Add 95% ethanol and incubate for 3 min. Centrifuge to form pellet and decant supernatant. Repeat the process with 75% and 50% ethanol.

6. Homogenize the pellet in a small amount (e.g., 50 μL) of 50% ethanol for 2 min by rotating up and down in the tube using the end of a plastic universal spatula (Alpha Laboratories Ltd).

7. Add PBS to each tube before centrifugation at 2,016 × g for 10 min. Decant supernatant and re-suspend in PBS.

8. Filter through a 50 μm Filicon nylon mesh and centrifuge again before removing the supernatant.

9. Depending on the size of the pellet, re-suspend in 100–300 μL PBS (see Note 9).

10. Make circular rings (about 1.0 cm diameter) on X-tra™ adhesive microscopic slides with a PAP pen.

11. Mix the nuclear suspensions with 4 μL of 10% poly-l-lysine before pipetting it onto the rings. Let the slides dry at room temperature.

12. Heat the slides with isolated nuclei at 60°C for 30 min. Upon cooling to room temperature, the slides are ready for immunoFISH. If desired, slides can be wrapped in foil for long-term storage at −20°C (see Note 2).

3.2. Immunofluorescence

1. Pretreatment is performed by boiling in 0.01 M Tris-EDTA pH 9 in a Biocare decloaking chamber (Biocare) for 2–5 min at operating pressure (see Note 3). Cool to room temperature and rinse twice in PBS for 2 min.

2. Add Protein Block for 10–15 min. Tap off Protein Block and without rinsing, add primary antibody and incubate for 1 h (see Note 4).

3. Wash in PBS for 5 min. Incubate with biotinylated goat anti-mouse or rabbit immunoglobulin for 30 min.

4. Wash in PBS for 5 min. Incubate for 30 min with a mixture of Alexa Fluor 350 Streptavidin (1:100) and the appropriate Alexa Fluor 350 conjugated secondary antibody, depending on the species and isotype of the primary antibody (Table 1).

5. Wash in PBS for 5 min. Incubate with Alexa Fluor 350 donkey anti-goat IgG for 30 min. Wash in PBS for 5 min.

6. Assess results of immunofluorescence under a fluorescence microscope before proceeding to FISH analysis (see Note 5).

3.3. FISH

1. Dehydrate tissue sections by passing the slides through a graded ethanol series: 70, 95, 100% (2 min for each step). Allow slides to air-dry completely.

2. Add probe mix to the slide and immediately cover with either a 22 × 22 mm square coverslip or 10 mm diameter circular coverslip (see Note 6). To perform FISH on a larger area using 22 × 22 mm coverslip, add 9 μL of the probe mix. If the 10 mm circular coverslip is used to FISH a smaller target area, just 1.33 μL of probe mix is sufficient.

3. Seal the edges of the coverslip preferably with the rubber sealant provided as part of the FISH Ancillary kit, but ordinary bicycle tire sealant will do as well.

4. Place the slides in a hybridizer (Dako Statspin Hybridizer, Dako A/S): 5 min denaturation step at 82°C and overnight hybridization at 37°C (see Note 7).

5. Upon completion of hybridization, prepare one Coplin jar of stringency buffer at room temperature and preheat another Coplin jar of stringency buffer (FISH ancillary kit, Dako A/S) to 65°C in a waterbath. Remove the rubber sealant from the slides and rinse off briefly (about 5 min) in the first jar of stringency buffer (at room temperature). Next, incubate the slides in the heated stringency wash for 10 min.

6. Remove the slides from the stringency wash and rinse off twice in diluted wash buffer (FISH Ancillary kit, Dako A/S) for 3 min each.

7. Apply 15 μL of fluorescence mounting medium without DAPI.

4. Notes

1. Often the cells of interest may be located only in a small area of a paraffin block. In such instances, we advocate using a tissue array needle to preselect cells of interest. On the other hand, with very small specimens, e.g., needle core biopsies, entire sections are cut from paraffin blocks instead and processed in a similar way.

2. For long-term storage of isolated nuclear suspension, add methanol/acetic acid (3:1 ratio, 100–300 μL) to each sample before storage at –70°C. To avoid clumping of nuclei when placed on slides, replace the methanol/acetic acid with PBS after removal from storage before transferring to microscope slides as described.

3. Depending on the age of the tissue and the state of antigenic preservation, a longer heating time may be required. Instead

of the Biocare decloaking chamber, one can use an ordinary domestic pressure cooker (e.g., Clipso Control Pressure Cooker, Tefal). Fill the pressure cooker with 2 L of Tris-EDTA buffer and bring to the boil. Place slides (within a metal rack) in the pressure cooker and heat at operating pressure for 20 min (with the Clipso Control Pressure Cooker) and then cool to room temperature with running tap water.

4. The incubation period may have to vary according to the antibody. Apart from increasing the antibody concentration (which tends to increase background staining), stronger immunofluorescence signal may be obtained by overnight incubation. In cases where the immunofluorescence signal is weak, one may improve signal-to-noise ratio significantly by using tyramide-based signal amplification. The CSA1 kit from Dako may be used for this purpose (with murine primary antibodies): Following the incubation with the primary antibody, the slides are washed in PBS, and then incubated with peroxidase block for 10 min.Wash in PBS for 5 min. Incubate with biotinylated goat anti-mouse/rabbit immunoglobulin (Dako A/S) for 15 min.Wash in PBS for 5 min. Incubate with streptavidin-biotin-horseradish peroxidase complex for 15 min.Wash in PBS for 5 min. Incubate with the amplification reagent (containing tyramide-biotin) for 15 min.Wash in PBS for 5 min. Incubate for 30 min with a mixture of Alexa Fluor 350 Streptavidin (1:100) and the appropriate Alexa Fluor 350 goat anti-mouse immunoglobulin, depending on the species and isotype of the primary antibody.Wash in PBS for 5 min. Incubate with Alexa Fluor 350 donkey anti-goat immunoglobulin for 30 min. Wash in PBS for 5 min. Assess results of immunofluorescence under a fluorescence microscope before proceeding to FISH analysis.

5. The success of immunofluorescence staining will depend on the type of primary antibody and degree of antigenic preservation (the latter may reflect the age of the paraffin block and quality of initial processing). We obtained good immunostaining of isolated nuclei from lymphoid tissue biopsies for a range of cytoplasmic (CD3, CD5, CD20, CD34, CD79a, Bcl-2, SLP-76, LAT, and vWF) and nuclear markers (Ki67, Oct2). If an antibody produces poor immunostaining with ordinary DAB chromogen, it is unlikely that it will yield good results with immunoFISH.

6. When applying the FISH probe mix (or chromosome paint), allow the probe to spread evenly beneath the coverslip and tap carefully with a pair of forceps to expel air bubbles.

7. Increasing the period of hybridization to 48 h and more will increase the signal strength, although the immunofluorescence signal may diminish somewhat.

8. This technique can also be used to highlight entire chromosomes in cells from paraffin biopsies using whole chromosome paints. These signals are larger in size and less discrete than those obtained using centromeric or locus-specific probes. However, numerical abnormalities can be demonstrated in some lymphomas by this approach and may be useful in instances where a centromeric probe is not suitable due to its tendency to cross-hybridize with alpha-satellite sequences of a different chromosome.

9. As extracted cell suspensions can be stored for long periods prior to use, custom nuclear arrays can be prepared as and when needed to allow a genetic anomaly to be studied by the immunoFISH procedure in many samples. It is also possible, provided the "spotted" nuclear suspensions are adequately separated, to apply a number of different antibody/probe combinations to multiple samples on a single slide.

Acknowledgement

We would like to give our immense gratitude to the late Prof David Mason, Leukaemia Research Fund Immunodiagnostics Unit, Nuffield Department of Clinical Laboratory Sciences, University of Oxford, without whom this chapter would not have been possible.

References

1. Paternoster SF, Brockman SR, McClure RF, Remstein ED, Kurtin PJ, Dewald GW. (2002). A new method to extract nuclei from paraffin-embedded tissue to study lymphomas using interphase fluorescence in situ hybridization. *Am J Pathol* **160**, 1967–1972.

2. Rowe LR, Willmore-Payne C, Tripp SR, Perkins SL, Bentz JS. (2006). Tumor cell nuclei extraction from paraffin-embedded lymphoid tissue for fluorescence in situ hybridization. *Appl Immunohistochem Mol Morphol* **14**, 220–224.

3. Martínez-Ramírez A, Cigudosa JC, Maestre L, Rodríguez-Perales S, Haralambieva E, Benítez J, Roncador G. (2004). Simultaneous detection of the immunophenotypic markers and genetic aberrations on routinely processed paraffin sections of lymphoma samples by means of the FICTION technique. *Leukemia* **18**, 348–353.

4. Korac P, Jones M, Dominis M, Kusec R, Mason DY, Banham AH, Ventura RA. (2005). Application of the FICTION technique for the simultaneous detection of immunophenotype and chromosomal abnormalities in routinely fixed, paraffin wax embedded bone marrow trephines. *J Clin Pathol* **58**, 1336–1338.

5. Mattsson G, Tan SY, Ferguson DJ, Erber W, Turner SH, Marafioti T, Mason DY. (2007). Detection of genetic alterations by immunoFISH analysis of whole cells extracted from routine biopsy material. *J Mol Diagn* **9**, 479–489.

Chapter 25

Fluorescence In Situ Hybridization on 3D Cultures of Tumor Cells

Karen J. Meaburn

Abstract

Genomes are spatially highly organized within interphase nuclei. Spatial genome organization is increasingly linked to genome function. Fluorescence in situ hybridization (FISH) allows the visualization of specific regions of the genome for spatial mapping. While most gene localization studies have been performed on cultured cells, genome organization is likely to be different in the context of tissues. Three-dimensional (3D) culture model systems provide a powerful tool to study the contribution of tissue organization to gene expression and organization. However, FISH on 3D cultures is technically more challenging than on monocultures. Here, we describe an optimized protocol for interphase DNA FISH on 3D cultures of the breast epithelial cell line MCF-10A.B2, which forms breast acini and can be used as a model for early breast cancer.

Key words: FISH, 3D cell culture, Acini, Interphase, Nuclear architecture, Tissue models, Genome organization, Early tumorigenesis

1. Introduction

Genes occupy preferred spatial positions within interphase nuclei. There is an emerging realization that the spatial positioning of DNA is critical for correct genome function (1–4). Moreover, the spatial positioning of the genome alters during diseases such as cancer (5–7). While most gene localization studies have been performed on cultured cells, genome organization is likely different in the context of tissues. Three-dimensional (3D) culture model systems provide a powerful tool to study the contribution of tissue organization to gene expression and organization. Many cell types have been grown under various 3D culture conditions, including liver, prostate, lung, and bone cells, and the cultures

Joanna M. Bridger and Emanuela V. Volpi (eds.), *Fluorescence in situ Hybridization (FISH):*
Protocols and Applications, Methods in Molecular Biology, vol. 659,
DOI 10.1007/978-1-60761-789-1_25, © Springer Science+Business Media, LLC 2010

closely recapitulate both tissue morphology and function (8). Perhaps, the most extensively studied 3D culture tissue models, however, are the mammary gland acinar models (9,10). In 3D cultures, normal mammary epithelial cells grown in contact with reconstituted basement membrane form growth-arrested spheroid structures, termed acini, which consist of a single layer of polarized cells surrounding a hollow lumen (11). The cultured acini not only resemble in vivo mammary gland morphology, but they recapitulate glandular functions, such as milk protein expression, making them far better models of tissue than monocultured cells (8,10). 3D cultured breast epithelial cells can be used to address various questions, particularly in the field of cancer (8,9). The effects of activating oncogenes in the context of tissue can be explored by experimental manipulations of normal cells grown in 3D cultures, as such manipulations are far harder and more time consuming in actual tissues (9). For example, MCF-10A.B2 cells are normal mammary epithelial cells that contain a construct for an inducible variant of ErbB2, an oncogene commonly overexpressed in breast cancer (12). MCF-10A.B2 acini in which ErbB2 is overactivated, by addition of a synthetic ligand, mimic the structural morphology and phenotypes of early tumorigenesis (12). Alternatively, the opposite can be done as cancerous breast cells can be grown in 3D cultures and manipulated with the aim of reverting the malignant phenotype and restoring the normal acini structure (10). The cellular processes and pathways affected can then be probed in these cancer models.

Fluorescence in situ hybridization (FISH) allows visualization of specific regions of the genome and ultimately mapping of genomic loci within interphase cells (13). The FISH procedure consists of sample fixation, permeabilization, and DNA denaturation, followed by probe hybridization to the target. For detection, samples are washed to remove unbound probe and probe that is bound to nuclear DNA with low homology, and the specifically bound probes are visualized using fluorochrome-conjugated reagents. Performing FISH on 3D cultures is technically more challenging than standard monocultures. 3D cultures are thicker than standard monocultures and they contain an extracellular matrix making probe penetration more difficult. In addition, 3D cultures are more fragile, more easily lost from the coverslip, and more difficult to handle and manipulate. To overcome these problems, several modifications have been made including development of permeabilization conditions to ensure that the biotin- and digoxigenin-labelled gene probes can penetrate the nuclei, yet the 3D structure morphology and attachment of acini to the slide are retained; the time and temperature for DNA denaturation have also been increased to account for the thicker structures; post-hybridization washes and detection of the probes are essentially carried out by conventional methods. We have successfully used this FISH method on breast epithelial MCF-10A.B2 acini grown under both normal and tumorigenic 3D culture growth conditions. This enabled us to compare the radial position of 11 genes and the identification of loci-specific repositioning during early tumorigenesis (6).

2. Materials

2.1. Probe Labelling by Nick Translation

1. 10× nick translation buffer: 0.5 M Tris–HCl pH 8.0, 50 mM MgCl$_2$, and 0.05 mg/mL bovine serum album (BSA, Fraction V, Sigma-Aldrich). Store at –20°C.

2. Nucleotide stock: 0.5 mM dATP (Invitrogen), 0.5 mM dGTP (Invitrogen), 0.5 mM dCTP (Invitrogen), 0.5 mM nucleotides containing dUTP conjugate of choice (e.g., biotin-11-dUTP (Roche Applied Science), digoxigenin-11-dUTP (Roche Applied Science)), and 20 mM Tris–HCl pH 7.5. Store this stock at –20°C.

3. 0.01 M β-mercaptoethanol: Add 0.7 μL of β-mercaptoethanol (Sigma-Aldrich) to 999.3 μL of sterile water. Store at –20°C.

4. DNase I: 1 mg/mL DNase I (Roche Applied Science) and 50% (v/v) glycerol. Store this stock at –20°C. Dilute 1–2 μL of this stock in 100 μL of ice-cold water immediately before use; discard afterwards (see Note 1).

5. DNA Polymerase: *Escherichia coli* DNA polymerase (10,000 U/mL, recombinant; New England Biolabs).

6. Agarose (Invitrogen).

7. 10 mg/mL Ethidium bromide solution (Invitrogen). Ethidium bromide is toxic and mutagenic. Gloves must be worn when handling this reagent. Waste ethidium bromide solution and gels must be discarded into a hazardous waste container. Contact your waste management department for proper disposal.

8. 6× loading dye: 0.25% (w/v) Bromophenol Blue, 0.25% xylene cyanol FF (Sigma-Aldrich), and 15% Ficoll (type 400; GE Healthcare) in water. Store at 4°C; this is a stable solution.

9. DNA ladder (hi-lo, Minnesota Molecular).

10. 10× Tris Acetate Buffer (TAE, Quality Biological Inc.).

11. 0.5 M EDTA.

2.2. Probe Precipitation

1. Competitor DNA: Human C$_0$t-1 (Roche Applied Science). The C$_0$t-1 needs to be species-specific; if your cells are derived from mice, use mouse C$_0$t-1.

2. Carrier DNA: yeast transfer RNA (tRNA, type X, Sigma-Aldrich).

3. 3 M sodium acetate pH 5.2.

4. Ethanol.

5. Hybridization mix: 10% (w/v) dextran sulphate (Sigma-Aldrich), 50% (v/v) formamide (Sigma-Aldrich), 2× SSC, and 1% (w/v) Tween-20 (Sigma-Aldrich). Store at –20°C. It is convenient to use the following stock solutions: 50% (w/v) dextran sulphate, this can be stored at –20°C, 4°C, or room temperature (RT). 20× SSC (Quality Biological Inc) stored at RT. 20× SSC stock can be prepared with 3 M NaCl and 0.3 M tri-sodium citrate in water; adjust pH to 7.0 using HCl and store at RT.

2.3. Sample Preparation

1. Lab-Tek eight-well glass chamber slides (Nalge Nunc).

2. Phosphate buffered saline (PBS): 1× PBS (Quality Biological Inc) stored at RT.

3. Paraformaldehyde: prepare 4% (w/v) paraformaldehyde solution by performing a 1 in 4 dilution of 16% (w/v) paraformaldehyde (Electron Microscopy Sciences) in PBS. The 4% solution can be stored at RT for 2 weeks or longer at –20°C, but freeze only once. Paraformaldehyde is carcinogenic and toxic by inhalation. Gloves and chemical fume hood must be used when handling this reagent.

4. Saponin/Triton-X solution: 0.5% (w/v) saponin (Sigma-Aldrich) and 0.5% (v/v) Triton X-100 (Sigma-Aldrich) in PBS. Make up fresh for each experiment.

5. 0.1 N HCl: dilute 1 N HCl in water. Store at RT for 1 week.

6. 2× SSC: dilute 20× SSC stock with water. Make fresh for each experiment.

7. 50% formamide/2× SSC: Formamide is diluted in water and 20X SSC as 5 parts formamide to 1 part 20× SSC and 4 parts water. Make up fresh for each experiment. Formamide is toxic and mutagenic. Gloves and chemical fume hood must be used when handling this reagent.

2.4. Washes and Detection

1. Wash buffer A: 50% formamide/2× SSC. Preheat to 45°C. Make up fresh for each experiment.

2. Wash buffer B: 1× SSC. Make up this dilution from a 20× SSC stock freshly for each experiment. Preheat at 60°C.

3. Wash buffer C: 0.05% (v/v) Tween-20 in 4× SSC. Make up fresh for each experiment. Wash buffer C is required at both RT and 42°C.

4. Blocking solution: make up a 3% BSA (Fraction V, Sigma-Aldrich) solution in wash buffer C, diluting from a 30% (w/v)

BSA stock. Make up fresh for each experiment, and keep on ice until used. Aliquots of 30% (w/v) BSA made in PBS can be stored at 4°C for short-term storage or –20°C for long-term storage.

5. Detection solution: detection antibody anti-digoxigenin-rhodamine (Roche Applied Science) and fluorescein-avidin DN (Vector Laboratories, Burlingame, CA, USA) are diluted 1:200 in blocking solution. A "starry night" background can be caused by an aggregation of detection reagents. To avoid this, antibodies should be spun briefly in a microcentrifuge, and only the supernatant should be used. Make up the dilution fresh before use, and keep on ice until used. Protect from light to avoid a reduction in fluorescent signal.

6. Mounting medium: 4,6,-diamidino-2-phenylindole (DAPI) containing Vectashield mounting medium (Vector Laboratories, Inc). DAPI is a known mutagen; avoid contact with eyes and skin. Protect DAPI solution from light.

3. Methods

3.1. Probe Labelling by Nick Translation

For DNA FISH, we generate probes from BACs (see Note 2). To label our probes we routinely use nick translation (see Note 3) using dUTP conjugated with biotin or digoxigenin, using a method adapted from Langer et al. (14).

1. Add the following chilled reagents to a microcentrifuge tube on ice and make up the volume with sterile water to 200 µL: 20 µL 10× nick translation buffer, 20 µL of either the biotin- or digoxigenin-nucleotide stock, 20 µL 0.01 M β-mercaptoethanol, 4 µg probe DNA, 2 µL diluted DNase I, 2 µL DNA polymerase .

2. Air bubbles should be avoided. After mixing, briefly spin the tube in a microcentrifuge at $10,000 \times g$ at 4°C.

3. Place the tube at 16°C for 1.5–2 h. If you do not have a cooling dry incubator or a chilled water bath, set up a standard water bath in a 4°C cold room. Following this incubation, place the tubes on ice while checking the length of the probe.

4. To check the probe length, mix 5 µL of nick translation product with 1 µL of 6× loading dye and run it on 1% (w/v) agarose in 1× TAE gel containing a 1:10,000 dilution of ethidium bromide solution in 1× TAE running buffer at 90 V for ~30 min. At the same time, run a DNA ladder. The probe should be between 100 and 600 bps in size (see Note 4). If the probe is too long, return it to 16°C for further digestion

by adding fresh enzymes. Check the probe length after 30 min, repeat if necessary.

5. Once satisfied with the length of the probe, stop the reaction by inactivating the enzymes by adding 1–2 μL of 0.5 M EDTA and heat at 65°C for 10 min. These products are then stored at –20°C until use (see Note 5).

3.2. Probe Precipitation

1. For each well of an eight-well glass chamber slide, mix 1,065 ng probe DNA (53.25 μL of nick translation product), 15 μg C_0t-1, and 71 μg tRNA into a microcentrifuge tube. If two probes are used for simultaneous detection of multiple genes, add 1,065 ng of both the probes, which must be labelled in alternative conjugated-dUTPs. The amount of competitor and carrier DNA remains the same regardless of the number of gene probes in each hybridization reaction. We commonly perform both single and dual gene FISH on acini structures.

2. Add 1/10th the volume of 3 M sodium acetate pH 5.2 (volume of step 1 of subheading 3.2) and two-times volume of ice-cold 100% ethanol.

3. Microcentrifuge at 10,000g for 20 min at 4°C. Decant the ethanol solution and leave the pellet, which should be white, to dry at RT or dry in a speed-vac (see Note 6). The pellet will become clear as it dries.

4. Resuspend the pellet in 50 μL of hybridization mix. Mix well and briefly microcentrifuge to remove air bubbles. Leave at RT for a minimum of 30 min before use (see Note 7).

5. An optional step is to predenature the probes at 95°C for 5 min just prior to use. Some people suggest that this step gives brighter FISH signals; however, in our hands there is no difference in signal intensity if this step is used or omitted.

3.3. Sample Preparation

Care must be taken to ensure the cells do not dry out at any point during the FISH procedure, detection, and washes.

1. MCF10A.B2 cells are grown for 20 days under 3D culture conditions (see Note 8) in eight-well glass chamber slides coated with basement membrane extract (see Note 9). Remove the growing media and wash the cultures with addition of PBS for 5 min (see Note 10). 400 μL is a suitable volume of solution to add to each well.

2. The cells are fixed by incubation in the paraformaldehyde solution for 10 min. Discard the paraformaldehyde solution into a hazardous waste container and rinse the cells with three changes of PBS (see Note 11). Paraformaldehyde fixation is used as it preserves much of the in vitro nuclear morphology.

3. For probes to enter the nuclei, permeabilization steps must be performed. For this, incubate the 3D cultures in the saponin/Triton-X solution for 40 min (see Note 12). Wash the cells in three changes of PBS.

4. Incubate the culture in 0.1 N HCl for 30 min (see Notes 12–13).

5. Wash the culture for 10 min in 2× SSC.

6. Equilibrate cells in 50% formamide/2× SSC for at least 30 min (see Note 14).

7. To codenature the probe DNA and nuclear DNA, first remove the 50% formamide/2× SSC solution. Waste formamide solutions must be discarded into a hazardous waste container. Contact your waste management department for proper disposal. Next, add the 50 μL of prepared probe in hybridization mix (from step 4 of subheading 3.2), remove air bubbles with a pipette tip, and ensure that the probe has spread over the entire well. Denature by placing the eight-well chamber slide at 85°C for 10 min (see Notes 15–16). For this, use a preheated slide moat or dry bath incubator. If your dry bath incubator has only blocks to accommodate tubes, simply turn the block over before heating for an even temperature over the slide.

8. Leave to hybridize overnight at 37°C (minimum of 18 h). To prevent the cells from drying, seal the lid of the chamber slide with parafilm and put the slide into a prewarmed humid container. A plastic Tupperware box with damp tissue inside makes an ideal humidified chamber. This box can then be either placed in a prewarmed incubator or floated in a water bath.

3.4. Washes and Detection

1. The next day, make wash buffers A–C and preheat to the appropriate temperatures.

2. Wash the cells three times, 5 min each, in wash buffer A (see Note 17). Do not remove the probe mix at the start of this procedure, simply add wash buffer A. Perform the washes in a 45°C water bath. The eight-well chamber slides are not very stable in floating when the wells are full with wash solutions. To increase its stability, place a tube rack in the water bath, and adjust the water level so it comes slightly above the rack. Place the chamber slide on this.

3. The washes are continued with three washes in wash buffer B. Although this solution is preheated to 60°C, perform the washes in the water bath set to 45°C (see Note 18). Each wash is for 5 min.

4. To cool down, incubate the cells in wash buffer C (RT) for 5 min (see Note 19).

Table 1
Troubleshooting guide

Problem	Possible solutions
Nick translated probes are too long	Add fresh enzymes and return the probes to 16°C for 30 min; repeat if necessary. If EDTA has already been added, repeat the nick translation, and increase the incubation time (see Note 2).
Probes will not shorten during nick translation	Check DNase I was added. Add fresh enzymes and incubate at 16°C again. The enzyme's activity may be compromised. Check the expiration dates of the enzymes and storage conditions. Order fresh enzymes or try a different aliquot. The DNA should be resuspended in water and not TE (see Note 2). Repeat DNA purification and nick translation.
Loss of acini during the protocol	Add and remove solutions gently (see Note 10). Do not use proteinases as they will also digest the proteinous substrates the cells are grown on (see Note 12). Check the chamber slides through a phase microscope periodically throughout the procedure to determine if a specific step is resulting in the loss of acini and modify this step accordingly.
No/weak signals	The probe may be of the wrong size (see Note 4). Check on a 1% agarose gel. Repeat the nick translation. The permeabilization could be inadequate. Modify permeabilization steps (see Notes 12–13). The time and temperature of denaturation may be inadequate. Increase time and/or temperature (see Note 15). Check the actual temperature of the dry incubator or slide moat. Wash buffer B may be too stringent, reduce the stringency by decreasing the temperature or increasing the concentration of SSC (see Note 18). The hybridization time may be inadequate; try leaving at 37°C over 2–3 nights. Ensure the solutions are DNase free. Try amplifying the signal, e.g., by adding anti-avidin-antibody (and an appropriate secondary antibody if this is not directly conjugated). The out of focus light from the thick structures may make some gene signals hard to see by looking down the microscope alone, image to check that there really is no signal (see Note 21). Check the bulb on the microscope; it may need changing or to be aligned.
High background	The probe may be too short. Check on a 1% agarose gel. If <100 bp, repeat the nick translation (see Note 4). Remember to add species-specific C_0t-1 and carrier DNA. They block nonspecific binding of the probe to repetitive sequences and increase the specificity of the probe binding. The wash conditions may not be stringent enough, check temperatures and see Note 18. Try increasing the concentration of Tween-20 in wash buffer C. The blocking step may be insufficient, increase the time or try other blocking reagents (such as serum).
"Starry night" background	The probe may be too long. Check on a 1% agarose gel. If >600 bp, repeat the nick translation (see Note 4). The DNA pellet may not have completely dried before adding hybridization mix. Always check before adding the hybridization mix (see Note 6). The detection antibodies commonly aggregate, spin briefly in a microcentrifuge and use only the supernatant.
Altered nuclear morphology	The time and temperature of denaturation may be too long/high causing over-denaturation (see Note 15). Reduce time and temperature. If proteinase treatments are used, the temperature or time may be too harsh for the cell type. Reduce time and temperature.

5. Since the probes are not directly labelled, detection steps are required. To block nonspecific binding, incubate the cells in blocking solution for 20 min in a humidified chamber. For the block and detection steps, use 200 µL per well (see Note 20).

6. To detect the gene probes, incubate cells with detection solution in a dark humid container for 2 h at 37°C.

7. Wash off the antibodies or avidin with 3 changes of wash buffer C at 42°C, 5 min each. Protect the cultures from the light as much as possible.

8. After the last wash solution is removed, add ~250 µL of DAPI-containing mounting medium to adequately cover the cells. Due to the thickness of the cell cultures, it is recommended to leave the slides at least 30 min at RT, in the dark, to allow good DAPI staining before viewing on a microscope (see Note 21). The slides can be stored at 4°C for a week or –20°C for longer. If the chamber slide is left intact, seal the lid with parafilm to reduce evaporation. For longer storage, it is worth checking the chamber slides every now and then to ensure they are not drying out. Alternatively, the chamber part can be removed and the glass bottom part mounted on a clean glass slide. Dismantling the chamber slide without breaking the glass or damaging the basement membrane extract can be challenging. Table 1 provides a troubleshooting guide. See Fig. 1.

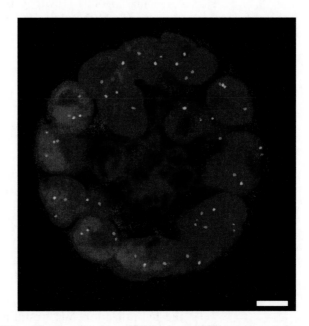

Fig.1. FISH on 3D culture structures. *PTEN* (*red*) and *VEGF* (*green*) gene loci were visualized by FISH in MCF10A.B2 cells grown for 20 days under 3D growth conditions without oncogenic activation. The nuclei were counterstained with DAPI (*blue*). A mid-section of an acinus is shown. Since not all loci from an individual cell are visible in the same plane of focus, a maximum intensity projection of an image stack is shown; the Z-stack was approximately one cell thick. The scale bar equals 10 µm.

4. Notes

1. The amount of DNase I stock to be used must be tested for each new batch.

2. We commonly purchase our BACs from BACPAC Resources center, C.H.O.R.I., CA, USA (http://bacpac.chori.org/) and purify DNA from bacteria cultures using the NucleoBond BAC100 kit (Macherey-Nagel Inc), following the manufacturer's instructions. The purified DNA stock can be stored for extended periods of time at –20°C. It is important to resuspend the purified DNA in water. Other solutions commonly used to resuspend purified DNA, such as TE, contain EDTA. EDTA contains chelating ions and inhibits the enzymes required for nick translation.

3. There are other effective methods to label DNA, e.g., degenerate oligonucleotide primer (DOP)-PCR. DOP-PCR has the advantage of amplifying DNA from a small amount of template (15,16). For probes generated from BACs, the quantity of DNA is not an issue, as one can easily amplify the DNA by growing bacteria cultures. Nucleotides directly conjugated to fluorochromes are available for direct labelling probes (such as FITC-dUTP from companies such as Molecular probes (Invitrogen) or Vysis (Abbot Molecular Inc, Illinois, USA)); however, we have not tried these for 3D cultured cells. Ready to use probes for your gene of interest may also be available commercially; however, it is cheaper to make your own which gives the user greater flexibility in designing experiments.

4. The probe DNA will run as a smear and not as a discrete band. Most of this smear should be within the 100–600 bp range. Probe length is critical for a good quality probe. If the probes are too long they give an increased background (a large number of bright spots, making the background look like a "starry night") and a decreased signal at the loci of interest. An increased background also occurs if the probe is too short. In this case, there is a homogenous, nonspecific staining of the DNA.

5. Nick translation products can be stored at –20°C for a long period of time (years) without a reduction in quality.

6. If the DNA becomes too dry, it is difficult to resuspend it into the hybridization mix, resulting in no/weak signals. If the DNA pellet still has ethanol, it is also difficult to resuspend, and the resulting probe will produce a high background.

7. Probes in hybridization mix can be stored at –20°C for extended periods of time (years).

8. There are several published methods to grow breast epithelial 3D cultures; the most commonly used involves growing acini on or in basement membrane extract (17, 18). In the "3D embedded" method, cells are embedded as single cells in basement membrane extract at the start of the 3D culture (18). In the "3D on-top" method, acini are grown on top of a thin layer of basement membrane extract in media containing a dilute solution of basement membrane extract (17). The "3D on-top" method has the advantage that it requires less basement membrane extract and is thus more cost-effective. We have achieved favorable results in FISH experiments using the "3D on-top" method. The acini grown using the "3D on-top" method still have a layer of basement membrane covering them, which needs to be permeabilized for the probes to penetrate, thus similar permeabilization conditions are likely required for acini grown using the "3D embedded" method. See Notes 12 and 13 for suggestions for optimizing the permeabilization conditions.

9. Cells can also be grown under 3D culture conditions on coverslips. This greatly reduces the amount of probe required. For a 12 mm round coverslip, 150–300 ng labelled DNA per probe, 3 µg C_0t-1 DNA, and 20 µg tRNA are resuspended in 5 µL of hybridization mix. For a 22×22 mm coverslip, 400–600 ng labelled DNA per probe, 10 µg C_0t-1 DNA, and 40 µg tRNA are resuspended in 10 µL of hybridization mix. However, in case of 3D tissue models in which cells are grown on gels, such as basement membrane extract for the breast acini cultures, the gels can be delicate and are easily dislodged from the glass coverslips. In our hands, performing the FISH procedure on cells grown in eight-well glass chamber slides gave better results.

10. Particular care must be taken when changing solutions on the 3D cultures. Acini are easily detached from the basement membrane on which they are grown. Moreover, the basement membrane extract itself is easily detached from the slide, resulting in loss of acini. Removing solutions is best done using a 1,000 µL pipette rather than aspirating using a vacuum suction system. Solutions should be replaced gently.

11. The cells can be stored at this point in 70% ethanol at –20°C for long-term storage (at least 1 year) unless RNA FISH is planned. To help stop evaporation, wrap in Parafilm.

12. These times are double the optimal times for the same cell line grown in monocultures. These times were optimized for acini grown in cultures for 20 days. During the culturing procedure, additional basement membrane accumulates around the acini. Thus, the optimum times for the permeabilization steps may be shorter for cultures grown for reduced periods of time. Prolonged permeabilization incubations can increase

the background. For example, in our hands, increasing the times in 0.5% saponin/0.5% Triton X-100 and/or the 0.1 N HCl step did not increase the signal intensity or quality but did significantly increase the background. On the other hand, inadequate permeabilization results in no or weak signals for the probes. For some 3D cultures, increased permeabilization may be required. Apart from optimizing the incubation times, increasing the concentration of Triton X-100 (to 1%) or HCl (to 0.2 N) can be beneficial. The addition of protein digestion steps by proteinases, such as pepsin or proteinase K, can also aid probe penetration and thus signal quality. However, in the case of the breast epithelial cell 3D cultures, a proteinase step is not appropriate because the basement membrane the cells are grown on will also be digested, resulting in loss of the structures when the solution is changed. Therefore, if proteinase digestion is required, FISH must be performed in suspension.

13. This step aids in deproteination. Consequently, it may need to be omitted to perform immuo-FISH, i.e., FISH performed in combination with immunofluorescence to allow for the simultaneous detection of genomic loci and cellular proteins. It is not easy to predict which antigens will be affected, and it must be empirically determined. In our hands, the 0.1 N HCl step was required to get FISH signals. Thus, if this step is removed other permeabilization steps must replace it, e.g., repeated freeze/thaw cycles in liquid nitrogen. For this, cells must be incubated in 20% (v/v) glycerol in PBS for at least 30 min prior to freeze/thaws and incubated in PBS for at least 1 h after them.

14. Cells can now be stored for at least 1 month at 4°C in this solution.

15. As a general rule, 3D cultures require longer times and higher temperatures than for the same cell line grown in standard monoculture conditions. For example, MCF-10A.B2 cells require 5 min at 75°C for monocultured cells, but 10 min at 85°C when grown under 3D culture conditions. Too short a time/low a temperature for denaturation will result in no/weak signals, whereas too high a temperature/too long a time will overdenature the nuclei, altering the nuclear morphology. The optimal time and temperature of a cell type/culture condition needs to be empirically determined. Generally, optimal times will be between 5–15 min and temperatures between 70–95°C. Routinely, use the shortest time/lowest temperature combination that gives good signals.

16. For coverslips, place the prepared probe mix on a clean slide. Remove the coverslip from the 50% formamide/2× SSC solution; gently tap it on tissue to remove the excess solution.

Place the coverslip cell side down onto the probe. Remove the excess solution from the back of the coverslip by gently patting the tissue over it. Any air bubbles need to be removed before hybridization (forceps or a pipette tip can be used to gently tap the coverslip). Seal the coverslip to the slide with rubber solution. The denaturation time and temperature will be identical to the eight-well glass chamber slides.

17. For coverslips, after the overnight hybridization, remove the rubber solution and place the slides in 2× SSC. This helps remove the coverslip from the slide without damaging the cells, structures, and the basement membrane bed. The washes can be performed in dishes floating in the water baths (12 mm round coverslips fit well in 24 well cell culture dishes and 22 × 22 mm coverslips in 6-well dishes). It is convenient to turn the coverslip cell side up at this point.

18. The stringency of the washes can be altered to help reduce background. Increasing the temperature and reducing the concentration of salts in wash B increases the stringency (e.g., 0.1 × SSC at 60°C). If the wash is too stringent, however, the signal intensity will be reduced.

19. If the probes are directly labelled, the cells are ready for mounting, and steps 5–7 of subheading 3.4 are omitted.

20. 20 μL of blocking and detection solution is adequate for a 12 mm round coverslip. To stop the coverslips drying out, perform the incubation in a humidified chamber.

21. Due to the thickness of the cell structures produced by 3D culture conditions (typically 50–100 μm for the normal acini and the tumor acini can be 10–100 times larger), confocal imaging is required. Imaging can be tricky due to the out of focus light from these thick structures. The solutions for improving image quality such as line averaging and decreasing the pin hole size, which require increasing the exposure times to compensate for decreasing signal, can lead to increased bleaching of the signals, even when using a neutral density filter. This is especially true when imaging throughout the structure with a fine Z-step series, which is optimal of spatial analysis of the genome. To get around this, we simply image sections of ~15–20 μm in thickness rather than the whole acini. This has the added advantage of increasing the number of acini analysed. Another alternative is to try other antifades.

Acknowledgements

The author would like to thank Dr. T. Misteli for critical reading of this chapter and for his continued advice and encouragement;

Dr S. Muthuswamy CSHL, NY, USA) for providing the MCF10. B2 cell line; and ARIAD Pharmaceuticals (http://www.ariad. com/regulationkits) for providing the synthetic ligand AP1510. This research was supported by the Intramural Research Program of the NIH, National Cancer Institute, Center for Cancer Research.

References

1. Fraser, P., and Bickmore, W. (2007) Nuclear organization of the genome and the potential for gene regulation. *Nature* **447**, 413–417.

2. Lanctot, C., Cheutin, T., Cremer, M., Cavalli, G., and Cremer, T. (2007) Dynamic genome architecture in the nuclear space: regulation of gene expression in three dimensions. *Nat Rev Genet* **8**, 104–115.

3. Meaburn, K. J., and Misteli, T. (2007) Cell biology: chromosome territories. *Nature* **445**, 379–781.

4. Misteli, T. (2007) Beyond the sequence: cellular organization of genome function. *Cell* **128**, 787–800.

5. Cremer, M., Kupper, K., Wagler, B., Wizelman, L., Hase Jv, J., Weiland, Y., Kreja, L., Diebold, J., Speicher, M. R., and Cremer, T. (2003) Inheritance of gene density-related higher order chromatin arrangements in normal and tumor cell nuclei. *J Cell Biol* **162**, 809–820.

6. Meaburn, K. J., and Misteli, T. (2008) Locus-specific and activity-independent gene repositioning during early tumorigenesis. *J Cell Biol* **180**, 39–50.

7. Wiech, T., Timme, S., Riede, F., Stein, S., Schuricke, M., Cremer, C., Werner, M., Hausmann, M., and Walch, A. (2005) Human archival tissues provide a valuable source for the analysis of spatial genome organization. *Histochem Cell Biol* **123**, 229–238.

8. Schmeichel, K. L., and Bissell, M. J. (2003) Modeling tissue-specific signaling and organ function in three dimensions. *J Cell Sci* **116**, 2377–2388.

9. Debnath, J., and Brugge, J. S. (2005) Modelling glandular epithelial cancers in three-dimensional cultures. *Nat Rev Cancer* **5**, 675–688.

10. Nelson, C. M., and Bissell, M. J. (2005) Modeling dynamic reciprocity: engineering three-dimensional culture models of breast architecture, function, and neoplastic transformation. *Semin Cancer Biol* **15**, 342–352.

11. Petersen, O. W., Ronnov-Jessen, L., Howlett, A. R., and Bissell, M. J. (1992) Interaction with basement membrane serves to rapidly distinguish growth and differentiation pattern of normal and malignant human breast epithelial cells. *Proc Natl Acad Sci U S A* **89**, 9064–9068.

12. Muthuswamy, S. K., Li, D., Lelievre, S., Bissell, M. J., and Brugge, J. S. (2001) ErbB2, but not ErbB1, reinitiates proliferation and induces luminal repopulation in epithelial acini. *Nat Cell Biol* **3**, 785–792.

13. Cremer, T., and Cremer, C. (2006) Rise, fall and resurrection of chromosome territories: a historical perspective. Part II. Fall and resurrection of chromosome territories during the 1950s to 1980s. Part III. Chromosome territories and the functional nuclear architecture: experiments and models from the 1990s to the present. *Eur J Histochem* **50**, 223–272.

14. Langer, P. R., Waldrop, A. A., and Ward, D. C. (1981) Enzymatic synthesis of biotin-labeled polynucleotides: novel nucleic acid affinity probes. *Proc Natl Acad Sci U S A* **78**, 6633–6637.

15. Fiegler, H., Carr, P., Douglas, E. J., Burford, D. C., Hunt, S., Scott, C. E., Smith, J., Vetrie, D., Gorman, P., Tomlinson, I. P., and Carter, N. P. (2003) DNA microarrays for comparative genomic hybridization based on DOP-PCR amplification of BAC and PAC clones. *Genes Chromosomes Cancer* **36**, 361–374.

16. Telenius, H., Pelmear, A. H., Tunnacliffe, A., Carter, N. P., Behmel, A., Ferguson-Smith, M. A., Nordenskjold, M., Pfragner, R., and Ponder, B. A. (1992) Cytogenetic analysis by chromosome painting using DOP-PCR amplified flow-sorted chromosomes. *Genes Chromosomes Cancer* **4**, 257–263.

17. Debnath, J., Muthuswamy, S. K., and Brugge, J. S. (2003) Morphogenesis and oncogenesis of MCF-10A mammary epithelial acini grown in three-dimensional basement membrane cultures. *Methods* **30**, 256–268.

18. Lee, G. Y., Kenny, P. A., Lee, E. H., and Bissell, M. J. (2007) Three-dimensional culture models of normal and malignant breast epithelial cells. *Nat Methods* **4**, 359–365.

Chapter 26

Simultaneous Ultrasensitive Subpopulation Staining/ Hybridization In Situ (SUSHI) in HIV-1 Disease Monitoring

Bruce K. Patterson

Abstract

The field of virology is undergoing a revolution as diagnostic tests and new therapies are allowing clinicians to treat, monitor, and predict outcomes of viral diseases. The majority of these techniques, however, destroy the factory of viral production and the information inherent in the reservoir – the cell. In this chapter, we describe a technique that combines cell surface immunophenotyping (to unequivocally identify cell types) and ultrasensitive fluorescence in situ hybridization (U-FISH) for HIV-1 to detect productively infected cells. Identification of virus and host (cells) allows earlier detection of changes in viral production and viral suppression but most importantly allows clinicians to monitor response to anti-viral therapy on a cell-by-cell and tissue-by-tissue basis taking into account the fact that the human body consists of very different, distinct compartments with unique selection pressures exerted on the viral life cycle.

Key words: Virus, HIV, Tropism, T-cells, Monocytes, Flow cytometry, Laser confocal image analysis, Oligonucleotides

1. Introduction

Failure to suppress Human Immunodeficiency Virus Type 1 (HIV-1) disease in some individuals has been attributed to host factors, viral factors, pharmacologic factors, and presence of viral reservoirs. Management of HIV-1 disease typically involves the use of plasma viral load, CD4 count, and gene sequencing for drug resistance mutations. Several studies indicated that replicating HIV-1 can persist in cells despite undetectable plasma viral loads in some individuals (1, 2). The persistence of replicating virus leads to continued mutation which in turn can lead to drug resistance and immune escape. The goal of therapy, therefore, should not only be to achieve undetectable plasma viral load but

Joanna M. Bridger and Emanuela V. Volpi (eds.), *Fluorescence in situ Hybridization (FISH): Protocols and Applications*, Methods in Molecular Biology, vol. 659,
DOI 10.1007/978-1-60761-789-1_26, © Springer Science+Business Media, LLC 2010

also suppression of viral replication within cellular reservoirs (Fig. 1). Understanding the cellular dynamics of HIV disease in each patient during various stages of HIV disease, as well as differential activities of Highly Active Antiretroviral Therapy (HAART) within the cellular compartments, is therefore critical for optimal management of HIV-1 disease.

Dronda and colleagues demonstrated in a Spanish cohort that 16.5% of patients had failed to demonstrate increases in CD4+ cell count in the year after HAART initiation and HIV-1 viral load (VL) suppression (3). Analysis of persistent HIV-1 replication within CD4 subsets (memory/naïve, activated/quiescent) would have greatly improved the management of these patients by guiding therapies directed at this persistent reservoir. Similarly, macrophages are resistant to cytopathic effects of HIV, are less susceptible to the CTL response, and persist as reservoirs for HIV, even in patients with serum viral load suppression secondary to HAART (4, 5). In addition, they play a pivotal role in the neuropathogenesis of HIV infection (4). A study by Li et al. has demonstrated that 87% of HIV isolates in patients with late stage disease are M-tropic and utilize CCR5 in both blood and peripheral tissues and the replicative ability of the virus increases

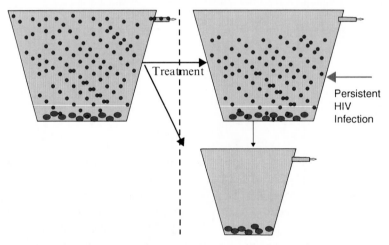

Fig.1. Schematic diagram of HIV viral dynamics in an individual following the treatment. Within an individual (*bucket*), HIV replicates in cells (*large ovals*) producing viral particles (*small circles*). A small amount (approx. 5 mL) of plasma is drawn from individuals (*spigot on buckets*) to perform plasma viral load. Prior to the therapy (*left of dashed line*), viral RNA is detected in the small amount of plasma taken from infected individuals. Following the treatment (*right of dashed line*), the dynamics of HIV changes, and most individuals achieve undetectable plasma viral load as represented by spigots without small circles. Prior to publications in 1999 (1, 2), it was assumed that individuals were represented by the lower right bucket, i.e., undetectable viral load was equivalent to complete viral suppression. We now know (12) that individuals can also exist in the top right bucket with undetectable plasma viral loads and persistence virus replication in cellular reservoirs (e.g., T-cells, macrophages). Using the technique described in this chapter, we can clearly distinguish between the three buckets, whereas plasma viral load only informs as to which side of the dashed line represents an infected individual. Using this information, individuals in the upper right bucket could undergo therapy intensification to achieve maximal suppression (*lower right bucket*). Achieving maximal suppression should be the goal of anti-retroviral therapy.

within the macrophages and MDM (6). Therefore, the progression of HIV disease to AIDS may be a result of replication within macrophages and MDM without the switch in tropism to X4 phenotypes. All of these observations may demonstrate that some patients may have a macrophage/MDM predominant disease, and others may have a T-cell predominant disease.

Certain antiretroviral drugs may provide suppression of HIV replication in specific cellular reservoirs such as macrophages. For example, HAART intensification with didanosine and hydroxyurea and stimulation and release of proviral HIV DNA using OKT3 and IL-2 have demonstrated depletion of residual HIV-1 reservoirs (7). It has also been demonstrated that tenofovir was more abundantly converted to its active metabolite in macrophages than AZT or 3TC (8). In summary, the goal for successful, long-term management of HIV-1 disease should require tailoring therapy based on HIV-1 reservoirs in an individual (8). Using an armamentarium of diagnostics such as ultrasensitive FISH (U-FISH) in addition to plasma viral load and CD4 counts would allow personalized tailoring of therapy based on an individual's HIV infected compartments. In addition, U-FISH could be used in the place of viral load in developing countries with access to flow cytometers for CD4 count but without the resources to develop a comprehensive HIV viral load monitoring program. The same sample that is used for CD4 counts (EDTA tubes) can be used for U-FISH and the assay can be set up at the same time as the protocols are not much different.

2. Materials

2.1. Reagents Included in This Test (ViroTect HIV-1 mRNA Kit, Invirion Diagnostics)

Label	Reagent	Preparatory Notes
Reagent 1	CellPerm	Add 15 mL of sterile water
Reagent 2	Prehybridization Buffer 1	
Reagent 3	Prehybridization Buffer 2	
Reagent 4	Hybridization Buffer	Add 3 mL formamide (Sigma Cat. # F7508 or equivalent)
Reagent 5	HIV Probe Cocktail	A mixture of hundreds of DNA oligonucleotides labeled with dyes similar to fluorescein and directed against gag-pol mRNA.
Reagent 6	Stringency Wash 1	
Reagent 7	Stringency Wash 2	

2.2. Materials and Equipment Required but not Provided in the Kit

1. Formamide, Molecular Biology Grade.
2. Sterile or nuclease-free water.
3. PBS, pH 7.4.
4. Disposable 12×75 mm Falcon tubes (polypropylene or polystyrene).
5. Pipetman or equivalent pipetors ($1–20$ µL, $20–200$ µL, $200–1,000$ µL range).
6. Vortex mixer.
7. Centrifuge.
8. Vacuum aspirator.
9. Waterbath or incubator at $43 \pm 1°C$.
10. Differential cell counter enumerating WBC's or hemocytometer.
11. Flow Cytometer or laser-based microscope.

2.3. Additional Supplies for Cell Culture (Optional)

1. 5% CO_2 incubator at 37°C.
2. Sterile culture hood.
3. ACH-2 cell line.
4. Appropriate cell culture media.

2.4. Appropriate Samples

Culture cell lines:
May include infected and uninfected CEM, H9, U937.
Recommended

- A3.01 A derivative of the CEM human T cell line (9).
- ACH-2 An HIV+ subclone of A3.01, can be induced with phorbol mynstate acetate (PMA) or TNF-α to secrete high levels of HIV-1 virus. See Special Techniques section for stimulation protocol (10).

2.4.1. Peripheral Blood Mononuclear Cells

1. In vitro-infected PBMCs (this is an excellent control for use with antibodies (e.g., CD4). In vitro-infected PBMCs should be harvested at 5 days post-infection for maximal intracellular replication.
2. Ficoll-separated cells from acid citrate dextrose (ACD) tubes (yellow top) or EDTA (purple top) tubes.
3. Cells collected in EDTA, ACD, or citrate CPT tubes (blue marble top).

2.4.2. Lymphoid Tissue Cell Suspensions

Isolate using a wire or 37 µm cloth mesh

1. Lymph node
2. Tonsil
3. Thymus

2.4.3. Control

The most appropriate controls are commercially available negative cell types such as CD Chex Plus (Streck Laboratories, Omaha, NE). The ACH-2 cell line before and after stimulation with TNF-α or PMA is an excellent negative and positive control, respectively.

3. Methods

3.1. Cell Fixation

1. Transfer 100 µL of whole blood (from ACD or EDTA tubes) or 1×10^6 prepared cells (in 100 µL) into a 1.5-mL microcentrifuge tube.

2. Add appropriate dilutions of antibodies (e.g., CD4, CD45RO, CCR5, CXCR4, HLA-DR) and incubate at ambient temperature for 20 min. Note: The only antibody conjugation that does not survive hybridization is PerCP although it can be added to a biotinylated antibody following the hybridization and post-washes.

3. Add 1 mL of 1× PBS, pH 7.4 and centrifuge at $400 \times g$ for 5 min at room temperature. Aspirate supernatant, being careful not to disturb cell pellet.

4. Add 300 µL of Reagent 1 to each sample tube and vortex the cells gently to prevent cell clumping. Let the reaction take place at room temperature for 1–2 h. If whole blood is used, then incubate for an additional 30 min in the water bath at $43 \pm 1°C$ to lyse the red blood cells.

3.2. Hybridization

1. Prepare a water bath at $43 \pm 1°C$ and place Reagent 6 and Reagent 7 in the preheated 43°C water bath.

2. Wash the cells in 1 mL of Reagent 2 (add directly to fixative without prior centrifugation), vortex the cells gently, and centrifuge at $400 \times g$ for 5 min at room temperature, full brake. Aspirate supernatant, being careful not to disturb the cell pellet.

3. Vortex the cells gently in residual fluid.

4. Add 1 ml Reagent 3, vortex the cells gently, and centrifuge at $400 \times g$ for 5 min at room temperature, full brake. Aspirate supernatant, being careful not to disturb the cell pellet. Remove as much residual volume as possible without disturbing the pellet.

5. Vortex the cells gently in residual fluid.

6. Prepare hybridization cocktail by mixing 100 µL of Reagent 4 with 3 µL of Reagent 5 per sample in a 15 mL polypropylene tube. Add 103 µL of the mixture to each sample tube. Vortex the cells gently.

7. Let the reaction to incubate in a preheated $43 \pm 1°C$ water bath for 30 min (up to 2 h is acceptable).

3.3. Cell Stringency Washes

1. Add 1 mL of prewarmed Reagent 6 to each sample tube and vortex the cells gently.

2. Centrifuge at $400 \times g$ for 5 min at room temperature, full brake. Aspirate supernatant, being careful not to disturb the cell pellet. Vortex gently in residual fluid.

3. Add 1 ml prewarmed Reagent 7 to each sample tube, vortex the cells gently.

4. Let the reaction to incubate in a 43°C water bath for 15 min.

5. Centrifuge at $400 \times g$ for 5 min at room temperature, full brake. Aspirate supernatant, being careful not to disturb the cell pellet. Vortex the cells gently in residual fluid.

6. Resuspend the cells in 400 μL of 1× PBS, pH 7.4 with 2% fetal calf serum.

7. Samples are ready for flow cytometric analysis.

3.4. Cell Analysis by Flow Cytometry

1. Prepare an analysis protocol similar to the ones that appear in Figs. 2 and 3 (see Note 1). Do not use forward scatter versus side scatter gating as HIV-1-infected lymphocytes have a bizarre morphology that alters scatter characteristics. Gating on cell surface staining is highly recommended (e.g., CD4 vs. side scatter or CD3 vs. side scatter) (see Note 2).

2. Collect a minimum of 5,000 gated events.

3. Display the samples as two parameter histograms and set quadrant cursors using either a negative control sample or an internal negative control cell population within the actual run sample.

3.5. Special Procedures (Controls)

3.5.1. Stimulation of ACH-2 Cells in Culture

In addition to PBMCs from HIV-1 seronegative individuals and commercially available control lymphocytes, stimulated and unstimulated ACH-2 cells are excellent positive and negative controls, respectively. Several published methods are available to stimulate ACH-2 cells to actively produce HIV-1 virus. The protocol included in this procedure is only one example. It is important for each laboratory to assess which assays work best for their individual needs and to determine the success of the stimulation reaction to produce HIV-1 virus by an established independent method.

1. Prepare exponentially growing cell cultures of ACH-2 cells (see Note 3).

   (a) Sterile cell culture growth media

   (b) Sterile 24-well cell culture plate(s)

   (c) Sterile cell culture-grade phorbol myristate acetate at 0.1 mg/mL in DMSO

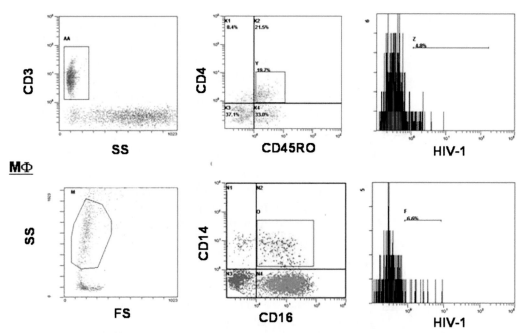

T-Cell

MΦ

Fig. 2. Representative SUSHI panels demonstrating HIV-1 replication in T-cells (*top row*) and macrophages (*bottom row*). HIV-1 replicates predominantly in CD3+, CD4+, CD45RO+ T-cells (12) and in CD14+, CD16+ circulating macrophages.

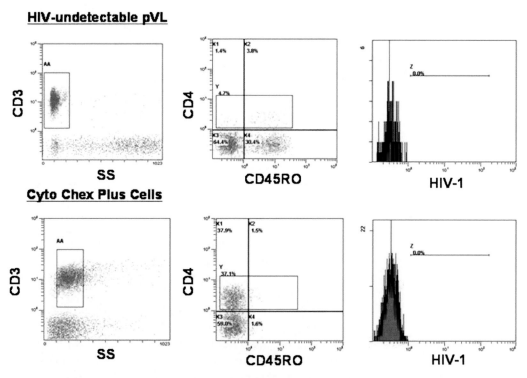

HIV-undetectable pVL

Cyto Chex Plus Cells

Fig. 3. Representative SUSHI panels demonstrating the lack of HIV-1 replication in T-cells from an individual with undetectable plasma viral load (*top row*) and in negative control Cyto Chex Plus cells (Streck Laboratories, Omaha, NE).

2. Perform a cell count on the exponentially growing cell cultures.

3. Remove 1×10^6 cells of A3.01 and 2×10^6 cells of ACH-2 for each experiment to be performed.

4. Centrifuge at $400 \times g$ for 5 min at room temperature, full brake. Aspirate supernatant, being careful not to disturb the cell pellet.

5. Vortex the cells gently in residual fluid.

6. Wash the cells in 1 mL of fresh culture medium typically used to grow the cells. Add 1 mL of fresh media, vortex the cells gently, and centrifuge at $400 \times g$ for 5 min at room temperature, full brake. Aspirate supernatant, being careful not to disturb the cell pellet.

7. Vortex the cells gently in residual fluid.

8. Resuspend the cells to a final concentration of 1×10^6 cells/mL.

9. Aliquot 1 mL of each cell sample into a separate, labeled well of a sterile 24-well culture plate.

10. Add 500 µL of 0.1 mg/mL of phorbol myristate acetate (PMA) in DMSO to culture. Pipette sample up and down several times to ensure adequate mixing.

11. Place the 24-well culture plate to a 37°C incubator. Let the cells get stimulated for 6–18 h. (At this dosage of phorbol myristate acetate, incubation for significantly less than 6 h will result in the cells not becoming fully activated).

12. After 6–18 h, remove the culture plate from the incubator. Pipette the cells gently up and down in the well to thoroughly mix, remove the cells from the plate, and deposit them into labeled 12×75 mL Falcon tubes.

13. Proceed with step 2 of Subheading 3.1.

4. Notes

1. Instrument Setup: Instrument standardization and quality control (Q/C) should be performed prior to any experiment using the methods previously described (11). First, this three-step process involves establishing the instrumentdetector settings using QC windows and Calibright APC beads (for appropriate instruments). The QC Windows kit, for example, contains reference standards labeled with FITC/PE/PE-Cy5. The kit contains a certificate of analysis with initial target channels for establishing PMT voltages and

amplifier settings for the fluorescence detectors. Following warm up, the instrument should be set at a flow rate to match that used with the samples of choice. The standard beads should be appropriately diluted and run with the FL-1, FL-2, FL-3, and FL-4 (if available) detector settings adjusted to match the manufacturer's specified target. Second, compensation for spectral overlap should be performed using antibody-labeled leukocytes stained individually with fluorochrome conjugates to match the fluorochromes to be used in the experiment. Compensation should be set according to the instrument using singly labeled FITC-, PE-, ECD-, PE/Cy5-, and APC-conjugated antibodies on leukocytes. Third, the daily target channels for the QC3 beads, Full Spectrum beads, and/or Calibrite APC beads should be established by running the beads with the compensation set for the individual fluorochromes to be used. Last, a negative control (CytoChex Plus, Streck Laboratories, Omaha, NE USA) should be hybridized as per the following protocol and set on scale (the entire negative population should be in the first or second decade) in a standard FL-1 log channel. The positive/negative cutoff should be based on the negative control or HIV-1 seronegative sample and/or the natural cutoff seen in internal HIV-1 negative cells and HIV-1 positive cells. Cells should not be gated based on scatter due to the blast-like morphology of the HIV-1 infected cell that places HIV-1 infected T-lymphocytes outside the normal lymphocyte scatter gate. Alternatively, a scatter gate excluding only debris should be used with additional gates based on cell surface markers (e.g., CD4, CD45RO, HLA-DR, CD62L). Representative histograms are shown in Figs. 2 and 3.

2. Due to the reagents and heating of the cell samples, cellular scatter characteristics may appear different from typical light scatter displays. In most cases, you will have to minimally change your forward- and side-scatter settings to put the cell populations on scale.

3. Low passage (<15) ACH-2 cells must be used since stimulation decreases with later passages, and the cells become aneuploid.

Acknowledgements

The author would like to acknowledge and thank Keith Shults for his collaborative efforts and unyielding support in the development of these techniques.

References

1. Patterson, B.K., Czerniewski, M.A., Pottage, J., Agnoli, M., Kessler, H., Landay, A. (1999) Monitoring HIV therapy in immune cell subsets using ultrasensitive fluorescence in situ hybridization. *Lancet* **353**, 211–212.

2. Furtado, M.R., Callaway, D.S., Phair, J.P., Kunstman, K.J., Stanton, J.L., Macken, C.A., Perelson, A.S., Wolinsky, S.M. (1999) Persistence of HIV-1 transcription in peripheral-blood mononuclear cells in patients receiving potent antiretroviral therapy. *N Engl J Med* **340**, 1614–1622

3. Dronda, F., Moreno, S., Moreno A., Casado, JL., Perez-Elias, M.J., and Antela, A. (2002) Long-term outcomes among antiretroviral-naïve human immunodeficiency virus-infected patients with small increases in CD4+ counts after successful virologic suppression. *Clin Infect Dis* **35**, 1005–1009.

4. Kedzierska, K. and Crowe, S. (2002) The role of monocytes and macrophages in the pathogenesis of HIV-1 infection. *Curr Med Chem* **9**, 1893–1903.

5. Schutten, M., van Baalen, C., Guillon, C., Huisman, R., Boers P., Sintnicolaas, K., Gruters, R., and Osterhaus, A. (2001) Macrophage tropism of human immunodeficiency virus type 1 facilitates in vivo escape from cytotoxic T-lymphocyte pressure. *J Virol* **75**, 2706–2709.

6. Li, S., Juarez, J., Alali, M., Dwyer, D., Collman, R., Cunningham, A., and Naif, H. (1999) Persistent CCR5 utilization and enhanced macrophage tropism by primary blood human immunodeficiency virus type 1 isolates from advanced stages of disease and comparison to tissue-derived isolates. *J Virol* **73**, 9741–9755.

7. Kulkosky, J., Nunnari, G., Otero, M., Calarota, S., Dornadula, G., Zhang, H., Malin, A., Sullivan, J., Xu, Y., DeSimone, J., Babinchak, T., Stern, J., Cavert, W., Haase, A., Pomerantz, R. (2002) Intensification and stimulation therapy for human immunodeficiency virus type 1 reservoirs in infected persons receiving virally suppressive highly active antiretroviral therapy. *J Infect Dis* **186**, 1403–1411

8. Balzarini, J., Van Herrewege, Y, and Vanham G. (2002) Metabolic activation of nucleoside and nucleotide reverse transcriptase inhibitors in dendritic and Langerhans cells. *AIDS* **16**, 2159–2163.

9. Folks T, Benn S, Rabson A, Theodore T, Hoggan MD, Martin M, Lightfoote M, Sell K. (1985). Characterization of a continuous T-cell line susceptible to the cytopathic effects of the acquired immunodeficiency syndrome (AIDS)-associated retrovirus. *Proc Natl Acad Sci U S A* **82**, 4539–4543.

10. Folks TM, Clouse KA, Justement J, Rabson A, Duh E, Kehrl JH, Fauci AS. (1989) Tumor necrosis factor alpha induces expression of human immunodeficiency virus in a chronically infected T-cell clone. *Proc Natl Acad Sci U S A* **86**, 2365–2368.

11. Purvis, N. and Stelzer, G. (1998) Multi-platform, multi-site instrumentation and reagent standardization. *Cytometry* **33**, 156–165.

12. Patterson, B.K., McCallister, S., Schutz, M., Siegel, J.N., Shults, K., Flener, Z., Landay, A. (2001) Persistence of intracellular HIV-1 mRNA correlates with HIV-1-specific immune responses in HIV-1-infected subjects on stable HAART therapy. *AIDS* **15**, 1635–1641.

Part IV

Protocols for Model Organisms

Part II

Chapter 27

Detection of Prokaryotic Cells with Fluorescence In Situ Hybridization

Katrin Zwirglmaier

Abstract

Fluorescence in situ hybridization with rRNA targeted oligonucleotide probes is nowadays one of the core techniques in microbial ecology, allowing the identification and quantification of microbial cells in environmental samples in situ. Next to the classic FISH protocol, which uses fluorescently monolabelled probes, the more sensitive CARD-FISH (also known as TSA-FISH), which involves an enzyme catalyzed signal amplification step, is becoming increasingly popular. This chapter describes protocols for both methods. While classic FISH has the advantage of being relatively cheap and easy to do on morphologically diverse samples, CARD-FISH offers a significantly higher sensitivity, allowing the detection of slow growing or metabolically inactive cells, which are below the detection limit of classic FISH. The drawback here is the considerably higher price for the probes and advanced cell fixation and permeabilization requirements that have to be optimized for different target cells.

Key words: Fluorescence in situ hybridization, Bacteria, Oligonucleotide probes, 16S rRNA, Cell fixation, Tyramide signal amplification, Catalyzed reporter deposition

1. Introduction

The development of FISH for bacterial cells some 20 years ago (1) has been hailed as a breakthrough for microbial ecologists, since it allowed both identification and quantification of microbial cells in situ within an ecosystem. The basic concept of microbial FISH is to use short oligonucleotide probes that are end-labelled with a fluorescent dye and target the ribosomal RNA of a microorganism. The rRNA presents an ideal target, since (1) it is a phylogenetic marker and (2) it is present in high copy numbers, within a metabolically active cell. The 16S rRNA, for which large and continually growing sequence databases are available, is the most common target for FISH probes, although occasionally the

Joanna M. Bridger and Emanuela V. Volpi (eds.), *Fluorescence in situ Hybridization (FISH):*
Protocols and Applications, Methods in Molecular Biology, vol. 659,
DOI 10.1007/978-1-60761-789-1_27, © Springer Science+Business Media, LLC 2010

23S rRNA is used, which has a higher information content than the 16S, but fewer sequences are available.

The classic FISH method is nowadays a standard technique in molecular biology and with more sequence data available and the price of fluorescently labelled probes falling, it is becoming ever more popular. There have been a number of improvements and technological advances in the last few years, many of which have their counterparts in eukaryotic cells or were in fact first developed for eukaryotes and later adapted for prokaryotes. Examples are (1) improved signal intensity with tyramide signal amplification (TSA-) FISH, also known as catalyzed reporter deposition (CARD-) FISH (2, 3), (2) detection of functional genes encoded on chromosomes or plasmids with RING-FISH (recognition of individual genes) (4), and (3) analysis of metabolic activity of a cell by uptake of isotope-labelled substrate with MAR-FISH (micro-autoradiography) (5, 6).

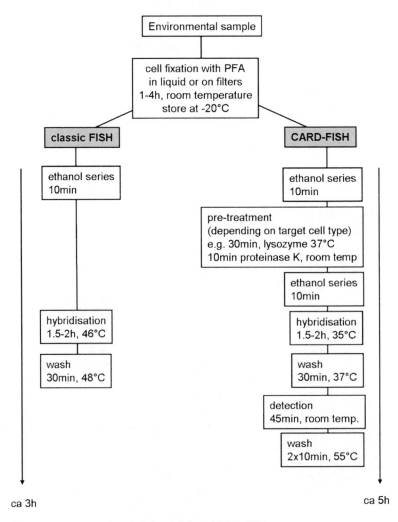

Fig. 1. Flow chart showing steps for classic and CARD-FISH.

Here, the protocols for both classic as well as CARD-FISH are presented. Classic FISH is recommended wherever possible, since it is relatively easy and cheap. The major drawback of classic FISH, however, is its poor detection of slow growing or metabolically inactive cells, which have a low ribosome content (and which can form a substantial part of the microbial community in some environments). If signal intensity with classic FISH proves to be too low, CARD-FISH, which provides a strong signal amplification, can be used as an alternative. The difference between classic and CARD-FISH is the probes – monolabelled with a fluorescent dye for classic FISH and conjugated with horse radish peroxidase (HRP) for CARD-FISH. The HRP catalyzes the deposition of fluorochrome labelled tyramide inside the target cells, thus leading to a significant signal amplification. The drawbacks of CARD-FISH are requirements for additional cell wall treatments (which have to be optimized for different types of target cells) to make the cells permeable for the enzyme as well as increased costs for HRP-labelled probes compared to fluorochrome-labelled probes.

The main steps, including the time requirements involved in both classic and CARD-FISH are summarized in Fig. 1.

2. Materials

2.1. Equipment and Consumables

1. Epifluorescence microscope with filter sets appropriate for the fluorescent label of the FISH probes and the counterstain.

2. Hybridization oven.

3. Multiwell microscope slides (Marienfeld) or

4. Membrane filters (0.2 μm pore size GTTP Millipore, 25 or 47 mm diameter).

5. Nail varnish to seal the slides.

6. Parafilm.

7. Plastic 50 mL tubes.

8. Plastic Petri dishes.

2.2. Cell Fixation

1. Phosphate buffered saline (PBS): 130 mM NaCl, 1.5 mM K_2HPO_4, 8.0 mM Na_2HPO_4, 2.7 mM KCl, pH 7.0.

2. Paraformaldehyde (PFA), 4% (w/v) in PBS, pH 7.0 (see Note 1), store as aliquots at −20°C.

2.3. Classic FISH

1. Hybridization buffer (900 mM NaCl, 20 mM Tris–HCl, pH 8.0, 0.01% SDS [w/v], x% formamide [concentration depending on probe]), prepare fresh on the day, 2 mL per slide.

2. 500 mM Na_2-EDTA.

3. Wash buffer (20 mM Tris–HCl, pH 8.0, 0.01% SDS [w/v], xmM NaCl [concentration dependent on the formamide concentration used for the hybridization, see Table 1]). Prepare fresh on the day, 50 mL per slide.

4. Ethanol solutions 50% (v/v), 80% (v/v), 100% (v/v).

5. Formamide, store at 4°C, hazardous.

6. Anti-fading slide mounting solution, e.g., Vectashield with 4,6-diamino-2-phenylindole-dihydrocloride (DAPI) (Vector Laboratories), store at 4°C in the dark.

7. Oligonucleotide probes, end-labelled with fluorochrome, dilute to 10 pmol/μL, and store aliquots at –20°C.

Table 1
The table lists the molar concentration of Na$^+$ in the washing buffer corresponding to different formamide concentrations in the hybridisation buffer

% Formamide in hybridisation buffer	Washing buffer for washing at 37°C – CARD-FISH (HRP probes)			Washing buffer for washing at 48°C – classic FISH (fluorescently labelled probes)		
	Na$^+$ conc. [M]	μL 5 M NaCl/50 mL	μL 0.5 M Na$_2$-EDTA/50 mL	Na$^+$ conc. [M]	μL 5 M NaCl/50 mL	μL 0.5 M Na$_2$-EDTA/50 mL
0	–	–	–	0.900	9,000	0
5	–	–	–	0.636	6,360	0
10	–	–	–	0.450	4,500	0
15	–	–	–	0.318	3,180	0
20	0.145	1,350	500	0.225	2,150	500
25	0.105	950	500	0.159	1,490	500
30	0.074	640	500	0.112	1,020	500
35	0.052	420	500	0.080	700	500
40	0.037	270	500	0.056	460	500
45	0.026	160	500	0.040	300	500
50	0.019	90	500	0.028	180	500
55	0.013	30	500	0.020	100	500
60	0.009	0	450	0.014	40	500
65	0.007	0	350	0.010	0	500
70	0.005	0	250	0.007	0	350

In the washing buffer stringency is adjusted through the salt concentration instead of formamide. The required salt concentration depends on the formamide concentration used in the hybridisation buffer

2.4. CARD-FISH

1. Blocking reagent (Roche), stock solution 10% (w/v) in maleic acid buffer (100 mM maleic acid, 150 mM NaCl, pH 7.5) autoclave and store in 2 mL aliquots at –20°C.

2. Proteinase K, 1 mg/mL in H_2O.

3. Lysozyme, 5 mg/mL in lysozyme buffer (100 mM Tris pH 8.0, 50 mM EDTA).

4. Hybridisation buffer (900 mM NaCl, 20 mM Tris–HCl pH 8.0, 0.01% SDS [w/v], 2% blocking reagent [v/v], x% formamide [concentration depending on probe]), prepare fresh on the day, 2 mL per slide.

5. 500 mM Na_2-EDTA.

6. Wash buffer (20 mM Tris–HCl, pH 8.0, 5 mM Na_2EDTA, 0.01% (w/v) SDS, x mM NaCl [concentration dependent on the formamide concentration used for the hybridisation, see Table 1]), prepare fresh on the day, 50 mL per slide.

7. Ethanol solutions 50% (v/v), 80% (v/v), 100% (v/v).

8. Formamide, store at 4°C, hazardous.

9. TSA-kit, containing fluorophore labelled tyramide and amplification diluent (PerkinElmer/NEN Life Sciences), reconstitute tyramide in H_2O or DMSO as stated in the product description and store in aliquots of 5 μL at –20°C, store amplification diluent at 4°C.

10. TSA working solution: amplification diluent: dextran sulphate: fluorophore labelled tyramide in a ratio of 25:25:1. Prepare fresh on the day (see Note 2).

11. TNT buffer (0.1 M Tris–HCl pH 7.5, 0.15 M NaCl, 0.05% [v/v] Tween-20).

12. Dextran sulphate, 40% (w/v) (see Note 3), store at 4°C.

13. Anti-fading slide mounting solution, e.g., Vectashield with DAPI, store at 4°C in the dark.

14. Oligonucleotide probes, end-labelled with HRP, store 10 μL aliquots (5 pmol/μL) at –20°C, once defrosted, do not refreeze, but keep at 4°C (up to 4 weeks).

3. Methods

3.1. Samples

Depending on the type of sample material and the concentration of bacterial cells in the sample, the hybridization will be carried out either on filters or on multiwell microscope slides. Typically, marine or freshwater samples will be collected on filters, whereas wastewater, soil, or clinical samples (due to their high cell concentration) can be hybridized directly on slides. The amount of sample needed again depends on the concentration of cells in the sample and has to be determined empirically. Ideally, when analyzing the hybridized cells under the microscope, cells should be evenly spread (avoiding cell clumps) and less than 1,000 cells per visual field. As a guideline, this will require ca 50 mL freshwater or coastal seawater (up to 150 mL for oligotrophic open ocean water) on a 25 mm diameter filter or 2–5 µL of wastewater sludge, soil, or clinical samples on a slide with 6 mm diameter wells.

3.2. Cell Fixation and Storage

3.2.1. On Filters

1. Filter the sample (volume required is dependent on cell concentration) onto a 0.2 µm pore size polycarbonate filters (Millipore GTTP) (see Note 4).
2. Air dry filters.
3. Place into container (Petri dish or similar) with 1:3 PFA (4%):PBS, fix for 1–4 h at room temperature.
4. Air dry filters.
5. Dehydrate cells by taking the filters through a series of increasing ethanol concentration of 50, 80, and 100%, 3 min in each solution.
6. Allow to air dry.
7. Filters can be stored at –20°C for several months, avoid repeated freezing–thawing.

3.2.2. In Solution

1. Spin down 2 mL of cells for 5 min, $10,000 \times g$ and remove supernatant.
2. Resuspend cells in 1.2 mL of PBS.
3. Add 400 µL of PFA (4%).
4. Incubate at room temperature for 1–4 h.
5. Spin down cells, 5 min, $10,000 \times g$ and remove supernatant.
6. Resuspend in 1 mL of PBS.
7. Spin down, 5 min, $10,000 \times g$, remove supernatant.
8. Resuspend in 200 µL of PBS.
9. Add 200 µL of ethanol.
10. Store at –20°C for up to several months.

3.3. Probes and Probe Design

Probes for both classic and CARD-FISH are usually about 18–24 nt long and monolabelled at one end with a fluorescent dye (for classic FISH) or HRP (for CARD-FISH). The most commonly used fluorescent dyes are fluorescein (excitation at 492 nm, emission at 520 nm [green]), Cyanine 3 (Cy3, excitation at 550 nm, emission at 570 nm [orange]), and Cyanine 5 (Cy5, excitation at 650 nm, emission at 670 nm [dark red]). Out of these, Cy3 has the brightest fluorescence. When choosing a fluorescent dye, the target cells also have to be considered, since some cells exhibit a strong autofluorescence, which can overlap with the emission wavelengths of some dyes and make hybridization signals difficult to observe.

Probe design and testing is a complex subject and a detailed discussion would go beyond the scope of this chapter. However, I have included an overview of the critical steps and available online resources that should enable readers to design and evaluate their probes.

For many major taxonomic groups, probes have already been developed and sequences can easily be found in the literature. Additionally, the online database probeBase (http://www.microbial-ecology.net/probebase/) hosted by the University of Vienna, Austria, currently contains about 1,500 rRNA targeted oligonucleotide probes, including hybridization conditions and references to the original publication (7).

To design new probes, various resources are available on the web. The three main rRNA sequence databases, SILVA (http://www.arb-silva.de), (8), greengenes (http://greengenes.lbl.gov/), (9) and RDP II (http://rdp.cme.msu.edu/), (10) each contain several hundred thousand published rRNA sequences and are regularly updated and curated by experts. Probes can be designed with the probe design tool within the freely available phylogenetic software package Arb (http://www.arb-home.de/, (11)), and then checked against other sequences in the database with the Arb probe match tool. The web interface probeCheck (http://www.microbial-ecology.net/probecheck, (12)), which is based on the Arb probe match tool, allows the user to check a probe sequence against the entire SILVA, greengenes, and RDP II databases.

Ideally, a probe should match the target sequence perfectly and have more than one mismatch to other non-target sequences, although discrimination with just one mismatch is possible. Mismatches in the middle of the probe sequence are more discriminating than near the end. The GC content of the probe should be in the range of 40–60%. Ribosomal RNA in situ forms a strong secondary structure and is complexed with various ribosomal proteins, which makes some regions of the rRNA sequence less accessible than others. This results in weak hybridization signals for some probes, even if the probe matches the target perfectly. A map of the 16S rRNA, showing five classes of target accessibility based on the brightness of the hybridization signal, has been developed for *E. coli* (13) and subsequently for other prokaryotes, archaea, and eukaryotes (14). Although the accessibility in individual target species may differ, the map can serve as a valuable guideline to choose suitable probe target regions.

Once a probe has been designed and checked against other sequences in a database, it still has to be tested experimentally in order to determine the optimal hybridization conditions, i.e. conditions that result in the brightest possible hybridization signals in target and no signals in non-target cells. Parameters that influence probe binding and stability (and therefore hybridization signal) are temperature, salt content, and formamide concentration in the hybridization buffer, as well as probe length and GC content. Due to differences in e.g. target accessibility or possible hairpin formation within the probe, optimal hybridization conditions for two probes with the same length and GC content are not necessarily the same and therefore have to be determined empirically.

It is recommended to keep the temperature constant for all probes, to allow simultaneous hybridization with several probes and instead adjust the stringency of the hybridization through the formamide content in the buffer. Optimal formamide concentration for a probe should be evaluated in steps of 5% with target and non-target cells. As a guideline, an 18 nt probe with a GC content of 50% will require approximately 30–40% formamide if hybridized with the buffer specified in this chapter at 46°C.

3.4. Classic FISH

3.4.1. Hybridization (on Slides or Filters)

3.4.1.1. On Slides

1. Spread 2–5 μL of PFA fixed cells evenly in a well on the lide.

2. Air dry (at room temperature or in an oven up to 50°C). It is important that the sample has dried completely, otherwise cells will wash off.

3. Take slide through ethanol series of 50, 80, and 100% ethanol, 3 min in each solution (see Note 5), then air dry.

4. Mix 9 μL hybridisation buffer + 1 μL probe (10 pmol/μL) and place on the sample. Make sure it is spread evenly across the well.

5. Use the remainder of the 2 mL hybridisation buffer to moisten a piece of tissue. Place tissue together with slide in a 50 mL tube, close the lid, and place the tube in a hybridization oven (horizontally in rack or similar) (see Note 6).

6. Hybridize for 1.5–2 h at 46°C.

7. While hybridizing: pre-warm the wash buffer to 48°C.

3.4.1.2. On Filters

1. Cut filters and label filter pieces with pencil (see Note 7).

2. Take filters through an ethanol series, 50, 80, and 100% ethanol, 3 min in each solution, then air dry.

3. Mix 27 μL of hybridization buffer and 3 μL (10 pmol/μL) probe and place on a piece of parafilm. Place the filter sample side down in the liquid, making sure that the whole area of the filter is in contact with the buffer/probe.

4. Use the remaining hybridization buffer to moisten a piece of tissue. Place tissue together with the filter on parafilm in a 50 mL tube or similar container, close the lid, and put the tube in a hybridization oven (horizontally in a rack or similar) (see Note 6).

5. Hybridize for 1.5–2 h at 46°C.

6. Whilst hybridizing: pre-warm wash buffer to 48°C.

3.4.2. Washing

1. Place slides/filters in pre-warmed wash buffer (see Note 8). If using 50 mL tubes, two slides fit in back to back.

2. Incubate for 30 min at 48°C.

3. Dip slides/filters briefly in H_2O.

4. Optional: take slides/filters through an ethanol series, 50, 80, and 100%, 2 min each (see Note 9).

5. Allow to air dry.

6. Mount slides in anti-fading slide mounting solution, e.g., Vectashield or Vectashield-DAPI (handle with gloves, since DAPI is a mutagen). For filters, put a drop of mounting solution on the cover glass, place the filter sample side down on the drop, and then place the cover glass on the slide. This minimises bubbles on the filter.

7. Seal the slides with nail varnish around the edges of the cover slip. This prevents drying out of the slide and leaking of slide mounting solution.

8. Slides can be stored like this in the dark for 1–2 weeks, although it is recommended to analyse them as soon as possible, since the signal fades with time.

3.5. CARD-FISH

3.5.1. Pre-treatment of Cells

Bacterial cell walls are generally quite permeable for small oligoucleotide probes and fluorescent dyes, which is why classic FISH does not require any special pre-hybridization treatment of the fixed cells. The enzyme-labelled probes used for CARD-FISH, however, are much larger and therefore the cell wall has to be permeabilized to allow the probes access to the cell. There is a wide range of permeabilization treatments and the optimal protocol depends on the cell wall properties of the target cells. The most common treatments are lysozyme and proteinase K. Other options are mutalysin, staphylolysin, antibiotics, and microwave fixation. There is a fine balance between hybridization signal intensity and cell integrity and harsh and excessive treatments can result in significant cell loss. This can be a problem in complex environmental samples if the aim is to determine the percentage of target cells relative to the total number of all DAPI stained cells. Here, the risk of getting false negatives due to insufficient cell wall permeability has to be weighed against possible loss of non-target cells. The optimal cell pre-treatment should be determined empirically for each type of target cell. Generally, Gram positive bacteria are more likely to require pre-treatment than Gram negative ones, although there are exceptions, e.g. *Synechococcus*, which, although Gram negative, does have a rather thick cell wall and requires pre-treatment with both lysozyme and proteinase K.

Below is the pre-treatment recommended for CARD-FISH with marine *Synechococcus* strains.

3.5.1.1. On Slides

1. Spread 2–5 μL of PFA fixed cells evenly in a well on the microwell slide.

2. Air dry (at room temperature or in an oven up to 50°C). It is important, that the sample has dried completely, otherwise cells will wash off.

3. Take slide through an ethanol series, 50, 80, and 100% ethanol, 3 min in each solution (see Note 5), then air dry.

4. Pipette 20 μL of lysozyme (5 mg/mL) in each well of the slide containing sample, incubate for 30 min at 37°C in a box or 50 mL tube containing moist tissue to avoid evaporation.

5. Dip slide in water to wash off lysozyme.

6. Take slide through an ethanol series, 50, 80, and 100%, 2 min in each solution, air dry at the end.

7. Pipette 20 μL of proteinase K (1 mg/mL) in each well containing sample, incubate at room temperature for a maximum of 10 min (see Note 10).

8. Dip the slide into the water to wash off the proteinase K.

9. Take the slide through an ethanol series, 50, 80, and 100%, 2 min each, air dry at the end.

10. Slides are now ready for hybridization (Subheading 3.5.2).

3.5.1.2. On Filters

1. Cut filters and label filter pieces with a pencil (see Note 7).

2. Take filters through an ethanol series, 50, 80, and 100% ethanol, 3 min each in each solution, then air dry.

3. Put 30 μL of lysozyme (5 mg/mL) on a piece of parafilm, place the filter sample side down into the lysozyme, incubate for 30 min at 37°C in a container with moist tissue to avoid evaporation.

4. Dip filters in water to wash off lysozyme.

5. Take filters through an ethanol series 50, 80, and 100%, 2 min each, air dry at the end.

6. Put 30 μL of proteinase K (1 mg/mL) on parafilm, place filter upside down into the proteinase K, incubate at room temperature for a maximum time of 10 min (see Note 10).

7. Dip filters in distilled water.

8. Take filters through an ethanol series 50, 80, and 100%, 2 min each, air dry.

9. Filters are now ready for hybridization (Subheading 3.5.2).

3.5.2. Hybridization

3.5.2.1. On Slides

1. Mix 9 μL hybridisation buffer + 1 μL HRP labelled probe (5 pmol/μL) and place on sample. Make sure it is spread evenly across the well.

2. Use the remaining hybridization buffer to moisten a piece of tissue. Place tissue together with the slide in a 50 mL tube, close the lid, and place the tube in a hybridization oven (horizontally in rack or similar) (see Note 6).

3. Hybridize for 1.5–2 h at 35°C.

4. While hybridizing: pre-warm wash buffer to 37°C.

3.5.2.2. On Filters

1. Mix 27 μL of hybridization buffer and 3 μL (5 pmol/μL) HRP labelled probe and put on a piece of parafilm. Place the filter upside down on the liquid, making sure that the whole area of the filter is in contact with the buffer/probe.

2. Use the remaining hybridization buffer to moisten a piece of tissue. Place tissue together with the filter on parafilm in a 50 mL tube or similar container, close the lid, and put the tube in a hybridization oven (horizontally in a rack or similar) (see Note 6).

3. Hybridize for 1.5–2 h at 35°C.

4. While hybridizing: pre-warm wash buffer to 37°C.

3.5.3. Washing

1. Place slides/filters in pre-warmed wash buffer (see Note 8). If using 50 mL tubes, two slides fit in back to back.

2. Incubate for 30 min at 37°C.

3. Dip slides/filters briefly into H_2O.

3.5.4. Detection – Catalysed Reporter Deposition

1. Place slides/filters in 50 mL of TNT buffer and equilibrate for 10–15 min at room temperature.

2. Use a piece of tissue to carefully dry the slide around the samples or for filters place filter briefly on tissue (face up). Do not let the slides/filters dry out completely.

3. For slides, pipette 10 μL of TSA working solution into each well.

4. For filters, pipette 30 μL of TSA working solution onto a piece of parafilm and place the filter sample side down in the liquid.

5. Incubate for 30–45 min at room temperature in the dark.

6. While incubating, pre-warm TNT buffer (2 × 50 mL per slide) to 55°C.

7. Wash slides/filters for 10 min in TNT at 55°C. Repeat in fresh buffer. This washes off excess substrate and deactivates the enzyme.

8. Dip slides briefly into H_2O.

9. Optional: take slides/filters through ethanol series, 50, 80, and 100%, 2 min each (see Note 9).

10. Air dry.

11. Mount slides in anti-fading slide mounting solution, e.g., Vectashield or Vectashield-DAPI (handle with gloves, as DAPI is a mutagen). For filters, put a drop of mounting solution on the cover glass, carefully place the filter sample side down on the drop, and then put the cover glass on the slide. This minimises bubbles on the filter.

12. Seal the slides with nail varnish around the edges of the cover slip. This prevents drying out of the slide and leaking of slide mounting solution.

13. Slides can be stored like this in the dark for 1–2 weeks, although it is recommended to analyze them as soon as possible, since the signal fades with time.

3.6. Analysis

Count at least several hundred DAPI stained cells under the microscope. Ideally, there should be less than 1,000 cells per visual field, evenly spread, rather than cell clumps. DAPI can fade quite fast under UV light, so it is best to count the DAPI stained cells first and probe-stained cells later.

4. Notes

1. PFA is toxic. Weigh out and dissolve in fume hood. It can be difficult to dissolve. Use heated PBS, add small amount of NaOH (it dissolves easily under alkaline pH), and titrate with HCl to pH 7.0. Store aliquots at $-20°C$. Once defrosted, PFA solution is stable for ca 1 week in the fridge. Avoid repeated freezing/thawing.

2. TSA working solution: One 5 μL aliquot of tyramide will give you 255 μL of working solution (125 μL amplification diluent + 125 μL dextran sulphate + 5 μL tyramide), which is enough for about 22 samples on slides or eight filters. Discard the rest of TSA working solution, avoid repeated freezing/thawing of tyramide. The dextran sulphate is very viscous and difficult to mix, therefore cut off the tip of a 1,000 μL pipette tip to mix the solution.

3. Dextran sulphate takes a very long time to dissolve completely (up to 1–2 days). It serves as a volume exclusion agent by reducing the volume of the buffer and thus indirectly concentrating the substrate.

4. Membrane filters generally have a matt and a shiny side. It doesn't matter which side is used for filtering. However, it is advisable to choose one side, e.g., the matt side, and then always filter cells onto this side to avoid confusion.

5. Ethanol series: This dehydrates cells and serves as additional cell fixation. It was also found to make cells stick better to the slide.

6. Formamide on tissues: It is essential that the same hybridization buffer (same formamide concentration) is used on the slide and for moistening the tissue to avoid evaporation and concentration effects. When using several probes which require different formamide concentrations, the hybridizations have to be carried out on separate slides.

7. Cut filters: A 25 mm diameter filter can be cut into up to eight sections, a 47 mm diameter filter into about 14–18 pieces.

8. Wash buffer: In the wash buffer, formamide is replaced with NaCl, to reduce the amount of toxic waste. Both formamide and Na^+ concentration influence the stability of nucleic acid double strands. The Na^+ concentration in the wash buffer is chosen so the resulting stringency is similar to the stringency based on the formamide concentration in the hybridization buffer (Table 1). The overall stringency of the washing step, however, is slightly higher than during the hybridisation due to the 2°C increase in temperature.

9. Final ethanol series: This washes off remnant salt, preventing formation of salt crystals which would increase background fluorescence and also speeds up the drying of slides/filters.

10. Proteinase K: The proteinase K step is critical. If incubated for too long, cells will completely lose integrity. Optimal length of proteinase K treatment will depend on the type of target cell.

References

1. DeLong, E.F., Wickham, G.S., and Pace, N.R. (1989) Phylogenetic stains: ribosomal RNA-based probes for the identification of single cells. *Science.* **243**, 1360–1363.

2. Pernthaler, A., Pernthaler, J., and Amann, R. (2002) Fluorescence in situ hybridization and catalyzed reporter deposition for the identification of marine bacteria. *Appl. Environ. Microbiol.* **68**, 3094–3101.

3. Schonhuber, W., Fuchs, B., Juretschko, S., and Amann, R. (1997) Improved sensitivity of whole-cell hybridization by the combination of horseradish peroxidase-labeled oligo-nucleotides and tyramide signal amplification. *Appl. Environ. Microbiol.* **63**, 3268–3273.

4. Zwirglmaier, K., Ludwig, W., and Schleifer, K.H. (2004) Recognition of individual genes in a single bacterial cell by fluorescence in situ hybridization – RING-FISH. *Mol. Microbiol.* **51**, 89–96.

5. Lee, N., Nielsen, P.H., Andreasen, K.H., Juretschko, S., Nielsen, J.L., Schleifer, K.-H., et al. (1999) Combination of fluorescent in situ hybridization and microautoradiography – a new tool for structure-function analyses in microbial ecology. *Appl. Environ. Microbiol.* **65**, 1289–1297.

6. Ouverney, C.C. and Fuhrman, J.A. (1999) Combined Microautoradiography-16S rRNA probe technique for determination of radioisotope uptake by specific microbial cell types in situ. *Appl. Environ. Microbiol.* **65**, 1746–1752.

7. Loy, A., Horn, M., and Wagner, M. (2003) probeBase: an online resource for rRNA-targeted oligonucleotide probes. *Nucleic Acids Res.* **31**, 514–516.

8. Pruesse, E., Quast, C., Knittel, K., Fuchs, B.M., Ludwig, W., Peplies, J., Glöckner, F.O. (2007) SILVA: a comprehensive online resource for quality checked and aligned ribosomal RNA sequence data compatible with ARB. *Nucleic Acids Res.* **35**, 7188–7196.

9. DeSantis, T.Z., Hugenholtz, P., Larsen, N., Rojas, M., Brodie, E.L., Keller, K., et al. (2006) Greengenes, a chimera-checked 16S rRNA gene database and workbench compatible with ARB. *Appl. Environ. Microbiol.* **72**, 5069–5072.

10. Cole, J.R., Chai, B., Farris, R.J., Wang, Q., Kulam-Syed-Mohideen, A.S., McGarrell, D.M., et al. (2007) The ribosomal database project (RDP-II): introducing myRDP space and quality controlled public data. *Nucleic Acids Res.* **35**(suppl 1), D169–D172.

11. Ludwig, W., Strunk, O., Westram, R., Richter, L., Meier, H., Yadhukumar, Buchner, A., Lai T., Steppi, S., Jobb, G., Förster, W., Brettske, I., Gerber, S., Ginhart, A.W., Gross, O., Grumann, S., Hermann, S., Jost, R., König, A., Liss, T., Lüssmann, R., May, M., Nonhoff, B., Reichel, B., Strehlow, R., Stamatakis, A., Stuckmann, N., Vilbig, A., Lenke, M., Ludwig, T., Bode, A., Schleifer, K.H. (2004) ARB: a software environment for sequence data. *Nucleic Acids Res.* **32**, 1363–1371.

12. Loy, A., Tischler, R.A.P., Rattei, T., Wagner, M., and Horn, M. (2008) probeCheck: a central resource for evaluating oligonucleotide probe coverage and specificity. *Environ. Microbiol.* **10**, 2894–2898.

13. Fuchs, B.M., Wallner, G., Beisker, W., Schwippl, I., Ludwig, W., and Amann, R. (1998) Flow cytometric analysis of the in situ accessibility of Escherichia coli 16S rRNA for fluorescently labeled oligonucleotide probes. *Appl. Environ. Microbiol.* **64**, 4973–4982.

14. Behrens, S., Rühland, C., Inácio, J., Huber, H., Fonseca, A., Spencer-Martins, I., Fuchs, B.M., Amann, R. (2003) In situ accessibility of small-subunit rRNA of members of the domains Bacteria, Archaea, and Eucarya to Cy3-labeled oligonucleotide probes. *Appl. Environ. Microbiol.* **69**, 1748–1758.

Chapter 28

FISH as a Tool to Investigate Chromosome Behavior in Budding Yeast

Harry Scherthan and Josef Loidl

Abstract

Fluorescence in situ hybridization (FISH) provides an effective means to delineate chromosomes and their subregions during all stages of the cell cycle. This makes FISH particularly useful for studying chromosome behavior in species with minute genomes and/or poor chromosome condensation at metaphase, which is the case in model organisms such as the budding yeast *Saccharomyces cerevisiae*. Since its introduction in 1992, FISH with composite whole chromosome or locus specific probes has become an indispensable tool in the analysis of chromosome behavior in metaphase and interphase cells, and especially of meiotic chromosome pairing of wild-type and mutant yeast strains.

Key words: DNA labeling, FISH, GISH, Immunofluorescence, *Saccharomyces cerevisiae*, Chromosome painting, Chromosome dynamics

1. Introduction

As compared to multicellular eukaryotes, the baker's yeast *Saccharomyces cerevisiae* ($n = 16$) challenges classical cytology, since metaphase occurs without nuclear envelope breakdown. Highly condensed chromosomes are absent and yeast nuclei lack a nuclear lamina, which makes them vulnerable to distortion during isolation procedures. Even though *S. cerevisiae* is extremely powerful for classical genetics, its cytology is further hampered by its size: diploid nuclei are only 2–4 μm in diameter and harbor 32 chromosomes with a total DNA content of a mere 24 Mbp of predominantly unique DNA sequences (1, 2). For these reasons, light and electron microscopic studies of yeast metaphase

Joanna M. Bridger and Emanuela V. Volpi (eds.), *Fluorescence in situ Hybridization (FISH):*
Protocols and Applications, Methods in Molecular Biology, vol. 659,
DOI 10.1007/978-1-60761-789-1_28, © Springer Science+Business Media, LLC 2010

chromosomes at best reveal tiny chromatin lumps that are barely reminiscent of the metaphase chromosomes of more complex eukaryotes (3, 4). However, a reasonable chromosome (bivalent) structure can be obtained by silver or fluorochrome staining of chromosome spreads of meiotic cells, in which homologues are connected by a joint protein zipper, the synaptonemal complex (SC) (5–7). Staining and tracking of particular chromosomes in cytological preparations from time course experiments became possible with the introduction of fluorescence in situ hybridization (FISH) for budding yeast (8). The combination of the above methods has since successfully facilitated cytological studies in this model organism (Fig. 1) (9).

Since FISH provides a powerful tool to delineate individual chromosomes and their subregions in chromosome spreads and nuclei, it has become a major tool for studying chromosome behavior in wild-type and mutant yeast strains (10–16). Furthermore, the delineation of specific chromosome regions by binding of fluorescent protein-tagged inducer or repressor molecules to chromosomally integrated LacO or TetO arrays has emerged as a valuable tool to study chromosomes and subregions, particularly in live cells (17–19). FISH, in comparison, is less laborious with respect to strain construction, since any chromosomal site can be delineated by the appropriate choice of probes. Moreover, FISH allows the size of the chromosome region(s) to be labeled to vary according to demand so that single-copy regions, chromosome arms, or even entire chromosomes can be painted by large (composite) DNA probes (Fig. 1). Furthermore, genomic in situ hybridization (GISH) can be used to delineate entire chromosomes or chromosome (sub)sets. GISH utilizes the preferential hybridization of homologous genomic DNA under suppression conditions to the chromosomes of one yeast species in a different species background (20) or a chromosome set of one species in a species hybrid (Fig. 1). Below we outline a protocol for yeast FISH that is routinely used in the analysis of budding yeast chromosome behavior in our laboratories.

2. Materials

2.1. Cell Growth and Preparation

1. Rich medium (YPD), liquid: 1% yeast extract, 2% peptone, 2% glucose in distilled water, autoclaved.

2. Presporulation medium (YPA), liquid: 1% yeast extract, 2% peptone, 1% potassium acetate in distilled water, autoclaved.

3. Sporulation medium (SPM), liquid: 2% potassium acetate in distilled water, autoclaved.

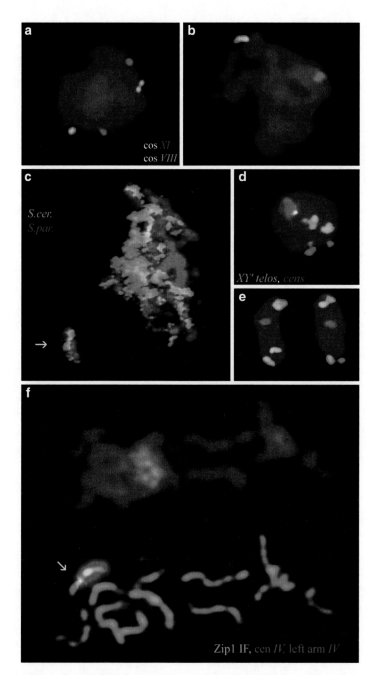

Fig. 1. (a) Spread meiocyte nucleus (DAPI, *blue*) hybridized with cosmids for chromosome *III* (*green*) and *XI* (*red*). Separation of signals of the same color (2 each) demonstrates absence of homologous pairing, while a split (closely spaced) signal for one locus indicates replicated DNA in prophase I. (b) Same experiment but pachytene nucleus with two large signals indicating homologous pairing. (c) Pachytene nucleus of a *S. cerevisiae* x *S. paradoxus* hybrid, differentially labeled by FISH with genomic DNAs (GISH) (*S. cerevisiae: green, S. paradoxus: red*). *Arrow*: A bivalent containing two synapsed homeologous chromosomes of different color. (d) Vegetative nucleus after FISH with a pan-centromere composite phage clone probe (*red*) and a XY' repeat pan-telomere probe (*green*) displaying Rabl orientation with centromeres clustered in a single signal region and few telomere clusters. (e) Two yeast bivalents (DNA, *blue*) after FISH with the same pan cen/telo probe as in detail (d) showing telomere (*green*) and centromere (*red*) labeling. (f) Spread bivalents of a pachytene nucleus displaying Zip1 immunofluorescence (SC protein, *green*) and FISH signals of a chromosome *IV* specific composite painting probe (450 kb, composed of 31 PCR products, *red*). The centromere is marked by a FISH signal (*blue, arrow*) of a centromeric PCR fragment probe (image (f) courtesy of J. Fuchs). For details *see* refs. 20, 24.

4. Zymolyase: Enzyme for dissolving the yeast cell wall. Prepare a 10 mg/mL stock solution of Zymolyase 100T (Seikagaku Co.) in distilled water (see Note 1).

5. Digestion solution: 0.8 M sorbitol with 10 mM dithiothreitol (prepared freshly from a frozen 1 M stock). Add 7 μL of Zymolyase stock solution (Subheading 2.1, item 4) per 500 μL.

6. Stop solution: 0.1 M 2-(N-morpholino) ethane sulfonic acid (MES), 1 mM EDTA, 0.5 mM $MgCl_2$, 1 M sorbitol, adjust pH to 6.4 using 0.1 M NaOH.

7. 2% N-lauroylsarcosine sodium salt.

8. 37% acid-free formaldehyde (Merck).

9. Fixative I: 4% formaldehyde solution made from an acid-free 37% formaldehyde stock solution (Merck) by diluting with distilled water. *Caution*: Formaldehyde is a hazardous chemical; handle always with care in a fume hood and wear protective equipment!

10. Detergent: Prepare a 1% solution of "Lipsol cleaning detergent" (Barloworld Scientific) in distilled water. The working solution can be stored for several months in the refrigerator (see Note 2).

11. Fixative II: 4% paraformaldehyde supplemented with 3.6% sucrose (see Note 3).

12. PBS: 130 mM NaCl, 7 mM Na_2HPO_4, 3 mM NaH_2PO_4 (pH 7.5).

13. Antifade agent, such as Vectashield (Vector Laboratories Inc.), is a useful tool to prevent fading of the fluorescent signals during microscopy.

2.2. Combined FISH and Immunostaining

1. Antibodies against various components of the yeast synaptonemal complex (SC), kinetochores, and several chromosomal proteins have been raised in various labs and several are commercially available. For instance, immuno labeling of microtubules with the YOL1/34 monoclonal rat anti yeast tubulin antibody (21, 22) produces good results. It can be obtained, e.g., from Serotec.

2. Secondary antibodies: obtain from commercial supplier like Jackson Laboratories, Sigma, or others. Always perform tests without primary antibodies to ensure specificity.

3. DAPI (4',6-diamidino-2-phenylindole) is used as a DNA-specific counterstain. It may be purchased ready-made in antifade solution (see Subheading 2.3).

4. Rubber cement for sealing cover slips; e.g., Fixogum (MarabuGmbH).

2.3. FISH

1. Probes for FISH can be obtained or generated in various ways: (1) Clones for the desired chromosomal loci can be selected from the *Saccharomyces* Genome Database (1) (http://www.stanford.edu/yeast) as cosmid or λ-phage clones and be purchased from the ATCC (Rockville, MD) (see Note 4). The probes can be combined to produce composite chromosome painting probes (see Note 5).

2. PCR fragments may be used as probes. Fragments of approximately 5–10 kb size are generated by long-range PCR. Suitable kits are available from various companies: e.g., Expand™ Long Template PCR System (Roche Applied Science); TaKaRa Ex Taq (TaKaRa Shuzo Co., Ltd.) and can be combined with appropriate primers chosen from the *Saccharomyces* Genome Database (1) (see Note 4).

3. Kits for nick translation can be purchased from commercial suppliers, either with a premade nucleotide labeling solution that contains a hapten-labeled nucleotide, like biotin-11-dUTP (Life technologies) or digoxigenin-11-dUTP (Roche Applied Science). If other labels have to be introduced, it is recommended to obtain nick translation kits without labeled nucleotides, which are then combined with the fluorophor-labeled nucleotide of choice (see Subheading 2.3, item 4) to generate probes suitable for multicolor FISH.

4. Labeled nucleotides: e.g., Cy3-dUTP, Cy5-dUTP (Amersham Pharmacia Biotech), fluorescein-dUTP, tetramethylrhodamine-dUTP, digoxigenin-dUTP, or biotin-dUTP (Roche Applied Science). Choose one specific nucleotide for each DNA probe. These can be combined in a single hybridization solution (see Subheading 2.3, item 9) and detected differentially.

5. RNase, DNase free (e.g., Sigma or Roche Applied Science).

6. BT buffer: 0.15 M $NaHCO_3$, pH 8.3, 0.1% Tween 20.

7. 20× SSC: 3.0 M NaCl, 0.3 M trisodium citrate, pH 7.0.

8. Tween-20.

9. Hybridization solution: 50% formamide, 2× SSC, 10% dextran sulfate, 1 μg/μL salmon sperm carrier DNA.

10. Rubber cement, e.g., Fixogum (Marabuwerke GmbH).

11. Heating block or a thermocycler capable of heating slides (e.g., Thermo Fisher Scientific Inc.; or AccuBlock).

12. Blocking buffer: 3% BSA, BT buffer (Subheading 2.3, item 6).

13. Detection buffer: 0.05% BSA, 0.1% Tween 20 in BT buffer.

14. Detection reagents: Avidin-FITC or -Cy3 conjugate (e.g., Extravidin® FITC conjugate, Extravidin® Cy3 conjugate; Sigma); Biotin-conjugated anti-avidin antibody (Vector

Labs); anti-digoxigenin-fluorophor conjugated antibodies (fluorescein, rhodamine, AMCA can be obtained, e.g., from Roche Applied Science).

15. Vectashield antifade solution (Vector laboratories). Can be obtained ready-made with DAPI for DNA costaining (see Subheading 2.2, item 3).

3. Methods

3.1. Cell Growth

1. For the study of mitotic cells, inoculate 5 mL YPD with a small colony from a plate and grow to a concentration of ~2×10^7 cells per mL, in a shaker at 30°C (usually overnight).

2. To obtain meiotic cells, inoculate 50 mL YPA with a small colony of a diploid strain from a plate and grow to a concentration of ~2×10^7 cells per mL in a shaker at 30°C (see Note 6). Centrifuge cell suspension for 4 min at $700 \times g$. Resuspend cells at a density of about 4×10^7 cells/mL in SPM and incubate until 2–4% 4-nucleate cells appear in DAPI-stained preparations (Subheading 3.2.1). At this point, the majority of cells will be at pachytene which is the most favorable stage for cytological examination. Sporulation time varies considerably between strains. Strain SK1 (23) which is widely used for meiotic studies shows the maximum number of pachytene cells ~4–5 h after transfer to SPM.

3.2. Cell Preparation

Here we describe the preparation procedures to study yeast chromosomes cytologically. Subheading 3.2.1 (Ethanol Fixation) is useful in combination with DAPI staining and inspection of GFP-labeled chromosomes and other cellular components. Formaldehyde fixation (Subheading 3.2.3) provides good preservation of cellular morphology and astral microtubules. It can be used in combination with GFP tagged protein detection or immunostaining (22). Nuclear spreading (Subheading 3.2.2) offers enhanced cytological resolution but it will disrupt cells. It is suitable for the visualization of SCs by Ag-staining or immuno-labeling of SC components and may also be used in combination with FISH (Fig. 1). Semi-spreading (Subheading 3.2.3) is a good compromise for obtaining a good spatial resolution of nuclear contents and a reasonable preservation of cell morphology for FISH.

3.2.1. Ethanol Fixation

1. Remove 1 mL of cells from the culture (Subheading 3.1).

2. Centrifuge the cell suspension for 10 s at $700 \times g$.

3. Discard the supernatant and resuspend pellet in PBS.

4. Repeat step 2.

5. Resuspend pellet in 1 mL 70% ethanol.

6. Repeat step 2.

7. Resuspend in 50 µL 70% ethanol.

8. Place 20 µL ethanol-fixed cells on a clean glass slide and streak out with the edge of a cover slip, without touching the surface of the slide.

9. Air-dry, add 18 µL antifade solution containing DAPI.

10. Cover with a glass coverslip.

11. Inspect under a phase contrast microscope.

3.2.2. Spreading

The spreading protocol described here is a modification by Loidl et al. 1991 (7) of the 1988 method of Dresser and Giroux (6).

1. Take 5 mL of a cell suspension obtained according to Subheading 3.1.

2. Spin the cell suspension for 20 s at $700 \times g$.

3. Resuspend the cell pellet in digestion solution (see Subheading 2.1, item 5).

4. Allow spheroplasting to proceed for 20 min at 37°C.

5. Put the cell suspension on ice.

6. Check the degree of cell wall degradation by placing 5 µL of the digested suspension on a glass slide and mix with an equal volume of 2% *N*-lauroylsarcosine.

7. Immediately place a glass coverslip onto the cell suspension and instantly observe the sample under a phase-contrast microscope at low power magnification.

8. After a few seconds, the cells should be bursting (initially bright cells become dark and then fragment). Approx. 80% of cells should rupture instantly, which is a good indicator for a degree of cell wall digestion suitable for FISH.

9. Stop the digest by adding an equal volume of ice-cold stop solution (Subheading 2.1, item 6) spin the cells down (Subheading 3.2.2, step 2), and resuspend them in 250 µL stop solution (see Note 7).

11. Place 20 µL of cell suspension on a slide, add sequentially: 40 µL fixative I (Subheading 2.1, item 9), 80 µL detergent (Subheading 2.1, item 10), 80 µL of fixative II (Subheading 2.1, item 11) (see Notes 8 and 9).

12. Disperse the mixture with a glass rod over the slide without touching its surface. Put slides in a chemical hood and let dry for >2 h – overnight (see Note 10).

13. Continue with one of the procedures described in Subheading 3.3.2. Slides may be stored at –20°C for months until further use.

3.2.3. Semi-spreading for Generation of Structurally Preserved Nuclei

This procedure is used when preservation of nuclear structure has to be combined with good access for the DNA probes (e.g., for combinatorial FISH and immunostaining (24)).

1. Transfer a sample from the sporulating culture (Subheading 3.1) at the desired time point immediately to 1/10th volume of acid-free ice-cold 37% formaldehyde (e.g., 1 mL culture to 0.1 mL 37% acid-free formaldehyde precooled on ice in an Eppendorf tube).

2. Incubate for 30 min on wet ice.

3. Sediment cells at $700 \times g$ for 2 min and wash them once in an excess of deionized H_2O.

4. Sediment cells at $700 \times g$ for 2 min and resuspend them in 1/10th of the sample volume of Solution I.

5. Spheroplast cells with Zymolyase by following Subheading 3.2.2, steps 2–4.

6. Drop 20 µL of the spheroplast suspension onto a slide and add sequentially 80 µL detergent and 80 µL of fixative. (Handle with care, the fixative contains formaldehyde; use protective equipment!)

7. Spread out the mixture with a glass rod and dry down the preparations in a chemical hood for >2 h – overnight (see Note 10).

8. Continue with the procedures described in Subheading 3.3. Alternatively, slides may be stored frozen at –20°C until use.

3.3. FISH

3.3.1. DNA Labeling and Probe Preparation

Nick-translation is the preferred method to label DNA probes for FISH. DNA probes of choice can be directly fluorophor or haptene labeled by the incorporation of fluorochrome-, biotin-, or digoxigenin-labeled deoxynucleotides (see Note 11) using commercial nick-translation kits. Nick-translation has the advantage that the fragment size of the labeled probe molecules is ~300 bp, which is imperative for access of the probe to the target molecules (see Note 12). Target regions as small as 2.4 kb can be successfully detected by yeast FISH (HS, unpublished observations).

1. Label DNA by nick translation according to the instructions of the supplier. (This typically requires 90 min of incubation at 16°C.)

2. Stop the reaction by adding 0.5 M EDTA to result in a final conc. of >30 mM and incubate for 10 min at 65°C.

3. Store the labeled DNA at –20°C until use.

4. Ethanol-precipitate labeled probe DNA with 1/10th volume 3 M Ammonium acetate and 2.5 volumes of ethanol for >30 min at −20°C.

5. Spin in microcentrifuge at >10,000 × g for 30 min.

6. Discard supernatant and briefly dry the DNA pellet by incubating the open tube a few minutes at 65°C, e.g. in a heating block.

7. Resuspend the DNA pellet in hybridization solution (Subheading 2.3, item 9) to result in a final concentration of 30 ng/μL.

8. Denature the DNA probe in hybridization solution at 95°C for 5 min.

9. Place tube on ice or in a fridge. Keep chilled until further processing.

3.3.2. Hybridization

1. Place a slide produced as described under Subheading 3.2.2 or 3.2.3 in a Coplin jar filled with deionized water until the sucrose-layer has dissolved.

2. Replace the deionized water once.

3. Drain and briefly air-dry the slide by standing it upright on a paper towel.

4. Apply 50 μL RNase (100 μg/mL in 2× SSC) to each slide, cover with a coverslip, and incubate for 60 min (up to 120 min) at 37°C in a moist chamber.

5. To denature chromosomal DNA, place 100 μL 70% formamide, 30% 2× SSC (pH 7.0) on top of the slides and cover with a 24×60 mm cover slip. Place slide on a hot plate or thermocycler block for 5–10 min at 80°C.

6. Add a few μL of the formamide, 2× SSC solution (Subheading 3.3.2) to the rim of the coverslip every 3 min, to replace the evaporating liquid.

7. Rinse off the cover slip with ice-cold deionized water, shake off water drops and let slide air dry, standing in an upright position on a paper towel.

8. Place a drop of antifade solution containing DAPI on the slide, cover with a large coverslip and locate a region with well-spread nuclei using the fluorescence microscope and mark reagion from below with a permanent pen.

9. Place a drop of BT buffer (Subheading 2.3, item 6) to the rim of the coverslip and let it sit for 2 min.

10. Carefully lift off the coverslip with forceps.

11. Wash briefly in deionized water and shake off the excess liquid from the slide, air-dry.

12. Apply 5 µL of denatured probe mix onto each slide (the amount depends on the size of the coverslip/region to be hybridized; 5 µL correspond to a 18×18 mm coverslip). Combine the probes in equimolar amounts, if two or more probes have to be hybridized simultaneously on the same slide.

13. Place a coverslip (18 mm×18 mm) over the region of interest containing the sample and seal with rubber cement (Fixogum).

14. Let the Fixogum dry until it appears clear.

15. Incubate at 37°C (e.g., in a humid cell culture incubator) for 24–48 h (see Note 13).

16. Peel off the rubber cement and gently rinse off the coverslip by placing the slides in a Coplin jar with 0.05× SSC at 37°C.

17. Wash slides 3× 5 min in 0.05× SSC at 37°C (see Note 14).

18. Wash once with BT buffer.

19. After hybridization with directly fluorophore-labelled DNA, apply a drop of antifade solution with DAPI and seal under a coverslip.

20. When digoxigenin- or biotin-labeled DNA probe molecules have to be detected after hybridization, refer to Subheading 3.3.3.

3.3.3. Signal Detection of Digoxigenin- or Biotin-Labeled Probes

1. Put a large drop (100 µL) of blocking buffer onto each slide in a Coplin jar and incubate for >10 min at 37°C in a moist chamber. Alternatively, slides can be submersed in a Coplin jar with blocking buffer.

2. To detect biotinylated probes, float off coverslip in a coplin jar with detection buffer and add 100 µL FITC-conjugated avidin (diluted 1:400 in BT) to each slide. Cover with a coverslip or a piece of Parafilm and incubate for 45 min at 37°C in a moist chamber. Continue with step 4 or 8.

3. To detect digoxigenin-labeled probes, apply rhodamine-conjugated anti-digoxigenin antibody (Roche Applied Science) diluted 1:200 in BT buffer and incubate for 45 min at 37°C as under point 2. Anti-digoxigenin-rhodamine and avidin-FITC may be mixed if two probes have to be detected differentially on the same slide.

4. Wash 3× 3 min in BT buffer at 37°C.

5. If a biotin-labeled probe displays only weak signals on the specimen, amplify its fluorescence as described up to step 10: Rinse coverslip away with BT buffer.

6. Wash slides three times 3 min in BT buffer and drain off excess liquid.

7. Incubate preparation with a biotin-conjugated anti-avidin antibody (diluted 1:250 in BT) for 35 min at 37°C.

8. Rinse off coverslip with BT buffer, wash slides 3× 3 min in BT, and drain off excess liquid.

9. Incubate preparation with FITC-conjugated avidin (diluted 1:450) as in step 2.

10. Rinse coverslip away with BT buffer, wash slides 3× 3 min in BT buffer, and drain off excess liquid.

11. Mount preparation in antifade solution containing DAPI and inspect under the microscope (Fig. 1).

3.4. Combined Immunostaining and FISH

1. Wash the slides that have been obtained by spreading or semi-spreading (Subheading 3.2.3), twice for 5 min in PBS, 0.05% Tween 20 at RT.

2. Drain off excess liquid.

3. To the slide add 50 µL of primary antibody (diluted in 1× PBS, 0.05% Tween 20; the appropriate dilution, this has to be tested empirically).

4. Cover with a coverslip or an appropriately sized piece of Parafilm and incubate at 4°C (e.g., in a fridge) overnight.

5. Float coverslip or Parafilm off with 1× PBS, 0.05% Tween 20.

6. Wash slides twice for 5 min in PBS, 0.05% Tween 20 and drain off excess liquid.

7. Dilute fluorochrome-conjugated secondary antibody 1× PBS, 0.05% Tween 20, according to the instructions of the supplier. Usually 1:250–400 works well for FITC labeled antibodies, and Cy3-labeled antibodies are diluted 1:500–1,500.

8. Add 100 µL of antibody solution under a coverslip or slide-sized piece of Parafilm and incubate for 90 min at room temperature.

9. Rinse coverslip or Parafilm off with 1× PBS, 0.05% Tween 20.

10. Wash slides twice for 5 min in 1× PBS, 0.05% Tween 20, and drain off excess liquid.

11. At this point, the preparation can be mounted in Vectashield + DAPI and images can be recorded using a fluorescence microscope.

12. After recording of images and cell coordinates, rinse off the coverslip and antifade solution with 1× PBS, 0.05% Tween 20 (see Note 15).

13. Fix cells for 1 min in 1% formaldehyde in 1× PBS.

14. Rinse the cells twice in PBS, 0.05% glycine to quench unsaturated aldehyde groups and apply the standard FISH procedure (Subheadings 3.3.2 and 3.3.3).

15. If the immunofluorescence was destroyed by the denaturation steps or has faded, relocate the cells from step 11 and record the FISH images of the same cells (see Note 14). Merge the two images digitally (25).

3.5. Microscopic
Evaluation
of Fluorescent Signals

An epifluorescence microscope equipped with appropriate filter sets for the excitation and emission of fluorescence spectra characteristic for the fluorochromes used, is necessary to visualize signals. The optimal combinations of excitation filter, beam splitter, and emission filter have to be chosen according to the experimental requirements. For high specificity, narrow bandwidth filters should be selected, which also minimize the bleed-through of the fluorescence of other fluorochromes in multicolor FISH experiments. Images are best recorded with a cooled CCD camera with high sensitivity to a wide spectrum of wavelengths, including far-red as emitted, e.g., by Cy5.

4. Notes

1. The Zymolyase stock can be stored at –20°C for several months and repeatedly refrozen. The powder does not dissolve completely; therefore, there is usually a pellet after thawing, which should be stirred up before use.

2. Lipsol is a laboratory cleaning agent, a mixture of nonionic and anionic detergents plus a chelating agent and builders (information provided by the manufacturer). Several standard laboratory detergents (Nonidet, Triton X-100, sodium dodecyl sulfate, *N*-lauroyl-sarcosine) failed to produce comparable results in our hands.

3. Add 4 g paraformaldehyde to 90 mL distilled water and heat the suspension on a magnetic stirrer to 80°C (*Caution*: Heating creates hazardous formaldehyde vapors! Work in a fume hood and wear protective equipment!). After 20–30 min, the solution should become clear. If it stays opaque, add 5 M NaOH until it becomes clear. After cooling, add 3.4 g sucrose to the solution. If NaOH has been added, the solution has to be titrated back to pH 8.5 using HCl. Thereafter, fill up volume to 100 mL. If the fixative is not completely clear, it may be filtered. It can be stored for several months in the refrigerator.

4. FISH probes should be carefully selected and their sequence be checked against the database (e.g., http://genome-www. stanford.edu) to avoid them containing repetitive genomic

elements (such as the Ty1 transposon), which would result in nonspecific speckled background staining (unpublished observations). Make sure that the probes are located outside duplicated genomic regions.

5. Attempts to isolate uncontaminated chromosome-specific DNA from pulsed field gels have failed in several labs. Hence, it is recommended to use contiguous sets of long-range PCR products and/or cosmid or P1 clones to paint extended chromosome regions. Pooled probe DNAs can be labeled by nick translation.

6. Presporulation growth in YPA (26) improves the synchrony of sporulation.

7. This suspension can be kept on ice for up to one day, during which it should be used for the preparation of slides.

8. Fixative is added to the slide before and after the detergent. A small amount of fixative present during detergent spreading prevents harsh disruption of spheroplasts but does not interfere too much with spreading. The relative amounts and order of application of nuclear suspension, detergent, and fixative should be optimized by testing, since the optimal spreading depends on the density of nuclei in the suspension, the degree of spheroplasting, and the age of solutions. The process of spreading can be watched under the phase-contrast microscope at low power magnification without a coverslip. Spheroplasts should swell slowly and gradually turn from white to black and then to gray. They should not disintegrate instantly!

9. The presence of sucrose in the fixative has the advantage that the mixture is hygroscopic and does not dry out completely. Therefore, this kind of preparation can be used for immunostaining even after prolonged storage in the refrigerator or in the freezer at -20°C or below.

10. During the spreading or semi-spreading procedures (Subheadings 3.2.2 and 3.2.3), chromatin tightly adheres to the surface of the slides owing to intimate charge interactions of DNA and glass. Coating of slides (e.g., poly-L-lysine, amino silane) is not necessary.

11. Fluorochrome-conjugated deoxynucleotides (e.g., Cy3-dUTP, Cy5-dUTP, fluorescein-dUTP/-dATP) in conjunction with nick-translation generates directly fluorescing DNA probes, which render FISH signals of sufficient intensity for most applications. However, when several probes are to be combined for multicolor (m)FISH applications, or if fluorescence signal intensity is too weak (in the case of single copy probes of a few kilobases), biotin or digoxigenin labeling of

probe DNAs followed by indirect immuno-detection is recommended (8, 27), since this renders stronger signals.

12. Labeling reactions need be optimized to result in labeled products of 100–500 bp length, which should be monitored by agarose gel electrophoresis. If the DNA fragments should still be too long after 90 min of nick-translation, they can be reduced in length, e.g., by adding 1 µL of a dilute (2 U/µL) DNAse I solution followed by a 10 min incubation. The effectiveness of this treatment should be first checked with test DNA and agarose gel electrophoresis. Commercial nick-translation kits (e.g., BioNick kit of Life technologies or DIG-Nick Translation Mix of Roche Applied Science), have performed well in our hands without further manipulation and are recommended for beginners.

13. In some circumstances FISH may fail, which is often due to closed chromatin state due to formaldehyde fixation. In such cases a second denaturation step, e.g., for 5 min at 75°C with the probe sealed under a coverslip is sufficient to bring about a successful hybridization.

14. It has been found that 0.05× SSC efficiently replaces the 70% formamide, 2× SSC solution usually used in standard procedures for posthybridization washes, since it delivers the same stringency (8, 24). This replacement also reduces the potential exposure to the fumes of the teratogenic chemical formamide.

15. If the signal from immunofluorescence (IF) staining has vanished after the FISH procedure previous recording of IF images and position of IF-stained cells with the help of the coordinate system of the microscope stage is recommended. After FISH, the same cells are relocated and the FISH signals are recorded. Finally, IF and FISH images are merged digitally (25). If the IF signal endures the FISH procedure (24) (which is often the case for SC proteins, but has to be tested empirically) IF-stained slides are immediately subjected to FISH as noted in the protocol and both types of signals are recorded together at the end of the procedure.

Acknowledgments

We thank A. Lorenz, J. Fuchs, and E. Trelles-Sticken for insightful comments and stimulating discussions. The work in the lab of HS was partly supported by the DFG (SCHE 350/10-1, SPP 1384) and H.-H. Ropers, Max-Planck-Inst. for Molecular Genetics, Berlin, FRG.

References

1. Cherry, J. M., Ball, C., Chervitz, S., Dolinski, K., Dwight, S., Harris, M., Hester, E., Juvik, G., Malekian, A., Roe, T., Weng, S., and Botstein, D. (1998) Saccharomyces Genome Database. http://genome-www.stanford.edu/Saccharomyces/

2. The Yeast Genome Directory. (1997). *Nature* **387**, 1–105.

3. Kater, J. M. (1927) Cytology of Saccharomyces cerevisiae with special reference to nuclear division. *Biol Bull* **52**, 436–449.

4. Wintersberger, U., Binder, M., and Fischer, P. (1975) Cytogenetic demonstration of mitotic chromosomes in the yeast *Saccharomyces cerevisiae*. *Mol Gen Genet* **142**, 13–17.

5. Kuroiwa, T., Kojima, H., Miyakawa, I., and Sando, N. (1984) Meiotic karyotype of the yeast *Saccharomyces cerevisiae*. *Exp Cell Res* **153**, 259–265.

6. Dresser, M. and Giroux, C. (1988) Meiotic chromosome behavior in spread preparations of yeast. *J Cell Biol* **106**, 567–573.

7. Loidl, J., Nairz, K., and Klein, F. (1991) Meiotic chromosome synapsis in a haploid yeast. *Chromosoma* **100**, 221–228.

8. Scherthan, H., Loidl, J., Schuster, T., and Schweizer, D. (1992) Meiotic chromosome condensation and pairing in Saccharomyces cerevisiae studied by chromosome painting. *Chromosoma* **101**, 590–595.

9. Loidl, J. (2003) Chromosomes of the budding yeast *Saccharomyces cerevisiae*. *Int Rev Cytol* **222**, 141–196.

10. Guacci, V., Hogan, E., and Koshland, D. (1994) Chromosome condensation and sister chromatid pairing in budding yeast. *J Cell Biol* **125**, 517–530.

11. Loidl, J., Klein, F., and Scherthan, H. (1994) Homologous pairing is reduced but not abolished in asynaptic mutants of yeast. *J Cell Biol* **125**, 1191–1200.

12. Weiner, B. M. and Kleckner, N. (1994) Chromosome pairing via multiple interstitial interactions before and during meiosis in yeast. *Cell* **77**, 977–991.

13. Rockmill, B., Sym, M., Scherthan, H., and Roeder, G. S. (1995) Roles for two RecA homologs in promoting meiotic chromosome synapsis. *Genes Dev* **9**, 2684–2695.

14. Gotta, M., Laroche, T., and Gasser, S. M. (1999) Analysis of nuclear organization in *Saccharomyces cerevisiae*. *Methods Enzymol* **304**, 663–672.

15. Trelles-Sticken, E., Adelfalk, C., Loidl, J., and Scherthan, H. (2005) Meiotic telomere clustering requires actin for its formation and cohesin for its resolution. *J Cell Biol* **170**, 213–223.

16. Admire, A., Shanks, L., Danzl, N., Wang, M., Weier, U., Stevens, W., Hunt, E., and Weinert, T. (2006) Cycles of chromosome instability are associated with a fragile site and are increased by defects in DNA replication and checkpoint controls in yeast. *Genes Dev* **20**, 159–173.

17. Straight, A. F., Belmont, A. S., Robinett, C. C., and Murray, A. W. (1996) GFP tagging of budding yeast chromosomes reveals that protein-protein interactions can mediate sister chromatid cohesion. *Curr Biol* **6**, 1599–1608.

18. Michaelis, C., Ciosk, R., and Nasmyth, K. (1997) Cohesins: chromosomal proteins that prevent premature separation of sister chromatids. *Cell* **91**, 35–45.

19. Aragon-Alcaide, L. and Strunnikov, A. V. (2000) Functional dissection of in vivo interchromosome association in *Saccharomyces cerevisiae*. *Nat Cell Biol* **2**, 812–818.

20. Lorenz, A., Fuchs, J., Trelles-Sticken, E., Scherthan, H., and Loidl, J. (2002) Spatial organisation and behaviour of the parental chromosome sets in the nuclei of *Saccharomyces cerevisiae* x *S. paradoxus* hybrids. *J Cell Sci* **115**, 3829–3835.

21. Kilmartin, J. V., Wright, B., and Milstein, C. (1982) Rat monoclonal antitubulin antibodies derived by using a new nonsecreting rat cell line. *J Cell Biol* **93**, 576–582.

22. Pringle, J. R., Adams, A. E., Drubin, D. G., and Haarer, B. K. (1991) Immunofluorescence methods for yeast. *Methods Enzymol* **194**, 565–602.

23. Kane, S. M. and Roth, R. (1974) Carbohydrate metabolism during ascospore development in yeast. *J Bacteriol* **118**, 8–14.

24. Trelles-Sticken, E., Dresser, M. E., and Scherthan, H. (2000) Meiotic telomere protein Ndj1p is required for meiosis-specific telomere distribution, bouquet formation and efficient homologue pairing. *J Cell Biol* **151**, 95–106.

25. Loidl, J., Jin, Q.-W., and Jantsch, M. (1998) Meiotic pairing and segregation of translocation quadrivalents in yeast. *Chromosoma* **107**, 247–254.

26. Roth, R. and Halvorson, H. O. (1969) Sporulation of yeast harvested during logarithmic growth. *J Bacteriol* **98**, 831–832.

27. Scherthan, H. and Trelles-Sticken, E. (2002) Yeast FISH: delineation of chromosomal targets in vegetative and meiotic yeast cells, in *FISH Technology, Springer Lab Manual* (Rautenstrauss, B., and Liehr, T., eds.), Springer, Heidelberg, New York, pp. 329–346.

Chapter 29

FISH on Chromosomes Derived from the Snail Model Organism *Biomphalaria glabrata*

Edwin C. Odoemelam, Nithya Raghavan, Wannaporn Ittiprasert, Andre Miller, Joanna M. Bridger, and Matty Knight

Abstract

The application of fluorescence in situ hybridization (FISH) for the mapping of single copy genes onto homologous chromosome has been integral to vast number genome sequencing projects, such as that of mouse and human. The chromosomes of these organisms are well-studied and are the staple resource of most of the early studies conducted in cytogenetics. However, there are now protocols for analyzing FISH probes in a number of different organisms on both metaphase and interphase chromosomes.

Here, we describe the methodologies for the chromosomal mapping of nonrepetitive (single-copy) genes of the snail *Biomphalaria glabrata* onto metaphase chromosomes derived from the only molluscan cell-line in existence. The technique described in this chapter was developed for the *B. glabrata* genome sequencing project through troubleshooting experimental procedures established for other organisms so that both the optimum resolution of metaphase chromosome and the effective hybridization of genes were achieved.

Key words: Fluorescence in situ hybridization, Snail chromosomes, *Biomphalaria glabrata*, Bge embryonic cell line, Molluscan, Planorbid gastropod

1. Introduction

The fluorescent mapping of genetic sequences onto their chromosomes of origin is an important tool in genomic analysis and often leads to a better understanding of the structure and organization of an organism's genome. Fluorescence in situ hybridization (FISH) as a technique is used frequently in mammals, birds, and plants but is underutilized in gastropods. Previous studies using FISH in gastropods have shown the mapping of 18S–28S,

Joanna M. Bridger and Emanuela V. Volpi (eds.), *Fluorescence in situ Hybridization (FISH): Protocols and Applications*, Methods in Molecular Biology, vol. 659, DOI 10.1007/978-1-60761-789-1_29, © Springer Science+Business Media, LLC 2010

5S rDNA, TTAGGG$_n$ telomeric repeats in periwinkle *Melarhaphe neritoides* (1) and *Cantareus aspersus* and *C. mazzullii* (2) and the nonlong terminal repeat (LTR)-retrotransposons, *nimbus, in Biomphalaria glabrata* (3). The latter organism is the subject of this chapter.

The fresh water snail *B. glabrata* ($2n = 36$) belongs to the taxonomic class Gastropoda (family Planorbidae) and is integral to the spread of the human parasitic disease schistosomiasis. The importance of this mollusc is such that it has been selected as a model molluscan organism for whole genome sequencing by the genome sequencing center (GSC) at Washington University, St. Louis, MO (4). Indeed, its significance as a model molluscan organism is underpinned by the fact that the only molluscan cell line in existence, to date, is derived from this snail (5).

The *B. glabrata* chromosomes are extremely small, and good spreading of metaphase chromosomes is therefore essential in enabling the recognition of specific chromosomes. Indeed, there is a large amount of aneuploidy in the Bge cell-line, (6) but so far the gene mapping has identified pairs of homologous chromosomes (6).

The methods in this chapter describe how to effectively map single-copy genes onto *B. glabrata* chromosomes (see Fig. 1).

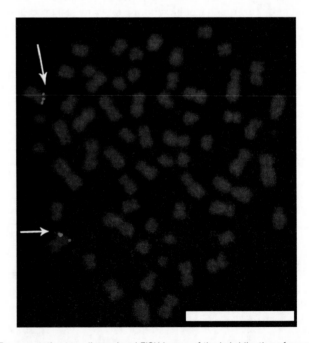

Fig. 1. Representative two-dimensional FISH image of the hybridisation of nonrepetitive (single copy) genes onto two homologous *B. glabrata* chromosomes. The fixed chromosomes were hybridised with a BAC probe (Fig. 1, *Red*) and counterstained with DAPI (*blue*). The *arrows* indicate the location of the gene on the chromosome arm. Scale bar = 10 μm.

2. Materials

2.1. Slide and Probe Preparation

2.1.1. Cell Culture of the Bge Cell-Line

1. The culture medium for growing the Bge cells comprises 22% (v/v) Schneider's *Drosophila* medium (Invitrogen), 0.13% (w/v) galactose (Invitrogen), 0.45% (w/v) lactalbumin hydrolysate (Invitrogen), and 14.1 μM phenol red (Invitrogen, Paisley, UK). The antibiotic gentamicin (Invitrogen) is required at a concentration of 20 μg/mL (see Note 1).

2. 0.22 μM pore filter units, 1,000 mL capacity (Fisher Scientific).

3. 10% inactivated characterized fetal bovine serum (FBS) (v/v), (Hyclone see Note 2).

4. An incubator that can be maintained at 27°C – no CO_2 is required – (see Note 3).

2.1.2. Bge Cellular Fixation

1. 10 μg/mL of colcemid dissolved in Hank's buffered salt solution (Invitrogen).

2. Potassium chloride (KCl) solution 0.05 M.

3. Methanol and acetic acid fixative (3:1) (v/v). Cooled on ice.

4. 20–200 μL pipetteman.

5. SuperFrost™ glass slides.

6. Phase-contrast microscope.

2.1.3. B. glabrata Genomic DNA Preparation

1. Bge cells, from the authors' laboratories or the American Tissue Culture Collection (ATCC).

2. Cell scraper.

3. Cellular digestion buffer, 100 mM Tris pH 8.0, 5 mM EDTA, 200 mM NaCl, 0.2% (w/v) SDS.

4. Proteinase K (stock 20 mg/mL, Invitrogen).

5. 70% (v/v) and 100% ethanol.

2.1.4. Probe Labeling and Preparation

1. 1 μg of *B. glabrata* BAC DNA containing genes from BS90 snail isolate BAC library (7); dissolved in 30 μL of sterile ddH$_2$O.

2. Nick translation kit (BioNick™, Invitrogen). Store at –20°C.

3. 1% agarose gel.

4. 3 μg of Herring sperm DNA.

5. 40 μg of sonicated *B. glabrata* genomic DNA – (see Note 4).

6. Stock solution of 20× sodium saline citrate (SSC); 3 M NaCl, 0.3 M tri-sodium citrate; pH 7.0.

7. Hybridization buffer (50% formamide, 10% dextran sulphate, 2× SSC, and 1% Tween 20). Make 100 μL aliquots.

2.2. Hybridization

2.2.1. Slide Denaturation

1. Ethanol solutions of 70, 90, and 100% (v/v).

2. Oven set to 70°C.

3. Stock solution of 20× SSC; 3 M NaCl, 0.3 M tri-sodium citrate; pH 7.0.

4. 70% (v/v) of formamide and 2× SSC; pH 7.0.

5. Glass Coplin jars.

6. Hot plate set to 37°C.

2.2.2. Probe Denaturation and Hybridization

1. Biotinylated probe.

2. 22 × 50 mm glass coverslips.

3. Rubber cement (Halfords).

4. Humidified chamber (see Note 5).

2.3. Washing and Counterstaining

2.3.1. Washing

1. Glass Coplin jars.

2. Stock solution of 20× SSC; 3 M NaCl, 0.3 M tri-sodium citrate; pH 7.0.

3. Formamide.

4. Buffer A: 50% formamide and 2× SSC; pH 7.0.

5. Buffer B: 0.1× SSC; pH 7.0.

6. 4× SSC; pH 7.0.

2.3.2. Blocking

1. Blocking solution; 4% (w/v) bovine serum albumin (BSA) in 4× SSC.

2. 22 × 50 mm coverslips.

2.3.3. Immunodetection and Washing

1. 100 μL of streptavidin conjugated to cyanine 3 (Amersham) in 1% BSA dissolved in 4× SSC (1:200 dilution).

2. 22 × 50 mm coverslips.

3. Humidified chamber.

4. 4× SSC with 0.1% Tween 20 (v/v).

5. Glass Coplin jars.

2.3.4. Counterstain

1. 4, 6-diamidino-2-phenylindole (DAPI) in Vectorshield antifade mountant (Vectorlaboratories).

2. 22 × 50 mm coverslips.

3. Methods

3.1. Slide and Probe Preparation

3.1.1. Cell Culture

1. The Bge cells are grown in the absence of CO_2, at 27°C and in Bge medium. The cells are very sensitive to slight changes to solute concentrations in the medium; it is thus important that Bge medium is prepared using precise laboratory balances.

2. Prepare 4.5 g of Lactalbumin hydrolysate, 1.3 g of Galactose, and 0.01 g of Phenol red. Add to 1,000 mL beaker.

3. Using a tissue culture flow hood, measure 220 mL of Schneider's *Drosophila* medium into a measuring cylinder. Add this to the 1,000 mL beaker and cover the top of the beaker with cling film (see Note 6).

4. Add 780 mL of ddH$_2$O to the beaker and mix using a magnetic stirrer. The lactalbumin hydrolysate and galactose tend to stick to the bottom of the beaker, in which case, gently move the beaker so that the magnetic stirrer can dislodge the solute. Try to keep the cling film on top of the beaker throughout this stage.

5. The medium is filter sterilized using 0.22 μm pore disposable filter unit of a 1,000 mL capacity and a sterile collection bottle. Using a tissue culture flow hood, remove cling film from the beaker of medium and decant the medium into the 0.22 μm pore filter unit. Attach a vacuum pump to the appropriate nozzle on the filter. The medium takes normally 1 min to pass through the filter.

6. After filtration, remove filter from bottle (within the tissue culture flow hood) and add 1 mL of 50 mg/mL gentamicin to medium. Screw lid onto bottle and gently swirl medium.

7. The medium is made complete by adding 10% inactivated FBS. The inactivation of the FBS can be carried out with 500 mL bottles of FBS, which can be aliquoted and frozen (see Note 7).

8. Prepare within a flow hood, complete Bge medium by pipetting 50 mL of FBS and 450 mL of Bge medium into a sterile bottle. Store both the 500 mL Bge medium with the FBS and the 500 mL complete Bge medium at 4°C.

3.1.2. Bge Cellular Fixation

1. The Bge cells are grown in T75 flasks, with 15 mL of complete Bge medium and passaged when they are 80% confluent. They are reseeded at a dilution of 1:9. This is usually done every 5 days.

2. Prepare 15 mL of complete Bge medium and 200 μL of colcemid in a conical tube. Invert to mix and store at 27°C.

3. Carefully remove the medium from the flask, without dislodging the cells from the substratum (see Note 8). Add the 15 mL complete Bge medium and 200 μL of colcemid to the flask. Incubate at 27°C for 1.5 h.

4. Dislodge the cells from the substratum by slapping the back of the flask sharply with an open-palmed hand; two slaps is normally enough to dislodge the mitotic cells. Centrifuge cells at $400 \times g$ for 5 min.

5. Resuspend pellet in 7 mL of hypotonic KCl solution. Begin this step by adding a few drops of KCl at a time and gently

increase the volume until 7 mL has been dispensed. Incubate at 27°C for 15 min. Invert the tube at 5 min intervals to prevent the cellular pellet reforming at the bottom of the tube.

6. Add two drops of fixative and invert tube. The addition of fixative at this stage prevents the cells from becoming too swollen. Centrifuge at $400 \times g$ for 5 min.

7. Add 7 mL of ice-cold fixative to the pellet as described in step 7. Incubate on ice for 30 min and then centrifuge at $400 \times g$ for 5 min.

8. Repeat step 7 but incubate on ice for 20 min. After centrifugation, resuspend pellet in 5–7 mL of ice-cold fixative.

9. Using a pipette, drop 20–40 µL of cell suspension onto a moist slide (see Note 9). Air-dry and use a phase contrast microscope to locate metaphases on the slide. Store the slide at room temperature for 2–5 days before hybridization.

10. Store the fixed cellular suspension at –20°C indefinitely (see Note 10).

3.1.3. B. glabrata Genomic DNA Preparation

FISH probes contain repetitive DNA that anneals to homologous regions found throughout the genome. In mammalian FISH studies, a fraction of species specific C_0t-1 DNA is used to suppress the repetitive elements that would then create signal all over the chromosomal preparation (8). We and others have not made a C_0t-1 DNA fraction for suppression but have used an excess of genomic sequences that contain within them the repetitive sequences. By performing a short reannealing step after denaturation of the probe mixture, including the genomic DNA, the repetitive sequences are suppressed. The genomic DNA used in this method for *Biomphalaria* is prepared from the Bge cells. The cells are cultured in Bge medium as described in Subheading 3.1.1. Two T75 flasks will produce enough DNA for a FISH experiment (40 µg).

1. Harvest 2× T75 flasks of Bge cells grown until 90% confluent. Remove the cells from substratum with a cell scraper and centrifuge at $400 \times g$ for 10 min.

2. Resuspend the pellet in 400 µL of cell digestion buffer and 40 µL of Proteinase K. Incubate this suspension at 55°C for 2 h.

3. Vortex vigorously the cellular suspension and spin in a cold microcentrifuge at $8,000 \times g$ for 5 min.

4. Precipitate the supernatant with 10% (v/v) 3 M sodium acetate and 2–2.5× (v/v) of ice cold ethanol. Incubate at –80°C for at least 30 min.

5. Spin again in a cold microcentrifuge for 30 min at 8,000×g; remove supernatant and wash pellet with 70% ethanol (v/v) equal to the amount of 100% ethanol added in step 4.

6. Spin in a cold microcentrifuge for 20 min at 8,000×g. Remove supernatant and air-dry the pellet.

7. Resuspend the pellet in 200 µL of ddH$_2$O and measure the concentration of DNA using a spectrophotometer (see Note 11).

8. Store at –20°C until ready to use.

3.1.4. Probe Labeling and Preparation

1. Label 1 µg of BAC DNA using the Nick translation kit and measure the size of the nick translated product on a 1% agarose gel. (see Note 12).

2. Precipitate together in a 1.5 mL tube for one slide: 500 ng of labeled BAC DNA, 40 µg of *B. glabrata* genomic DNA, and 3 µg of herring sperm DNA. Precipitate as described in Subheading 3.1.3.

3. Air-dry and resuspend the pellet in 12 µL of Hybridization mixture. Allow the probe to dissolve overnight at room temperature.

3.2. Hybridization

3.2.1. Slide Denaturation

The slides prepared in Subheading 3.1.2 should be aged at room temperature for 2–5 days before beginning denaturation. This section should be performed concomitantly with Subheading 3.1.4 so that both slide and probe are ready for hybridization at the same time.

1. Set two water baths at 70 and 37°C. Use the 70°C water bath to heat a Coplin jar full of 70% formamide solution. Additionally, set oven to 70°C and hot plate to 37°C.

2. The slides are dehydrated through a series of coplin jars filled with ethanol solutions of 70, 90, and 100% (5 min each). The slides are dried on the hot plate and then transferred to the 70°C water bath for 5 min. During these 5 min incubation periods, prepare a solution of ice cold 70% (v/v) ethanol.

3. Next, denature the slide in the 70% formamide solution for 1.5 min (see Note 13).

4. Transfer the slide to the solution of ice-cold ethanol for 5 min, before an additional 90 and 100% ethanol dehydration cycle.

5. Dry the slide on hot plate.

3.2.2. Probe Denaturation

1. Denature probe at 75°C for 5 min and then incubate at 37°C for 30 min (reannealing step for repetitive sequence suppression).

2. Add 10 µL of probe onto the denatured sample on the slide, cover with a 22×50 mm coverslip, and seal with rubber cement.

3. Hybridize denatured slide and probe in a humidified chamber at 37°C for 12–16 h.

3.3. Washing and Counterstaining

1. Carefully remove the rubber cement from the slide with forceps (see Note 14).

3.3.1. Washing

2. Wash slide three times for 5 min in Buffer A, prewarmed to 45°C.

3. Repeat procedure but with Buffer B instead.

4. After wash, immerse slide in 4× SSC at room temperature.

3.3.2. Blocking

1. Add 100 µL of blocking solution to slide. Place a coverslip over slide and incubate at room temperature for 10 min.

3.3.3. Immunodetection and Washing

1. Remove coverslip (see Note 15) and add 100 µL of streptavidin conjugated to cyanine 3 to the slide. Incubate in a humidified chamber at 37°C for 30 min in the dark.

2. Wash the slides thrice for 5 min in 4× SSC with 0.1% Tween 20 (v/v) at 42°C in the dark.

3. Complete the wash with a brief rinse in ddH_2O (see Note 16).

4. Mount in Vectashield containing DAPI to counterstain the chromosomes. The slides are now ready to be visualized with a fluorescence microscope.

4. Notes

1. There is no antibiotic substitute for gentamicin.

2. The Bge cells although easy enough to culture are very sensitive to changes in serum and will die if serum from other companies is substituted. Thus, you must use Hyclone-characterized serum.

3. The Bge cells do not require high concentrations of CO_2 as mammalian cells do and so can be grown without extra CO_2. We use flasks to grow our cells that have their lids tightly closed.

4. Whole genomic DNA can be used to substitute for the suppression abilities of C_0t-1 DNA since it contains the repetitive DNA that the C_0t-1 DNA fraction does. The reannealing step prior to adding the probe to the sample, immediately following denaturation of the probe allows the repetitive

fraction of the genomic DNA to anneal to the repetitive sequences in the probe and thus suppress them within the FISH experiment. We normally use 40× excess genomic DNA.

5. We constructed humid chambers out of sandwich boxes that are covered in foil to exclude light. The floor of the box has three to four layers of tissue that is moistened during a hybridization reaction. We cut plastic pipettes to size and create a platform for the hybridizing slides to rest upon. The chambers will float in a water bath or can be placed in an oven.

6. The cling film on top of the beaker prevents evaporation of the medium when it is being mixed in the open laboratory.

7. Prepare a bottle of 500 mL of water, which is equivalent in dimension to that of the FBS bottle. Place both bottles in a water bath set to 56°C and allow the temperature in both bottles to reach 56°C. The water in the bath must be of a sufficient volume to completely immerse the bottles. This can be monitored by periodically (every 5 min) placing a thermometer in the bottle filled with water. Once this bottle has reached 56°C, begin timing for 30 min. During the 30 min, routinely invert the FBS bottle (every 5 min) to prevent the serum proteins congealing at the bottom of the bottle. After 30 min, place the FBS on ice and wait for it to cool down to room temperature. Prepare sterile 50 mL aliquots of the FBS. Store at −20°C.

8. The cells do not adhere that well to the substratum and so it is important to treat the flasks carefully.

9. To produce evenly spread metaphases, it is best to drop the cellular suspension onto the top of a moist slide that is at an angle of 45°; then allow the cellular suspension to run gently down the slide.

10. The fixative should be replaced monthly.

11. If the amount of genomic DNA is less than 40 μg, then repeat the preparation. The volume of reagents such as the cellular digestion buffer and Proteinase K can be adjusted for more than two flasks.

12. For the optimal FISH results, with clear signals and a minimum of background, the probe length needs to be between 200 and 500 bps. This is achieved by the DNAse I activity within the nick translation reaction.

13. The optimal time for the denaturation of the Bge cell chromosomes is 1.5 min. This has been carefully calculated from a range of denaturation times and temperatures to maximize chromosome structural morphology and efficiency of the probe hybridization.

14. If the rubber cement cannot be removed initially from the slide with forceps, then place the slide in the Buffer A for 5 min; this will loosen the rubber cement.

15. Allow the coverslip to slip off the slide. There will be excess liquid on the slide and so the coverslip should not be stuck. The coverslip can be removed using gravity.

16. Run water very gently over the slide. This will remove salts present from the other buffers and prevent crystals forming.

Acknowledgement

This study was funded by a grant from NIH-NIAID, RO1-AI63480.

References

1. Colomba M.S., Vitturi R., Castriota, B.L., and Libertini A. (2002). FISH mapping of 18S-28S and 5S ribosomal DNA, $(GATA)_n$ and $(TTAGGG)_n$ telomeric repeats in the periwinkle *Melarhaphe neritoides* (Prosobranchia, Gastropoda, Caenogastropoda). *Heredity.* **88**, 381–384.

2. Vitturi R., Libertini A., Sineo L., Sparacio I., Lannino A., Gregorini A., and Colomba M. (2005). Cytogenetics of the land snails *Cantareus aspersus* and *C. Mazzulli* (Mollusca: Gastropoda: pulmonata). *Micron.* **36**, 351–357.

3. Knight M., Bridger J.M., Ittiprasert W., Odoemelam E.C., Masabanda J., Miller A., and Raghavan N. (2008). Endogenous retrotransposon sequences of the *Schistosoma mansoni* intermediate snail host, *Biomphalaria glabrata.* In: Brindley, P.J. (Ed.), Mobile Genetic Elements in Metazoan Parasites. Austin, Landes Bioscience. Available from: <http://550 www.eurekah.com/chapter/3457>.

4. Raghavan N., Knight M. (2006). The snail (*Biomphalaria glabrata*) genome project. *Trends Parasitol.* **22**, 148–151.

5. Hansen, E.L. (1976). A cell line from embryos of *Biomphalaria glabrata* (Pulmonata): Establishment and characteristics. In: Maramorosch, K. (Ed.), Invertebrate 532 Tissue Culture: Research Applications. Academic Press, New York, pp. 75–97.

6. Odoemelam E., Raghavan N., Miller A., Bridger J.M., and Knight M. (2009). Revised karyotyping and gene mapping of the *Biomphalaria glabrata* embryonic (Bge) cell line. *Int J Parasitol.* **39**, 675–681.

7. Raghavan N., Tettelin H., Miller A., Hostetler J., Tallon L., and Knight M. (2007). Nimbus (BgI): an active non-LTR retrotransposon of the Schistosoma mansoni snail host Biomphalaria glabrata. *Int J Parasitol.* **37**, 1307–1318.

8. Lichter, P., Cremer, T., Tang, C.J. Watkins, P.C., Manuelidis, L., and Ward, D.C. (1988). Rapid detection of human chromosome 21 aberrations by in situ hybridization. *Proc Natl Acad Sci USA.* **85**, 9664–9668.

Chapter 30

Fluorescence *in situ* Hybridization with Bacterial Artificial Chromosomes (BACs) to Mitotic Heterochromatin of *Drosophila*

Maria Carmela Accardo and Patrizio Dimitri

Abstract

The organization of eukaryotic chromosomes into euchromatin and heterochromatin represents an enigmatic aspect of genome evolution. Constitutive heterochromatin is a basic, yet still poorly understood component of eukaryotic genomes and its molecular characterization by means of standard genomic approaches is intrinsically difficult. *Drosophila melanogaster* polytene chromosomes do not seem to be particularly useful to map heterochromatin sequences because the typical features of heterochromatin, organized as it is into a chromocenter, limit cytogenetic analysis. In contrast, constitutive heterochromatin has been well-defined at the cytological level in mitotic chromosomes of neuroblasts and has been subdivided into several bands with differential staining properties. Fluorescence in situ hybridization (FISH) using Bacterial Artificial Chromosomes (BAC) probes that carry large genomic portions defined by sequence annotation has yielded a "revolution" in the field of cytogenetics because it has allowed the mapping of multiple genes at once, thus rendering constitutive heterochromatin amenable to easy and fast cytogenetics analyses. Indeed, BAC-based FISH approaches on *Drosophila* mitotic chromosomes have made it possible to correlate genomic sequences to their cytogenetic location, aiming to build an integrated map of the pericentric heterochromatin. This chapter presents our standard protocols for BAC-based FISH, aimed at mapping large chromosomal regions of mitotic heterochromatin in *Drosophila melanogaster*.

Key words: Bacterial artificial chromosomes, *Drosophila melanogaster*, Fluorescence in situ hybridization, Mitotic heterochromatin

1. Introduction

The introduction of the fluorescence in situ hybridization (FISH) in the 1980s (1) allowed the drawbacks of isotopic hybridization to be overcome by giving significant advances in resolution, speed, and safety. The methodological improvements introduced over the years have made FISH accessible to all and applicable to

Joanna M. Bridger and Emanuela V. Volpi (eds.), *Fluorescence in situ Hybridization (FISH): Protocols and Applications*, Methods in Molecular Biology, vol. 659,
DOI 10.1007/978-1-60761-789-1_30, © Springer Science+Business Media, LLC 2010

many investigations on nucleic acids (i.e., gene mapping, comparative genomic hybridization, characterization of chromosome structural abnormalities, aneuploidy detection, microdeletion syndromes, cancer cytogenetics).

Bacterial artificial chromosome (BAC) technology was developed in early 1990s (2) and immediately introduced technical advantages as compared to plasmid, cosmid, and Yeast Artificial Chromosomes (YAC) cloning technologies. BACs can hold much larger pieces of DNA than a plasmid or a cosmid can (with a range from 100 to 350 kb of DNA) and are more stable (chimerism is substantially if not completely absent) than YAC and allow cells that harbor them to grow much faster. In cytogenetics, for example, Bacterial Artificial Chromosomes (BAC)-based FISH can be used to map genomic DNA and groups of genes to specific chromosomes and subchromosomal regions. Further, genomic analyses such as large-scale physical mapping have been made possible which has aided the completion of a number of genome sequencing projects considerably. Altogether, these characteristics have made BACs the most broadly used resource to study larger genes, several genes at once, and to perform more efficient screening either by hybridization or PCR-based screening.

As a direct consequence of the human genome sequence assembly, BACs are now being widely used to screen the genome for genetic abnormalities and their contribution to the functional analysis of genes is becoming very important in modeling genetic disease with transgenic mice. Moreover, BACs have also been used in a variety of organisms to sequence the genome of agriculturally important plants such as corn and rice, and of animals such as the mouse.

In *Drosophila*, the first application of FISH was performed in 1981, when a fluorochrome-labeled RNA was utilized as a probe for identification of specific DNA sequences to salivary gland polytene chromosomes (3). However, polytene chromosomes seem to be not so useful in mapping single-copy genes located in heterochromatin because of its characteristics which limits cytogenetic analysis. In polytene chromosomes, the mitotic heterochromatin appears as a tiny compact mass located in the middle of the chromocenter (α-heterochromatin), and the rest of the chromocenter is probably a mesh-like material corresponding to the region found at the border between euchromatin and heterochromatin (β-heterochromatin) in mitotic chromosomes (4). On the other hand, in neuroblast mitotic chromosomes, heterochromatin has been well-defined at a cytological level and it is divided into 61 heterochromatic bands with differential staining properties (Fig. 1; (5, 6)). Therefore, performing BAC-based FISH analysis on *Drosophila* mitotic chromosomes allows correlation between the genomic sequence with the cytogenetic locations, thus permitting an integrated map of the pericentric

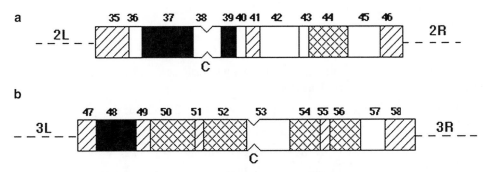

Fig. 1. Cytological map of mitotic heterochromatin Example showing schematic representation of mitotic heterochromatin of chromosomes 2 (**a**) and 3 (**b**) displaying several regions with different degree of fluorescence intensity visible after staining with 4′, 6-diamino-2-phenylindole (DAPI) (6). Each heterochromatin region of the mitotic map of chromosomes 2 and 3 is identified by a number; *filled areas* represent the DAPI-bright regions; the *shaded boxes* represent regions of intermediate fluorescence and the *open boxes* are regions of dull fluorescence. 2L= heterocromatin of the left arm of chromosome 2; 2R= heterocromatin of the right arm of chromosome 2; 3L= heterocromatin of the left arm of chromosome 3; 3R= heterocromatin of the right arm of chromosome 3. C= centromeric region.

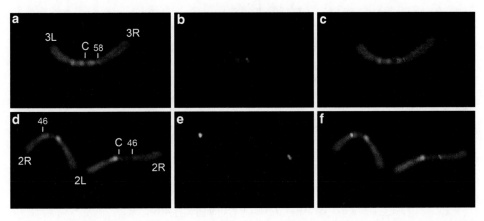

Fig. 2. Examples of FISH mapping of two BAC-DNA clones to the *Drosophila melanogaster* mitotic heterochromatin. BACR22A05 maps to chromosome 3 heterochromatin (**a-c**); BACR11B14 maps to chromosome 2 heterocromatin (**d-f**). For BAC22A05 the prominent signal maps to h58 of 3L heterochromatin (**c**), but there are additional weaker signals due to repetitive sequences present within the BAC. A single prominent signal mapping to region h46 of 2R heterochromatin is seen with BACR11B14 (**f**).

heterochromatin to be achieved ((7–9); Fig. 2). This methodology can support a further characterization of *Drosophila* heterochromatin at molecular level allowing the functional analysis of heterochromatic genes in order to prevent the obsolete classification of this ubiquitous component of eukaryotic genomes as a genomic "wasteland" (10). This chapter presents methods routinely used for mitotic chromosome preparations and FISH with BAC probes to mitotic heterochromatin in *D. melanogaster*.

2. Materials

2.1. Preparation of Mitotic Chromosome Squashes

1. Siliconized glass slides (only as support for droplets of dissection solutions).
2. Siliconized glass coverslips (20 × 20 mm or 22 × 22 mm).
3. Nonsiliconized glass slides.
4. Petri dishes (35 × 10 mm).
5. Microscope for dissection.
6. Dissecting forceps (Dumont #5) or needles.
7. Bibulous paper.
8. Razor blade.
9. Saline (0.7% NaCl in H_2O). Store at 4°C.
10. Hypotonic solution (0.5% sodium citrate in H_2O). Store at 4°C.
11. Fixatives: acetic acid, methanol, H_2O (11:11:2); acetic acid (45%). Make fresh.
12. Absolute ethanol, chilled to –20°C.
13. Liquid nitrogen or a block of dry ice.

2.2. BAC Extraction

1. LB medium with antibiotic (20 μg/mL chloramphenicol).
2. Tris–HCl 1 M (pH 8.5).
3. EDTA 0.5 M.
4. NaOH 2 N.
5. Sodium dodecyl sulphate (SDS) solution (10% w/v).
6. Ammonium acetate (CH_3COONH_4, solution 7.5 M).
7. 3-Methyl-1-butanol (isopropanol).
8. 70% ethanol.
9. RNase A (Roche).

2.3. FISH

1. Glass Coplin jars.
2. Glass coverslips (22 × 22 mm or 24 × 24 mm).
3. Moist chamber.
4. Dry, dust-free box for holding slides.
5. Rubber cement.
6. Formamide (J. T. Baker). Stored at 4°C.
7. Biotin-Nick translation Mix (Boehringer). Stored at –20°C.
8. Digoxygenin-Nick translation Mix (Boehringer). Stored at –20°C.
9. Rhodamine-Nick translation Mix (Boehringer). Stored at –20°C.

10. Fluorescein isothiocyanate (FITC)-conjugated avidin (DCS grade; Vector laboratories) for biotinylated probes. Store at 4°C.

11. Cy3-conjugated avidin (Boehringer) for biotinylated probes. Store at 4°C.

12. Rhodamine-conjugated antidigoxigenin sheep IgG, Fab fragments (Boehringer, Mannheim) for digoxigenin-labeled probes. Store at 4°C.

13. Sonicated salmon sperm DNA.

14. Sodium acetate (3 M, pH 4.5).

15. Ethanol (70, 90% and absolute) at room temperature.

16. Ethanol (70%) chilled at –20°C.

17. Ethanol (90% and absolute) chilled at 4°C or on ice.

18. 20× SSC.

19. Tween-20.

20. 4,6-diamino-2-phenylindole-dihydrocloride (DAPI; 0.2 μg/mL), dissolved in 2× SSC. Store at 4°C.

21. Vectashield H-1000 (Vector laboratories). Store at 4°C.

3. Methods

Making mitotic chromosome squashes is a crucial step for FISH on the constitutive heterochromatin of *D. melanogaster* diploid cells. The quality of chromosome morphology determines the ease of recognizing the heterochromatic landmarks produced by fluorochromes such as DAPI or Hoecst-33258 that are general indicators of AT-rich regions. The following protocols describe the preparation of mitotic chromosomes squashes from the neural ganglia of *D. melanogaster* larvae.

3.1. Preparation of Mitotic Chromosome Squashes from Larval Brains

1. Grow larvae in moderately crowded vials. Select large larvae that are climbing up the sides of the tube (see Note 1). At this stage, determine the sex of the larvae if chromosomes of a particular sex are required for analysis.

2. Collect and wash the larvae in a 35 × 10 mm petri dish with 0.7% saline at room temperature.

3. Transfer three drops of saline (50 μl each) onto a siliconized slide. Place one or two larvae in each drop.

4. Perform dissection in saline using sharp forceps (Dumont #5) or dissecting needles as follows: grasp the mouth hooks with one forcep, then grasp the body of the larva midway down with the other pair of forceps. Gently separate the mouth hooks from the rest of the larval body. The brain frequently

remains attached to the head portion together with imaginal discs (Fig. 2).

5. Remove the brains from the mouth hooks and gently clean off imaginal discs, then collect the brains in a fresh drop of saline.

6. Transfer the brains into a drop (50 μl) of hypotonic solution placed on a siliconized slide and incubate at room temperature for 10 min.

7. Move brains to a 35 × 10 mm petri dish containing a freshly prepared mixture of acetic acid/methanol/H_2O (11:11:2) and leave for approximately 30 s.

8. Transfer a single brain into a small drop (2 μl) of 45% acetic acid placed on a dust-free siliconized coverslip. One to four brains can be placed on the same coverslip. Leave the brains in the 45% acetic acid drops for 1–2 min.

9. Pick up coverslip carrying the brains using a dust-free nonsiliconized slide so that the coverslip will adhere to the slide. Try to avoid the formation of air bubbles.

10. Flip the slide over and gently press out excess acetic acid between two sheets of blotting paper, then squash hard using thumb pressure. When squashing avoid lateral moving of the coverslip.

11. Freeze slides either in liquid nitrogen or on dry ice for 5 min. Using a sharp razor blade, flip off coverslip with a quick motion, and immediately plunge slides in cold (–20°C) absolute ethanol.

12. Let them gradually reach room temperature (it usually takes about 30 min), remove from ethanol, and air-dry. Slides can be stored at 4°C for weeks in a dry, dust-free box.

13. Dry preparations can be checked without a coverslip under a phase-contrast microscope. Chromosomes suitable for FISH experiments should appear flat and gray with no refractivity.

3.2. Extraction of BAC DNA

1. Seed a bacterial culture in 10 mL of LB medium with antibiotic (20 μg/mL cloramphenicol). Transfer 2 mL of overnight culture to a 2 mL microcentrifuge tube and centrifuge for 50 s in a benchtop microcentrifuge at 15,700 × g.

2. Discard the supernatant and thoroughly resuspend the bacterial pellet in 100 μl of 50 mM Tris–HCl pH 8.5, 10 mM EDTA buffer.

3. Add 200 μl of fresh 0.2 M NaOH, 1% SDS solution. Invert the tubes to mix and immediately add 150 μl of 7.5 M ammonium acetate solution. Invert to mix. Centrifuge at 15,700 × g for 15 min in the microcentrifuge.

4. Transfer the supernatant to a 1.5 mL microcentrifuge tube with 300 µl of isopropanol. Invert to mix. Centrifuge at $15,700 \times g$ for 10 min in the microcentrifuge.

5. Remove the supernatant. Wash the pellet and the walls of the tube with 300 µl of 70% ethanol; centrifuge at $15,700 \times g$ for 5 min, remove the supernatant, and air-dry the pellet.

6. Resuspend the pellet in 50 µl of 10 mM Tris–HCl pH 8, 1 mM EDTA buffer containing 50 µg/mL RNase A. Incubate at 65°C for 10 min. Store DNA at 4°C for short-term use. Long-term storage of the minipreps at –20°C is possible but repeated freezing and thawing must be avoided.

3.3. Fluorescence In Situ Hybridization with BACs on Mitotic Chromosomes

The FISH technique coupled with DAPI staining and digital recording of images allows the mapping of BAC DNA. Digital images of FISH signals and DAPI staining can be captured separately by a CCD camera, pseudocolored and merged using specific softwares for image analysis, such as Adobe Photoshop. The subsequent co-visualization of the hybridization signals and DAPI banding pattern on the same chromosomes allow the mapping of a given BAC DNA to specific cytological regions of mitotic heterochromatin (Fig. 3; (7)).

It is important to point out that most of the BACs carrying genomic DNA derived from heterochromatin contain variable amounts of sequences related to transposable elements (TEs) which are known to be accumulated in pericentric regions of chromosomes. A priori, the intensity and number of hybridization signals may be expected to reflect several factors including the

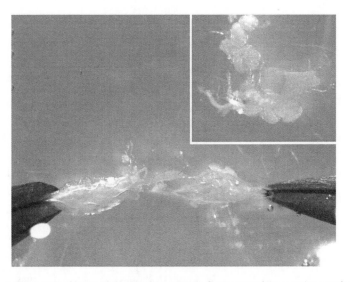

Fig. 3. Dissection of larval brains in saline solution. Separation of the mouth hooks from the rest of the larval body using forceps. The inset shows the brain together with the surrounding imaginal discs, separated from the head portion.

copy number and arrangement of TE-sequences within the genome, as well as their relative concentration within a given BAC. Experimentally, however, we found that most BACs tested produced a prominent fluorescent signal on 2R or 3L, the mapping of which was consistent with the BAC location on polytene chromosome. Only in a few cases, additional hybridization sites were detected in heterochromatin, but they usually show a lower fluorescence intensity compared to that of the master signal. It is also important to stress that the BACs tested gave similar mapping results in both of *y; cn bw sp* and Oregon-R strains of *Drosophila*.

FISH mapping of BACs to mitotic heterochromatin can be used to answer several kinds of questions. For example, it has proven successful in mapping to heterochromatin of *D. melanogaster* a large group genes, predicted by computational analyses (7–11).

3.3.1. Hybridization

1. Dehydrate 2–3 days old slides (prepared according to Subheading 3.1) by immersion in 70, 90, and 100% ethanol (3 min each time). Air-dry slides after denaturation at room temperature.

2. Immerse one to three slides in 50 mL of prewarmed denaturation solution (35 mL ultrapure formamide, 5 mL 20× SSC and 10 mL distillated water). Incubate for 2 min in a water bath at 70°C.

3. Quickly transfer the slides to 70% ethanol at –20°C, incubate for 3 min, and then dehydrate in ice-cooled 90 and 100% ethanol (3 min each time). Let slides air-dry at room temperature.

4. Label 1 μg of DNA probe (BAC DNA) by Nick-translation using biotin-11-dUTP or digoxigenin-1-dUTP. For DNA labeling, we routinely use biotin-nick translation mix or digoxigenin-nick translation mix (Roche) (see Note 2).

5. Remove unincorporated nucleotides by ethanol precipitation (see Note 3) and store the probe at –20°C.

6. Precipitate the labeled DNA (40–80 ng per slide; see Note 4) with the addition of sonicated salmon sperm DNA (3 μg per slide), 0.1 volumes of 3 M sodium acetate, pH 4.5, and 2 volumes of cold (–20°C) absolute ethanol. Place at –80°C for 15 min and spin at $15,700 \times g$ for 15 min. Dry the pellet in a Savant centrifuge (see Note 5).

7. Resuspend DNA in the hybridization mixture (10 μl per slide) by vortexing.

8. Heat the probe solution at 80°C for 8 min. Place tubes on ice for 5 min and centrifuge briefly to bring down any condensation. Leave on ice until required.

9. Put 10 µl of probe solution to denatured slides and cover with 24 × 24 mm dust-free clean coverslip so that no air bubbles are trapped. After spreading of the probe solution, seal the edges of coverslip with rubber cement.

10. Put slides in a moist chamber and incubate overnight at 37°C (see Note 6).

11. Roll off the rubber cement and gently remove the coverslip. If coverslip does not come off, rinse the slide once in the washing solution prewarmed to the temperature used for hybridization and try again (see Note 7).

12. Wash slides three times (5 min each) in 50% formamide, 2× SSC at 42°C.

13. Wash slides three times (5 min each) in 0.1× SSC at 60°C and remove excess liquid from the slide edges (see Note 8).

14. Apply 100 µl of blocking solution to each slide. Cover with 24 × 24 mm coverslip and incubate at 37°C for 30 min.

3.3.2. Detection of Biotin-Labeled DNA

1. Remove the coverslips and blot excess blocking solution from the edges of the slide.

2. Put onto each slide 50–100 mL of 3.3 µg/mL FITC-conjugated avidin (Vectorlaboratories) diluted in 4× SSC, 1% BSA, 0.1% Tween 20; cover with a 24 × 24 mm coverslip and incubate for 30 min at 37°C in a dark moist chamber.

3. Remove coverslip and wash three times (5 min each) in 4× SSC, 0.1% Tween 20, at 42°C. Remove slides from the washing solution and let them air-dry at room temperature (see Note 9).

4. Stain with 0.16 µg/mL 4,6-diamino-2-phenylindole-dihydrocloride (DAPI) dissolved in 2× SSC for 5 min at room temperature.

5. Rinse slides once in 2× SSC at room temperature, remove slides from 2× SSC, and air-dry.

6. Mount slides in 20 mM Tris–HCl, pH 8, 90% glycerol containing 2.3% of DABCO (1,4-diazo-bicyclo-(2,2,2) octane; Merck) antifade (see Note 10).

7. Seal coverslips with rubber cement and store at 4°C. Slides can be stored for weeks.

3.3.3. Detection of Digoxigenin-Labeled DNA

The procedure is identical to that for biotinylated probes described in Subheading 3.3.2, with the exception of step 2, which is modified as follows:

Put onto each slide 50–100 µl of 2 µg/mL rhodamine-conjugated anti-digoxigenin sheep IgG, Fab fragments (Boehringer), diluted in 4× SSC, 1% BSA, 0.1% Tween 20; cover with a 24 × 24 mm coverslip and incubate for 30 min at 37°C in a dark moist chamber.

3.3.4. Detection of Rhodamine-Labeled DNA (see Note 11)

After the posthybridization washes (see Subheading 3.3.1, steps 12 and 13), slides with probes directly labeled with tetramethylrhodamin-6-dUTP or other fluorophores must be treated as follows:

1. Wash slides once for 5 min in 2× SSC at room temperature.

2. Stain slides with DAPI and mount as described in Subheading 3.3.2, steps 4–7.

3.3.5. Double Labeling

1. For simultaneous in situ hybridization, mix the desired amount of biotin- and dig-labeled probes.

2. Probe preparation: As described in Subheading 3.3.1.

3. Hybridization: As described in Subheading 3.3.1.

4. Signal detection: Prepare a mixture of 3 μg/mL FITC-conjugated avidin, 2 μg/mL rhodamine-conjugated anti-Dig sheep IgG, Fab fragments diluted in 4× SSC, 1% BSA, 0.1% Tween 20.

5. Apply 80–100 μl per slide and cover with 22×22 mm or 24×24 mm coverslip and incubate at 37°C in a dark, humid chamber.

6. Wash slides, stain, and mount preparation as described in Subheading 3.3.2.

For toxicity and hazards see Note 12.

4. Notes

1. Making squashes of chromosome preparations, neural ganglia from female are preferred because they frequently have better chromosomes than male larvae.

2. To label FISH probes, direct incorporation of fluorescently conjugated nucleotides can also be performed. Fluorescein-labeled dNTP (green emission) or Cy3-labeled dUTP (red emission) are available from several suppliers. We routinely prepare TE probes labeled with tetramethylrhodamin-6 dUTP (red emission) using the rhodamin-nick translation mix from Roche.

3. Labeled DNA may be also recovered by centrifugation with Microcon centrifugal filter device (Millipore) according to the standard protocol.

4. Use 100–150 ng per slide for each BAC.

5. Alternatively, transfer the desired amount of labeled DNA in an eppendorf tube, add sonicated salmon sperm DNA (3 μg per slide), and dry in a savant centrifuge speed vacuum concentrator (Savant instruments Inc.).

6. In hybridization experiments aimed to test whether or not a given sequence is present within heterochromatin of mitotic chromosomes, it may be helpful to use a positive control for probe labeling. One possibility is to check the probe on polytene chromosome preparations; if the probe is labeled successfully, multiple euchromatic signals corresponding to the euchromatic copies of the element will be revealed.

7. Temperature used for middle repetitive probes is 37°C. For higher stringency, hybridization can be performed overnight at 42°C.

8. Do not let slides dry from step 1 of Subheading 3.3.5 to step 6.

9. Low-stringency washes can be performed in 2× SSC or 4× SSC at 35°C.

10. Commercial antifade such as Vectashield H-1000 (Vector laboratories) may also be used.

11. The use of DNA probes directly labeled with tetramethylrhodamin-6-dUTP or with other fluorescently conjugated nucleotides avoids the blocking and detection steps and thus is particularly useful because it reduces the background and shortens the procedure. The hybridization signal intensity obtained with tetramethylrhodamin-6 dUTP labeled probes is comparable to that observed using biotin-labeled or digoxigenin-labeled probes coupled with secondary detection.

12. Absolute acetic acid: may be harmful by inhalation, ingestion, or skin absorption. Wear gloves and use in a chemical fume hood.

    Ethanol: may be harmful by inhalation, ingestion, or skin absorption. Wear gloves and use under a chemical fume hood.

    Methanol: may be harmful by inhalation, ingestion, or skin absorption. Ventilation is necessary to limit exposure to vapors. Wear gloves and use under a chemical fume hood.

    Dry ice: handle with caution, can cause frostbite. Wear appropriate gloves.

    Liquid nitrogen: handle frozen samples with caution. Do not breath the vapors. Wear cryo-gloves and a face mask.

    3-Methyl-1-butanol: Flammable. Harmful by inhalation. Irritating to respiratory system. Repeated exposure may cause skin dryness or cracking. Target organ(s): Central nervous system. Causes eyes and skin irritation. Wear appropriate gloves and safety glasses and always use in a chemical fume hood. Keep solutions covered.

    Formamide: produce teratogenic effects. It may be harmful by inhalation, ingestion, or skin absorption. The vapors are irritating to the eyes, skin, mucous membranes, and upper respiratory tract. Wear appropriate gloves and safety glasses and always use in a chemical fume hood. Keep solutions covered.

DAPI: it may have carcinogenic effects. Harmful by inhalation, ingestion, or skin absorption. Wear appropriate gloves.

All the following reagents may be harmful by inhalation or ingestion and it must avoid contact with eyes, skin, or clothing. Wear appropriate gloves: SDS, Tris–HCl, Tween 20, FITC, Cy3-conjugated avidin, rhodamine-conjugated anti-digoxigenin sheep IgG, DABCO, Vectashield H-1000, Rubber cement.

Acknowledgments

M. C. Accardo and P. Dimitri are supported by grants from National Institute of health (NIH) and Istituto Pasteur-Fondazione Cenci-Bolognetti. Work published despite of dramatic reduction in financial support to public research by the Italian Government.

References

1. Bauman JG., Wiegant J., Borst P., Van Duijn P. (1980) A new method for fluorescence microscopical localization of specific DNA sequences by in situ hybridization of fluorochrome labelled RNA. *Exp. Cell Res.* **128**, 485–490.

2. Shizuya H., Birren B., Kim UJ., Mancino V., Slepak T., Tachiiri Y., Simon M. (1992) Cloning and stable maintenance of 300-kilobase-pair fragments of human DNA in Escherichia coli using an F-factor-based vector. *Proc. Natl. Acad. Sci. USA.* **89**, 8794–8797.

3. Bauman JG., Wiegant J., Van Duijn P., Lubsen NH., Sondermeijer PJ., Hennig W, Kubli E. (1981) Rapid and high resolution detection of in situ hybridisation to polytene chromosomes using fluorochrome-labeled RNA. *Chromosoma.* **84**, 1–18.

4. Miklos GL., Cotsell JN. (1990) Chromosome structure at interfaces between major chromatin types: alpha- and beta- heterochromatin. *Bioessays.* **12**, 1–6.

5. Gatti M., Pimpinelli S. (1992) Functional elements in *Drosophila melanogaster* heterochromatin. *Annu. Rev. Genet.* **26**, 239–275.

6. Gatti M, Bonaccorsi S, Pimpinelli S. (1994) Looking at *Drosophila* mitotic chromosomes. *Methods Cell Biol.* **44**, 371–391.

7. Corradini N., Rossi F., Vernì F., Dimitri P. (2003) FISH analysis of *Drosophila melanogaster* heterochromatin using BACs and P elements. *Chromosoma.* **112**, 26–37.

8. Rossi F., Moschetti R., Caizzi R., Corradini N., Dimitri P. (2007) Cytogenetic and molecular characterization of heterochromatin gene models in *Drosophila melanogaster. Genetics.* **175**, 595–607.

9. Hoskins RA., Carlson JW., Kennedy C., Acevedo D., Evans-Holm M., Frise E., Wan KH., Park S., Mendez-Lago M., Rossi F., Villasante A., Dimitri P., Karpen GH., Celniker SE. (2007). Sequence finishing and mapping of *Drosophila melanogaster* heterochromatin. *Science.* **316**, 1625–1628.

10. Dimitri P., Caizzi R., Giordano E., Accardo MC., Lattanzi G., Biamonti G. (2009) Constitutive heterochromatin: a surprising variety of expressed sequences. *Chromosoma.* **118**, 419–435.

11. Yasuhara JC., Marchetti M., Fanti L., Pimpinelli S., Wakimoto BT. (2003) A strategy for mapping the heterochromatin of chromosome 2 of *Drosophila melanogaster. Genetica.* **117**, 217–226.

Chapter 31

Three-Dimensional Fluorescence In Situ Hybridization in Mouse Embryos Using Repetitive Probe Sequences

Walid E. Maalouf, Tiphaine Aguirre-Lavin, Laetitia Herzog, Isabelle Bataillon, Pascale Debey, and Nathalie Beaujean

Abstract

A common problem in research laboratories that study the mammalian embryo is the limited supply of live material. For this reason, new methods are constantly being developed and existing methods for in vitro models using cells in culture are being adapted to represent embryogenesis. Three-dimensional fluorescence in situ hybridization (3D-FISH) is an important tool to study where genomic sequences are positioned within nuclei without interfering with this 3D organization. When used in the embryo, this technique provides vital information about the distribution of specific sequences in relation to embryonic nuclear substructures such as nucleolar precursor bodies and chromocenters. In this chapter, we will present a detailed description of FISH in order to perform 3D-FISH in the early preimplantation murine embryos.

Key words: 3-Dimensional, FISH, In situ, Embryo, Mouse, Development

1. Introduction

Fluorescence in situ hybridization (FISH) is a cytogenetic technique that provides researchers with a direct way to visualize and map the genetic material in individual cells, including specific genes or subchromosomal regions and whole chromosomes (1). This technique has become critical in preimplantation genetic diagnosis (PGD). Often families with known or suspected genetic diseases want to know more about their unborn child's conditions before proceeding with full-term pregnancy. These concerns can be addressed by FISH to diagnose a number of diseases such as Prader-Willi syndrome, Angelman syndrome, 22q13 deletion syndrome, chronic myelogenous leukemia, acute lymphoblastic leukemia, Cri-du-chat, Velocardiofacial syndrome, and Down

Joanna M. Bridger and Emanuela V. Volpi (eds.), *Fluorescence in situ Hybridization (FISH):*
Protocols and Applications, Methods in Molecular Biology, vol. 659,
DOI 10.1007/978-1-60761-789-1_31, © Springer Science+Business Media, LLC 2010

syndrome (2). This technique is also important for basic research with many researchers studying the nonrandom spatial distribution of chromosomes and genes within the nuclear space. The organization of the genome inside the interphase nucleus appears to be a central determinant of genome function and that just knowing the sequence of a genome is not sufficient to understand its physiological function (3). Using interphase nuclei with preserved 3D shape, one can indeed determine the three-dimensional arrangements of chromosomes, providing detailed information about the compartmentalization of chromosomes in discrete territories (4). It is also possible to locate gene-rich versus gene-poor regions and evaluate the dynamic interactions of specific genomic sequences with other nuclear components (5). For instance, we recently demonstrated using the 3D-FISH that the organization of constitutive heterochromatin in the early mouse embryo is correlated with their development to term (6).

According to the experimental design, probes used for FISH will bind only to parts of the chromosomes or to specific genomic sequences with which they show a high degree of sequence homology. In this chapter, continuous repetitions of short monomer probes will be discussed to study two repetitive sequences from the heterochromatic region of the nucleus, centromeric, and pericentromeric repeats, also known as minor and major satellites. The probes will be directly tagged with different fluorochromes for simultaneous detection. Thereby, using multiple probes labeled with different fluorochromes, scientists are able to label each chromosome in its own unique color (the result is known as a spectral karyotype) but are also able to delineate several genomic sequences at the same time, within the same nucleus (7). In all these scenarios, fluorescence microscopy will be used to locate the FISH signal in the sample.

Unlike cells in culture, mouse embryos are surrounded by glycoprotein rich membrane known as the zona pellucida (8) and therefore, they do not attach to the substratum when cultured in vitro. Moreover, the nuclei of early preimplantation embryos are much larger than those of differentiated cells which makes it more difficult to preserve the three-dimensional structure during the procedure. In this chapter, we will describe multicolor FISH with centromeric and pericentromeric probes in mouse embryos, using both direct and indirect detection.

2. Materials

2.1. Embryo Preparation

1. M2 culture medium embryo-tested (Sigma-Aldrich). Store at 4°C.

2. Phosphate Buffered Saline (PBS) tablets (Sigma-Aldrich) dissolved in distilled water according to manufacturer's instructions, autoclaved, and then stored at 4°C.

3. Glass dishes (Electron Microscopy Sciences, Euromedex).

4. Thin glass manipulation pipettes for embryos, usually home-made with glass capillaries or Pasteur pipettes.

5. Tyrode's solution (Sigma-Aldrich) stored in single use aliquots of 1 mL at –20°C.

6. Fixative: 20% paraformaldehyde (PFA) solution, EM grade (Electron Microscopy Sciences, Euromedex) with adjusted pH to 7.4, and stored at 4°C with light protection (e.g., covered with aluminium foil). Working solutions of 4% are prepared by dilution with PBS just before use.

7. SuperFrost Plus slides (VWR).

8. Hydrophobic barrier pen (Sigma-Aldrich).

2.2. Fish

1. Phosphate Buffered Saline (PBS) tablets dissolved in distilled water according to manufacturer's instructions and autoclaved. Store at 4°C.

2. Permeabilization stock solution: 10% (w/v in distilled water) Triton X-100 solution. Store at 4°C. The 0.5% working solution is prepared by dilution with PBS just before use.

3. 20× saline sodium citrate (SSC) buffer (Sigma-Aldrich) containing 0.3 M sodium citrate and 3 M NaCl at pH 7.0. The working solutions of 0.1× and 2× SSC at a pH 6.3 are prepared by diluting the 20× stock in PBS and lowering the pH with the appropriate volume of HCl.

4. Cy3-labeled probe: StarFISH pan-centromeric probe Cy3 conjugated (~200nt; Cambio).

5. Bio-labeled probe: consensus 80-nt forward sequence for murine minor satellites biotinylated at its 5' end (Invitrogen) (9).

6. RNase A stored at 1 M in single use aliquots of 100 µL at –20°C.

7. Heating block set at 85°C (Grant, VWR).

8. Deionized formamide (Molecular Biology Grade).

9. Hybridization buffer: 50% formamide, 10% dextran sulphate (from 50% w/v stock solution), 1× Denhardt, 40 mM NaH_2PO_4 in 2× SSC, pH 7 (all products from Sigma-Aldrich). Store in single use aliquots of 1mL at –20°C.

10. Hybridization incubator with humidified loading tray for slides, set at 37°C (Grant Boekel, VWR).

11. Humidified nontransparent box in which you can keep the slides during the different incubation steps.

12. Detection compound for the unlabeled probe: Streptavidin conjugated with Alexa 633 (Molecular Probes, Invitrogen).

13. Nuclear counterstain: 10 µM Yo-Pro-1 (Molecular Probes, Invitrogen).

14. Mounting medium: Citifluor (Biovalley).

15. Glass coverslips 22 × 40 mm (CML).

3. Methods

Preimplantation embryos can be collected in vivo just before 3D-FISH or cultured according to standard protocols. However, embryos are known to be very sensitive to culture conditions and this can affect their development (e.g., cleavage rates, fragmentation, and quality of postimplantation). It is entirely possible that this may impact on chromosome/gene positioning. We therefore advise to prepare all materials and equipments in advance and perform mounting of the embryos onto slides as quickly as possible.

3.1. Mounting Embryos onto Slides

1. Label the slide with name and date, and draw one, maximum two, squares (1 cm × 1 cm) on your slide with the hydrophobic pen (see Note 1). Each square should be allocated to one group of embryos (with 10–20 embryos/group) and therefore, several slides might be required.

2. Prepare culture plates with microdrops (25–50 µL) of M2 solution to rinse the embryos, and keep at room temperature in open air to equilibrate.

3. Defrost on the bench one aliquot of Tyrode's solution per group of embryos (see Note 2).

4. The first step is to denude the embryo by removing the zona pellucida in order to improve access of the probes within the embryos (see Note 3). Take the first group of embryos to be processed and transfer it from the collection/culture medium into the M2 medium. Load the manipulation pipette with fresh Tyrode's solution, and then transfer embryos into a glass dish containing 500 µL of Tyrode's solution. Immediately reload the manipulation pipette with fresh Tyrode and then transfer embryos into a second glass dish containing another 500 µL of Tyrode's solution. Incubate the embryos for no more than 1.5 min (see Notes 4 and 5).

5. Quickly rinse the embryos in multiple drops of the preequilibrated M2 and place them inside the drawn square on the glass slide.

6. Aspirate slowly excess M2 medium in order to fix the embryos onto the glass slide (see Note 6).

7. As soon as the embryos are fixed and M2 removed, add 10–15 µL of 4% PFA.

8. Keep slide(s) in a humidified box at 4°C, overnight (see Note 7).

3.2. Hybridization with the Probes

1. On day 2, gently aspirate the PFA and replace with PBS for 30 min to rinse the embryos.

2. Permeabilize the embryos for 30 min with 0.5% Triton X-100 (diluted in PBS just before use).

3. Rinse embryos in PBS and then in 2× SSC for 5 min each time.

4. To decrease nonspecific background, RNAs are digested by incubation in 200 µM of RNAse diluted in 2× SSC (pH 6.3) for 30 min at 37°C.

5. Rinse embryos in 2× SSC (pH 6.3) and then in PBS for 5 min each time.

6. Equilibrate in hybridization buffer for 1–2 h (see Note 8).

7. In a 0.5 mL tube, mix 1 µL of each of the probe solutions with 12 µL of hybridization buffer and denature at 85°C for 10 min, and transfer the tube into ice straight after.

8. At the same time, put your glass slides with the embryos (still in the hybridization buffer) on the heating block to denature the DNA at 85°C for 10 min (see Note 9).

9. Replace the buffer on the embryos with the hybridization mix and place the slides in the humidified loading tray of the hybridization incubator at 37°C for 24 h (see Note 10).

10. On day 3, several washing steps are required to remove the remaining unbound probe and decrease the nonspecific background. First, rinse embryos in prewarmed (37°C) 2× SSC for 3×5 min at 37°C. Then in prewarmed (60°C) 0.1× SSC for 3×5 min at 60°C.

11. Further rinse the embryos at RT in 0.1× SSC then in PBS for 5 min each time.

12. The DNA is counterstained with 10 µM of Yo-Pro-1 in PBS for 15 min (see Note 11).

13. Rinse the embryos for 2×5 min and postfix with 2% PFA for another 5 min (postfixation helps maintain the fluorescent signal until observation).

14. After removal of the PFA, slowly add one drop of mounting medium, and gently place the coverslip onto the slide.

15. Seal the borders of the coverslip with nail varnish and keep slide at 4°C in the dark until observation with a microscope (see Note 12).

16. An example of the results produced is shown in Fig. 1 (see Note 13).

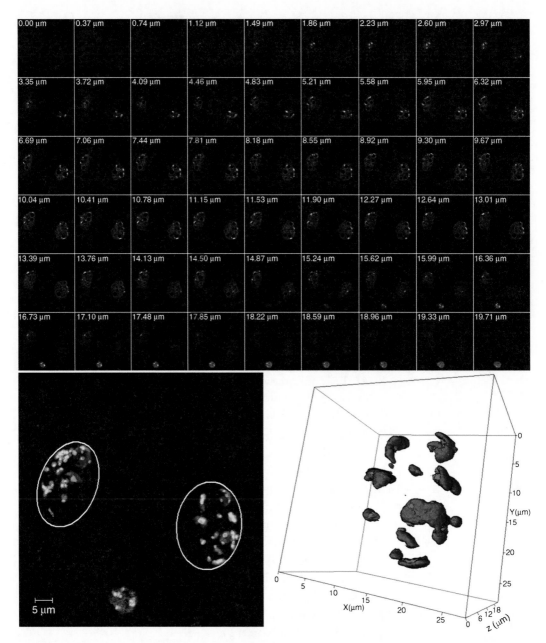

Fig. 1. 3D-FISH on a two-cell stage mouse embryo with the pan-centromeric and centromeric probes counterstained with Yo-Pro-1. The z-series (*upper panel*, z-step of 0.37 μm) was acquired on a Zeiss LSM510 confocal with a plan-apochromat 63×/1.4 oil DIC objective at 488, 543, and 633 nm wavelenghts sequentially. *Blue* color was assigned to DNA, *red* to the pan-centromeric signal, and *green* to the centromeric one to allow better signal vs. background contrast. With the LSM browser software, the z-series can be merged to give a single image with the two probes (*lower panel, left*) on which each nucleus is underlined by a *white circle*. 3D reconstruction of one nucleus from this embryo (*lower panel, right*) clearly shows the conserved shape of the nucleus and the so-called "chromocenters" formed by aggregations of major (pan-centromeric probe) and minor (centromeric probe) satellites.

4. Notes

1. Due to the nature of the solutions that are used in this method and that can easily dilute/erase ink (even permanent ones), a diamond pencil is usually used to label the slides.

2. Do not use Tyrode's solution more than 15 min after defrosting as it loses its activity and becomes ineffective on embryos.

3. When dealing with several groups, we advise to carry out the denudation process for one group at a time as it is a very critical step in this experiment. An extended incubation with the acid will damage the embryos and their nuclei, while a short incubation time might not be enough to remove the zona pellucida completely.

4. This step and all subsequent ones are carried out at room temperature (RT) unless otherwise stated.

5. Under the dissection microscope, you can observe the zona pellucida being degraded by the acid activity of the Tyrode's solution.

6. Even though this step is carried out at room temperature, the light from the microscope produces enough heat to dehydrate the embryos. This must be avoided to preserve the 3D structure of the nuclei. Rehydration can be accomplished by adding M2 medium if the embryos are not yet fixed on the slide.

7. Alternatively, incubate in 4% PFA for 30 min at room temperature and proceed to the next step.

8. It is recommended to switch-on the heating block when equilibration starts in order to reach the desired temperature for the subsequent steps.

9. To prevent the microdrops from evaporating during that step, filter paper humidified with formamide 50% should be placed in the bottom of the tray.

10. All subsequent steps are carried out in a nontransparent humidified chamber.

11. Other nuclear counterstains, such as propidium iodide, can be used according to the choice of fluorochromes used to label the probes.

12. Confocal or grid microscopy is necessary for three-dimensional imaging. The optimal parameters to obtain the best 3D reconstruction is to perform Z-sectioning at no more than 0.5 µm between sections and to take 2–3 additional sections on top and below the limits of the nucleus.

13. If the fluorescent signal is too weak, amplification of the signal may be necessary. Secondary components such as fluorescently tagged secondary antibodies can be used to provide a more pronounced signal.

Acknowledgments

We thank Pierre Adenot for technical assistance on the confocal microscope platform MIMA2 (Microscopie et Imagerie des Microorganismes, Animaux et Elements) and UEAR for animal care. We are also grateful to Eve Devinoy (INRA) and Claire Francastel (Institut Cochin) for their help. This work was supported by INRA grant "Crédits incitatifs PHASE".

References

1. Halling, K. C., Kipp, B. R. (2007) Fluorescence in situ hybridization in diagnostic cytology. *Hum Pathol* **38**, 1137–1144.

2. Sreekantaiah, C. (2007) FISH panels for hematologic malignancies. *Cytogenet Genome Res* **118**, 284–296.

3. Foster, H. A., Bridger, J. M. (2005) The genome and the nucleus: a marriage made by evolution. Genome organisation and nuclear architecture. *Chromosoma* **114**, 212–229.

4. Walter, J., Joffe, B., Bolzer, A., Albiez, H., Benedetti, P. A., Muller, S., Speicher, M. R., Cremer, T., Cremer, M., Solovei, I. 2006 Towards many colors in FISH on 3D-preserved interphase nuclei. *Cytogenet Genome Res* **114**, 367–378.

5. Kupper, K., Kolbl, A., Biener, D., Dittrich, S., von Hase, J., Thormeyer, T., Fiegler, H., Carter, N. P., Speicher, M. R., Cremer, T., Cremer, M. (2007) Radial chromatin positioning is shaped by local gene density, not by gene expression. *Chromosoma* **116**, 285–306.

6. Maalouf, W. E., Liu, Z., Brochard, V., Renard, J. P., Debey, P., Beaujean, N., Zink, D. (2009) Trichostatin A treatment of cloned mouse embryos improves constitutive heterochromatin remodeling as well as developmental potential to term. *BMC Dev Biol* **9**, 11.

7. Werner, M., Wilkens, L., Aubele, M., Nolte, M., Zitzelsberger, H., Komminoth, P. (1997) Interphase cytogenetics in pathology: principles, methods, and applications of fluorescence in situ hybridization (FISH). *Histochem Cell Biol* **108**, 381–390.

8. Wassarman, P. M. (2005) Contribution of mouse egg zona pellucida glycoproteins to gamete recognition during fertilization. *J Cell Physiol* **204**, 388–391.

9. Bouzinba-Segard, H., Guais, A., Francastel, C. (2006) Accumulation of small murine minor satellite transcripts leads to impaired centromeric architecture and function. *Proc Natl Acad Sci U S A* **103**, 8709–8714.

Chapter 32

Fluorescence *in situ* Hybridization (FISH) for Genomic Investigations in Rat

Andrew Jefferson and Emanuela V. Volpi

Abstract

This chapter concentrates on the use of fluorescence in situ hybridization (FISH) for genomic investigations in the laboratory rat (*Rattus norvegicus*). The selection of protocols included in the chapter has been inspired by a comprehensive range of previously published molecular cytogenetic studies on this model organism, reporting examples of how FISH can be applied for diverse investigative purposes, varying from comparative gene mapping to studies of chromosome structure and genome evolution, to characterization of chromosomes aberrations as well as transgenic insertions. The protocols, which include techniques for the preparation of mitotic chromosomes and DNA fibers from short-term cell cultures, have been gathered through the years and repeatedly tested in our laboratory, and all together aim at providing sufficient experimental versatility to cover a broad range of cytogenetic and cytogenomic applications.

Key words: Fluorescence in situ hybridization, FISH, ZOO-FISH, Molecular cytogenetics, Chromosomes, Rat

1. Introduction

Fluorescence in situ hybridization (FISH) is an accurate and sensitive cytological technique for the visualization of specific DNA sequences on chromosome preparations. As a physical mapping technique, FISH has proved a valuable tool for the completion of various genome projects. The technique can be used to map clones on metaphase chromosomes, and because of the correlation between the physical proximity of DNA sequences in interphase and their genomic distance, FISH can also be used to order clones on interphase nuclei (1–3). When needed, hybridization on stretched DNA fibers (FiberFISH) allows clone ordering at a significantly higher resolution (5–500 kb resolution range on fibers compared to the 100 kb to 1 Mb resolution range on

Joanna M. Bridger and Emanuela V. Volpi (eds.), *Fluorescence in situ Hybridization (FISH):*
Protocols and Applications, Methods in Molecular Biology, vol. 659,
DOI 10.1007/978-1-60761-789-1_32, © Springer Science+Business Media, LLC 2010

interphase nuclei), and it also makes estimating the size of gaps and overlaps in contig maps possible (4).

Most advantageously, FISH can also be used for the assessment of specific genetic changes on chromosome preparations. This particular application, by providing a link between standard karyotypic analysis and molecular genetic studies, has prompted the reinvention of classical cytogenetics into molecular cytogenetics. The arrival of molecular cytogenetics on the genetic research scene was particularly appropriate at a time when the strategies for the isolation and identification of human disease genes were being revised and new approaches to cloning, based solely on positional information such as positional cloning and candidate gene approach, were being developed (5).

The pessimistic prediction that the completion of the human genome sequence would have caused a rapid decline in the demand for FISH as a physical mapping technique has been proved wrong. In fact, in the so-called post-genomic era, FISH is still widely used, most frequently for the resolution of cytolocation discrepancies and as a validation technique for copy number variation (CNV). Furthermore, the wealth of resources, which have become available – especially chromosome specific probes – combined with the well-known versatility of the technique – has led to a diversified range of technically sophisticated applications for diagnostic purposes, clinical investigations and research on chromosome biology and genome organization (ref. (6, 7)).

FISH has also successfully contributed to the analysis of the mouse, rat, and other model organisms' genomes. Particularly relevant has been the deployment of FISH in support of gene transfer technology, more precisely for the characterization of transgenic insertion and assignment of the transgenes' chromosomal positions in mouse (8–10) and rat (11–13).

This chapter concentrates on the use of FISH for genomic and post-genomic investigations in rat and was inspired by a comprehensive range of previously published molecular cytogenetic studies on this model organism, reporting examples of how different FISH applications can be applied for diverse investigative purposes, including karyotype refinement (14), comparative gene mapping (15–17), studies on chromosome structure and genome evolution (18–20), characterization of mutagenically induced chromosome aberrations (21), and identification of oncogene amplification in tumours (22–25), as well as the aforementioned FISH application for detection and chromosomal mapping of transgenic loci. The selection of protocols presented in this chapter, including techniques for the preparation of mitotic chromosomes and DNA fibers from short-term fibroblast cultures and spleen cells, aims at providing sufficient technical details and experimental versatility to cover most of the FISH applications enlisted above.

Although, over the years there have been substantial efforts by single research groups to isolate rat chromosome-specific probes, particularly with the intent of developing a 24-colour karyotyping protocol for rat chromosomes (26–29), until recently the commercial availability of resources for molecular cytogenetics analysis in this animal species has been very limited. An experimental expedient to overcome this technical impasse can be provided by heterologous chromosome painting analysis or ZOO-FISH, wherein single chromosome "paints" (chromosome specific DNA libraries obtained either by microdissection or flow-sorting) from one animal species are applied to chromosomes from another species. The ZOO-FISH technique was originally devised for evolutionary studies (30), but the method, combined with the information available on chromosomal homology between mouse (*Mus musculus*) and rat (*Rattus norvegicus*) (19, 31, 32), by allowing mouse paints to be used in an indirect fashion for chromosome identification in rat, enables in some instances to by-pass the technical hindrance caused by limited access to rat-specific probes. Examples of heterologous painting for hybridization-based karyotype technology and other FISH applications in rat can be found in Fig. 1.

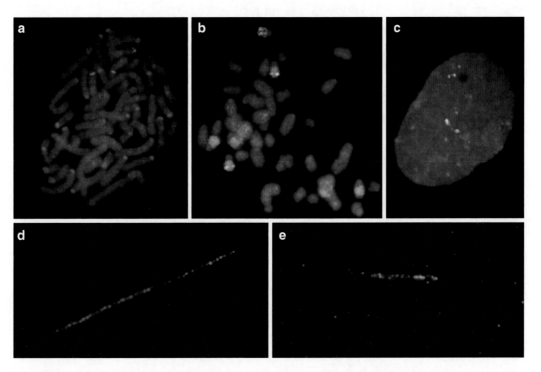

Fig. 1. Fluorescence in situ hybridization (FISH) on rat chromosomes and DNA fibres. (**a**) Example of metaphase mapping of a BAC probe (*red*) from rat chromosome 13; (**b**) Suspected trisomy of chromosome 13 is confirmed by heterologous hybridization (ZOO-FISH) using a mouse chromosome 1 paint (*green*); (**c**) Example of clone ordering by three-colour FISH on an interphase nucleus; Chromosomes and nuclei are counterstained with DAPI. (**d**) and (**e**) Examples of two- and three-colour FiberFISH experiments. The slides were examined and imaged on a CytoVision system (Genetix, formerly Applied Imaging), comprised of an Olympus BX-51 epifluorescence microscope coupled to a Sensys CCD camera (Photometrics) and software package Genus.

2. Materials

2.1. Cell Culture Procedures

2.1.1. Short-Term Fibroblast Culture from Ear Explants/Tail Tips

1. Ear explants or tail tips.
2. Absolute ethanol.
3. 1× Phosphate buffered saline (PBS).
4. Dulbecco's Modified Eagle's Medium (DMEM).
5. Foetal bovine serum (FBS).
6. L-glutamine 200 mM.
7. Penicillin at 10,000 U/mL and streptomycin at 10 mg/mL.
8. Sterile scalpels (Swann and Morton).
9. 10-cm Petri dish(es)
10. 25-cm^3 tissue culture flask(s).
11. Access to a Class II microbiological cabinet.
12. Access to an incubator at 37°C with 5% CO_2.

2.1.2. Short-Term Cell Culture from Spleens

1. Spleen(s).
2. 1× Phosphate buffered saline (PBS).
3. RPMI-1640 medium.
4. Foetal bovine serum (FBS).
5. Penicillin at 10,000 U/mL and streptomycin at 10 mg/mL.
6. Lipopolysaccharides (LPS) (5 mg/mL) (from *Salmonella enteritidis*) (Sigma).
7. 10-cm Petri dish(es).
8. 10-mL syringes and 26-gauge needles.
9. 15-mL centrifuge tubes.
10. 25-cm^3 tissue culture flask(s).
11. Access to a Class II microbiological cabinet.
12. Access to an incubator at 37°C with 5% CO_2.

2.1.3. Lymphocyte Culture from Peripheral Blood

1. Ten drops of peripheral blood from the tail vein collected in heparinized tube.
2. Vacuette sodium heparin tubes (Greiner Bio-One).
3. RPMI-1640 medium.
4. Foetal bovine serum (FBS).
5. Penicillin at 10,000 U/mL and streptomycin at 10 mg/mL.
6. Phytohaemagglutinin (PHA).
7. Lipopolysaccharides (LPS) (5 mg/mL) (from *S. enteritidis*) (Sigma).
8. 15-mL centrifuge tubes.

9. 7-mL Bijou container(s).

10. Access to a Class II microbiological cabinet.

11. Access to an incubator at 37°C with 5% CO_2.

2.1.4. EBV-Transformed B-Lymphoblastoid Cell Lines Culture

1. EBV-transformed B cells.

2. RPMI-1640 medium.

3. Foetal bovine serum (FBS).

4. Penicillin at 10,000 U/mL and streptomycin at 10 mg/mL.

5. 15-mL centrifuge tubes.

6. 25-cm³ tissue culture flask(s).

7. Access to a Class II microbiological cabinet.

8. Access to an incubator at 37°C with 5% CO_2.

2.2. Chromosome Preparation

1. A growing, semi-confluent culture of cells.

2. KaryoMAX Colcemid Solution (10 μg/mL) (Invitrogen).

3. Trypsin–EDTA solution (1×).

4. Hypotonic solution: 0.075 M KCl or alternatively 0.034 M KCl, 0.017 M trisodium citrate.

5. Carnoy's fixative: Three volumes methanol, one volume glacial acetic acid. Make fresh.

6. Clean microscope slides (VWR International).

7. 1-mL disposable plastic Pasteur pipettes.

8. Waterbath at 37°C.

9. A phase contrast microscope with a 10× objective.

10. Slide storage boxes (VWR International).

2.3. DNA Fibres Preparation

1. A growing culture of cells.

2. Shandon Sequenza staining rack and disposable coverplates (Thermo Electron Corporation).

3. 15-mL centrifuge tubes.

4. Haemocytometer.

5. A hotplate set to approximately 40°C.

6. Clean glass microscope slides (VWR International)

7. 1× Phosphate buffered saline (PBS)

8. Cell lysis solution: Five volumes 0.07 M NaOH, two volumes of 100% absolute ethanol.

9. Methanol.

10. 70, 90, and 100% ethanol series.

11. A phase contrast microscope with a 10× objective.

12. Slide storage boxes.

2.4. Labelling of DNA Probes by Nick-Translation

1. ~1 µg of DNA probe.

2. 0.2 mM labelled dUTP (Biotin-11-dUTP (Roche) or Digoxigenin-11-dUTP (Roche) or Alexa 594 – dUTP (Invitrogen)).

3. Vysis nick-translation kit (Abbott Laboratories), containing 0.1 mM dTTP, a 0.1-mM dATP, dCTP, dGTP mix, nick-translation buffer, and nick-translation enzyme mix.

4. 1.5-mL microcentrifuge tubes.

5. Waterbath at 16°C in a cold room.

6. Heating block at 68°C.

7. Agarose gel running apparatus.

8. 2% agarose gel (Sigma).

9. Small fragment DNA size marker (e.g. 100-bp ladder (Invitrogen) or PhiX174 *Hae*III digest (New England Biolabs)).

10. 1× TBE gel running buffer.

11. Precipitation mix: 1:1 mix of Salmon sperm DNA (Sigma) and *E. coli* tRNA (Roche).

12. 3 M sodium acetate.

13. Cold absolute ethanol.

14. Tris–EDTA (1×) (Sigma).

2.5. Hybridization Procedure

1. Labelled DNA probe(s).

2. Chromosome "Paints" (mouse or rat), available upon request from the following manufacturers: Cambio, Genetix, Spectral Imaging, and Metasystems.

3. 1.5-mL microcentrifuge tubes.

4. Chromosome spreads or DNA fibres on slide.

5. Sonicated genomic rat DNA.

6. Hybridization mix: 50% formamide (VWR International), 10% dextran sulphate (Sigma), and 2× SSC.

7. Denaturation solution: 70% formamide, 0.6× SSC. Make fresh.

8. Washing solution: 50% formamide, 1× SSC. Make fresh.

9. 2× SSC.

10. 70, 90, and 100% ethanol series.

11. Blocking solution: 5% non-fat dry milk in 1× PBS, 0.1% Tween-20 (VWR International). Make fresh.

12. Detection solution: 1% non-fat dry milk in 1× PBS. Make fresh.

13. Antibodies and/or affinity molecules (Table 1).

14. Vectashield mounting medium with 4′,6-diamidino-2-phenylindole (DAPI) (Vector Laboratories).

Table 1
Combination of antibodies/affinity molecules for the detection of hapten-labelled DNA probes

Probes labeling	Antibodies/affinity molecules	Final concentration (μg/mL)	Supplier
One-colour FISH: Biotin	Fluorescein Streptavidin	10	Vector Laboratories
One-colour FISH: Biotin	Texas Red Streptavidin	10	Invitrogen
One-colour FISH: Digoxigenin	Mouse anti-digoxigenin	0.4	Roche
	Goat anti-mouse Alexa Fluor 488	20	Invitrogen
Two-colour FISH: Biotin and Digoxigenin	Texas Red Streptavidin	10	Invitrogen
	Mouse anti-digoxigenin	0.4	Roche
	Goat anti-mouse Alexa Fluor 488	20	Invitrogen
Three-colour FISH: Biotin, Digoxigenin and Alexa Fluor 594 directly labelled probe	Fluorescein Streptavidin	10	Vector Laboratories
	Mouse anti-digoxigenin	0.4	Roche
	Rabbit anti-mouse Cyanine5	30	JacksonImmunoResearch
	Goat anti-rabbit Cyanine5	30	JacksonImmunoResearch

15. 22 mm × 22 mm glass coverslips, number 1½ thickness (VWR International).

16. 22 mm × 50 mm glass coverslips, number 1½ thickness (VWR International).

17. Vulcanizing solution (otherwise known as rubber cement, typically bicycle puncture repair glue).

18. Diamond tipped scribe.

19. Fine forceps.

20. Humid hybridization chamber (airtight plastic box with a piece of damp blotting paper in it).

21. Heating block at 60°C.

22. Heating block at 72°C.

23. Heating block at 37°C.

24. Waterbath at 70°C.

25. Waterbath at 42°C.

26. Incubator at 37°C.

27. 50-mL plastic and glass Coplin jars (TAAB).

2.6. Viewing the Slides by Fluorescence Microscopy

1. Epifluorescent light microscope with filters for DAPI, FITC, TRITC, and Cy5 excitation and emission. 10× dry and 60× or 100× oil immersion objective lenses fitted to above microscope.

2. Immersion oil.

3. Charge-coupled device (CCD) camera and appropriate operating software for digital image capturing and analysis.

3. Methods

3.1. Cell Culture and Chromosome Preparation Procedures

3.1.1. Short-Term Fibroblast Culture from Ear Explants/Tail Tips

1. Working in a 10-cm Petri dish, wash the ear explant/tail tip in 100% ethanol and twice in 1× PBS (see Note 1).

2. With a sterile scalpel finely cut the ear explant/tail tip into very thin slices.

3. Place the slices into a 25-cm^3 flask containing 5 mL of DMEM supplemented with 10% FBS, 1% L-glutamine, and penicillin at 100 U/mL and streptomycin at 0.1 mg/mL.

4. Incubate the flask in a 37°C incubator with 5% CO_2.

5. Inspect the flask for adherent cell growth daily. Fibroblasts should become established after 1–2 days.

6. Cultures that are 80–90% confluent are suitable for harvesting of metaphase chromosomes (see Note 2).

7. Add Colcemid to a final concentration of 0.2 µg/mL and incubate at 37°C for 1–2 h (see Note 3).

8. Transfer the medium to a 50-mL centrifuge tube. Add 5–10 mL of 1× PBS to the flask, swirl and then aspirate and transfer to the tube with the medium.

9. Add 2–3 mL of Trypsin–EDTA solution to the flask and incubate at 37°C for 3 min or until the cells start to detach from the flask. This can be checked on a phase contrast microscope. Cell detachment can be helped by gently tapping the flask.

10. Once the cells have detached, transfer the original medium into the flask, mix well, and return the cell suspension to the centrifuge tube.

11. Centrifuge at 280×g for 5 min and aspirate and discard the supernatant, being careful not to disturb the cell pellet.

12. Slowly add to the tube 10 mL of pre-warmed hypotonic solution, while gently flicking the bottom of the tube to resuspend the pellet.

13. Incubate at 37°C for 10 min.

14. Centrifuge at 280×g for 5 min. Discard supernatant, leaving some fluid to flick the pellet into suspension again.

15. Add Carnoy's fixative drop by drop (see Note 4). After few drops, fill the tube with 5–10 mL of fixative.

16. Centrifuge the cells at $280 \times g$ for 5 min and remove the supernatant. Resuspend the cells in 5–10 mL of fixative. Repeat this step two more times.

17. Resuspend the pellet in a small volume of fixative, so that the suspension appears cloudy.

18. With a 1-mL plastic Pasteur pipette, drop a small volume of suspension onto a glass microscope slide from approximately 20 cm above the slide and allow to air-dry overnight.

19. The mitotic index (or percentage of mitotic cells) of the harvest and the quality of chromosome spreading can be assessed using a phase contrast microscope with a 10× objective (see Note 5).

20. Slides should be made in batches and stored in a sealed container at −20°C prior to hybridization.

3.1.2. Short-Term Cell Culture from Spleens

1. Collect two or more spleens in 1× PBS or RPMI-1640 Medium (see Note 1).

2. Transfer the spleens to RPMI-1640 to a 10-cm Petri dish.

3. Using a 10-mL syringe with a 26-gauge needle, pierce the spleens all over. Holding the spleens with one needle, use another needle to wash medium gently through the spleens to remove as many cells as possible.

4. Transfer the cells to a 15-mL centrifuge tube and centrifuge at $195 \times g$ for 6 min.

5. Discard the supernatant and resuspend the cells in 5 mL of RPMI-1640 medium supplemented with 15% FBS, 1% L-glutamine, and penicillin at 100 U/mL and streptomycin at 0.1 mg/mL.

6. Set up two or three cultures (1–2 mL each) in 25-cm³ tissue culture flasks.

7. Add 50 μL of LPS (stock concentration of 5 mg/mL) to each culture (see Note 6).

8. Incubate cultures at 37°C for 44–46 h.

9. Add KaryoMAX Colcemid solution to a final concentration of 0.2 μg/mL and incubate at 37°C for 30 min (see Note 3).

10. Transfer the cells to a 15-mL centrifuge tube and centrifuge at $195 \times g$ for 5 min. Aspirate and discard supernatant.

11. Slowly add to the tube 10 mL of pre-warmed hypotonic solution, while gently flicking the bottom of the tube to resuspend the pellet.

12. Incubate at 37°C for 10 min.

13. Centrifuge at $280 \times g$ for 5 min. Aspirate supernatant, leaving some fluid.

14. Flick the pellet into suspension again, and add Carnoy's fixative drop by drop (see Note 4). After few drops, fill the tube with 5–10 mL of fixative.

15. Centrifuge the cells at $280 \times g$ for 5 min and remove the supernatant. Resuspend the cells in 5–10 mL of fixative. Repeat this step two more times.

16. Resuspend the pellet in a small volume of fixative so that the suspension appears cloudy. With a 1-mL plastic Pasteur pipette, drop a small volume of suspension onto a glass microscope slide from approximately 20 cm above the slide and allow to air-dry overnight.

17. The mitotic index (or percentage of mitotic cells) and the quality of chromosome spreading of the harvest can be assessed using a phase contrast microscope with a 10× objective (see Note 5).

18. Slides should be made in batches and stored in a sealed container at −20°C prior to hybridization. Unused cell suspension can also be stored at −20°C.

3.1.3. Lymphocyte Culture from Peripheral Blood

1. Add ten drops of fresh peripheral blood to a 25-cm³ flask containing 5 mL of RPMI-1640 Medium supplemented with 15% FBS, 1% L-glutamine, and 0.1-mL PHA and/or 50-µL LPS (see Note 6).

2. Incubate the culture for 72 h at 37°C without CO_2.

3. Proceed as from step 9 in Subheading 3.1.2 above.

3.1.4. EBV-Transformed B-Lymphoblastoid Cell Lines Culture

1. B-lymphoblastoid cells should be cultured in RPMI-1640 Medium supplemented with 10% FBS, 1% L-glutamine, and penicillin at 100 U/mL and streptomycin at 0.1 mg/mL.

2. Prepare metaphase chromosomes from rapidly dividing cultures, typically 24 h after splitting a cell culture.

3. Proceed as from step 9 in Subheading 3.1.2 above.

3.2. DNA Fibres Preparation

1. Take 5 mL of a growing cell culture and transfer to a 15-mL centrifuge tube.

2. Centrifuge at $195 \times g$ and remove the supernatant. Resuspend the cells in 1 mL of 1× PBS and assess the cell density using a haemocytometer.

3. Adjust the concentration of cells to 2×10^6 cells/mL with 1× PBS.

4. Pipette 10 µL of cell suspension onto a glass microscope slide (near the frosted end, but not on it).

5. Allow the cell suspension to dry by placing the slide on a hot-plate at approximately 40°C. Do not leave the slide at 40°C for longer than is necessary for the cell suspension to dry.

6. Place the slide onto a Shandon Sequenza coverplate and mount in the staining rack.

7. To each slide, add 150 µL of lysis solution and allow to drain for 30 s.

8. Then add 200 µL of methanol to fix the chromatin fibers and leave for 1 min.

9. Remove the slides from the coverplate and let them air-dry.

10. Assess the degree of chromatin release and length of fibers using a phase contrast microscope with a 10× objective.

11. Slides should be made in batches, taken through a 70, 90, and 100% (1 min each) ethanol series, and stored at –20°C prior to hybridization.

12. The remaining cell suspension can be centrifuged, the cells fixed with Carnoy's fixative (as described earlier) and stored in a sealed container at –20°C for future use (see Note 7).

3.3. Labelling of DNA Probes by Nick-Translation

1. Place ~1 µg of probe DNA into a 1.5-mL microcentrifuge tube (see Note 8). The maximum volume should not exceed 17.5 µL (concentration of DNA should be no less than 60 ng/µL). Add distilled water to make the volume up to 17.5 µL if necessary.

2. Add 2.5 µL of 0.2 mM labelled dUTP (either biotin-11-dUTP or digoxigenin-11-dUTP or directly labelled Alexa 594-dUTP), 5 µL of 0.1 mM dTTP (included in the Vysis nick-translation kit), 10 µL of 0.1 mM dNTP mix (dATP, dCTP, and dGTP included in kit), 5 µL of 10× reaction buffer (included in kit), and 10 µL of nick-translation enzyme mix (included in kit).

3. Gently mix the content of the tube, centrifuge briefly, and incubate at 16°C for 2 h. Terminate the reaction by incubating at 68°C for 10 min.

4. To check the probe has been cut to the optimal length, run a volume of nick-translation reaction corresponding to 100 ng of DNA on a 2% agarose gel using a small DNA marker such as a 100-bp ladder or PhiX174 *Hae*III digest. The DNA probe fragment's length should range from 200 to 600 bp (see Note 9)

5. The probe should be precipitated with the addition of a quarter volume of precipitation mix (1:1 mix of Salmon sperm DNA and *E. coli* tRNA), a tenth volume of 3 M sodium acetate and twice the volume of cold absolute ethanol. The probe should be left to precipitate at –80°C for 2 h or at –20°C overnight.

6. The probe should then be purified by centrifugation at 15,500×g for 20 min, followed by two washes in 70% ethanol and centrifugation at 15,500×g for 5 min.

7. Allow the probe pellet to air-dry and resuspend in 1× Tris–EDTA at a concentration of 20 ng/µL.

3.4. Hybridization Procedure

1. Place 10 µL (~200 ng) of labelled DNA probe in a 1.5-mL microcentrifuge tube (see Note 10). Add a 3× excess of sonicated rat genomic DNA ("competitor DNA") (see Note 11).

2. Place the tube on a heating block at 60°C to dry down the DNA.

3. Resuspend the probe in 10 µL of 1× hybridization solution (final DNA concentration of 20 ng/µL). Allow to resuspend for a minimum of 15 min at room temperature.

4. Denature the probe at 72°C for 5–10 min.

5. Following denaturation, the probe should be let to pre-anneal at 37°C for 30–60 min (see Note 11).

6. Dehydrate the slide in a 70, 90, and 100% ethanol series and allow to air-dry.

7. Denature slide in 70% formamide/2× SSC at 70°C for 1½–3 min (see Note 12).

8. Quench the slide in cold 2× SSC for 5 min.

9. Dehydrate the slide in a 70, 90, and 100% ethanol series and allow to air-dry.

10. Mark the area of the slide containing the cells by breathing on the slide and circling the area from underneath with a diamond tipped scribe.

11. Place the slide on a heating block set to 37°C.

12. Add 10 µL of probe mix to the hybridization area, carefully cover with a 22 mm × 22 mm coverslip and seal the edge of the coverslip with vulcanizing solution. Lower the coverslip very gently to avoid air bubbles.

13. Place the slide in a pre-warmed humid hybridization chamber at 37°C. Allow to hybridize for 16–48 h (see Note 13).

14. Following hybridization, carefully peel off the dried vulcanizing solution with fine forceps and place the slide in a Coplin jar filled with 2× SSC, allowing the coverslip to soak off (see Note 14).

15. Wash the slide twice for 5 min in 50% formamide, 1× SSC at 42°C.

16. Wash the slide for 5 min in 2× SSC at 42°C and transfer to 1× PBS at room temperature.

17. Blot excess liquid from the back and the sides of the slide. Place the slide horizontally in the humid hybridization chamber

and add 200 μL of blocking solution (5% dried non-fat milk in 1× PBS/0.1% Tween-20). Spread with a 22 mm × 50 mm coverslip. Incubate in humid hybridization chamber at 37°C for 60 min.

18. Briefly wash the slide in a Coplin jar with 1× PBS, allowing the coverslip to soak off. Briefly drain the slide (do not allow to dry out) and return to the humid hybridization chamber.

19. Add to the slide 100–200 μL of detection solution with appropriate concentration of primary antibody or affinity molecule (see Table 1), apply a 22 mm × 50 mm coverslip, and incubate at 37°C for 30–60 min (see Note 15).

20. Wash the slide in a Coplin jar containing 1× PBS three times for 5 min (ensuring the coverslip soaks off during the first wash). If secondary antibody is not necessary, proceed as from to step 24.

21. If secondary antibody is necessary, briefly drain the slide and return to the humid hybridization chamber.

22. Add 100–200 μL of detection solution with appropriate concentration of secondary antibody (see Table 1), apply a 22 mm × 50 mm glass coverslip, and incubate at 37°C for 30–60 min.

23. Wash the slide in a Coplin jar containing 1× PBS three times for 5 min (ensuring the coverslip is soaked off during the first wash).

24. Briefly drain the slide but do not allow it to dry out. Mount the slide with 25 μL of Vectashield containing DAPI and a 22 mm × 50 mm glass coverslip (thickness 1½).

25. If the probe(s) used in the hybridization was/were directly labelled with a fluorophore (Alexa 594 or similar), skip steps 18–24.

3.5. Viewing the Slides by Fluorescence Microscopy

1. Place the slide on the microscope stage.

2. Using the 10× objective and the DAPI fluorescence filter, focus on the cells using the coarse focusing wheel.

3. Identify a spread of metaphase chromosomes and centre it in the field of view.

4. Apply a drop of immersion oil to the coverslip and move the 60× or 100× objective into position. Refocus if necessary using the fine-focus wheel.

5. To verify whether the hybridization has worked, the metaphase chromosomes and interphase nuclei should be examined for signals with the FITC and/or TRITC filters (depending on the combination of secondary antibodies used). (For troubleshooting guide see Note 16).

4. Notes

1. Tissue should be immediately transferred to the medium and the cell culture set up as soon as possible.

2. A sub-confluent cell culture contains a proportion of cells undergoing cell division and that will increase the likelihood of harvesting a good number of metaphase spreads. Sub-confluence is usually reached in less than a week. Longer culturing can result in the increased chance of in vitro chromosomal abnormalities (e.g. tetraploidy).

3. Colcemid is a colchicine-related spindle inhibitor. A longer incubation will increase the percentage of mitotic cells but will also increase the degree of chromosome condensation. Colcemid is toxic and adequate precautions should be taken when handling it, particularly to avoid skin contact. For more detailed information, refer to hazard sheet.

4. After hypotonic treatment, the swollen cells are very fragile. Adding the fixative drop by drop prevents possible rupture of the cell membrane and subsequent loss of chromosomes.

5. Chromosomes should be well spread (there should be no crossovers) and there should be no visible cytoplasm surrounding them. Spreading can be improved by dropping the cells from a greater height and by leaving the slides over a hot water bath (i.e. in humid conditions) until the fixative has slowly evaporated. If no improvement is noticed, the cells should be reharvested and the hypotonic treatment adjusted accordingly by increasing the length of the incubation at 37°C. Do not heat the slides to speed up drying, but allow them to dry for few hours at room temperature (before storing them at −20°C) as this results in the chromosomes and nuclei adhering to the surface of the slide better.

6. Lipopolysaccharides (LPS) and PHA are mitogens used to stimulate lymphocyte proliferation. They can be used either separately or in conjunction to improve the metaphase yield.

7. DNA fibres can also be obtained from fixed cells following the same method (as from Subheading 3.2, step 5). The only difference is that the slides should be rehydrated with 1× PBS before the addition of the lysis solution.

8. Most suitable probes for FISH analysis are genomic clones with an insert ranging from few Kb to ~300 kb (cDNAs, BACs, PACs, YACs, cosmids, plasmids, etc.). The probe (vector plus insert) DNA does not need to be linearized before nick-translation.

9. It is critical to check that after the labelling the probe fragment range in size from 200 to 600 bp. Nick-translation

products that are too large will result in a bright, speckled background. Instead, products that are too small will result in a diffuse background. In both cases, the result will be a lower signal/noise ratio.

10. If multiple probes are to be used in the same hybridization, ensure that the concentration of each probe remains the same (final combined DNA concentration will increase according to the total number of probes). For recommended concentration of chromosome paints and other commercial probes, follow the manufacturer's instructions.

11. This short incubation is a necessary step to "compete" ubiquitous repetitive sequences. In this case, the sonicated genomic DNA, previously added to the probe, works as a "competitor". Overannealing of the probe will result in a significantly reduced amount of unique sequence, single-stranded DNA available for hybridization, and hybridization efficiency and intensity of the signal might be negatively affected.

12. The optimal length of the denaturation step – within the range given in the protocol – should be determined empirically for different batches of slides by increasing the time from 1½ min by 15 s at a time until the chromosome morphology becomes unacceptable. Overdenaturation of the slides will result in "puffy" chromosomes and can be remedied by reducing the denaturing time. Chromosomes can be easily overdenatured if they have not aged properly. Slides should be stored at –20°C for at least 1 week prior to hybridization. Alternatively, slides should be baked at 55–60°C for several hours before hybridization and hardened by dehydration in an ethanol series. Formamide is toxic (harmful by inhalation, ingestion, and skin contact) so it should be handled/used under a fume hood and gloves should be worn. For more detailed information, refer to hazard sheet.

13. The length of the hybridization time, within the range given in the protocol, should be decided on the basis of the copy number of the target sequence and the size of the probe insert. Repetitive targets need a shorter hybridization. Long insert probes need a longer hybridization.

14. To avoid scraping off the cells, handle the slides with care at all stages, especially when removing coverslips (such as after hybridization). To remove the coverslip, soak the slide in 2× SSC.

15. Fluorochromes are susceptible to photo-bleaching. When processing slides and working with fluorescence conjugated antibodies, try to limit their exposure to light as it can fade fluorochromes and make the analysis more difficult.

16. Troubleshooting guide: (1) If there is no hybridization signal on the slide (or the signal is very weak), check the following steps: (a) The concentration of the probe might have been too low. Recheck DNA concentration on gel. (b) The probe might not have been denatured properly. Check/adjust the temperature of waterbath/heating block. (c) The chromosomes might not have been denatured properly. Check/adjust accordingly temperature of waterbath and formamide concentration. (d) The post-hybridization washes were too stringent. Check/adjust accordingly temperature of waterbath and formamide concentration. (e) The inappropriate detection reagents have been used. The choice of antibodies/affinity molecules for the probes is determined by the reporter molecule used to label the probe. (f) Antibodies are too old. Antibodies should be stored in accordance with the manufacturer's instructions. Use fresh aliquots whenever possible and avoid freeze–thawing of antibodies. (2) If there is too much background and the signal/noise ratio is low, the steps to check are the following: (a) The chromosomes might have been surrounded by cytoplasm. Cytoplasm on the slide (visible under phase contrast) can trap probe and result in a high background. If necessary, prior to hybridization pre-treat the slide with RNAse A (Sigma) (20 mg/mL) and/or Proteinase K (Sigma) (0.25 mg/mL). (b) The probe has dried onto the slide. Ensure that the edges of the coverslip are sealed with glue to prevent the probe from drying on the hybridization area. The slide should also be incubated in a moist chamber at 37°C for hybridization. (c) Competitor DNA was not sufficient. Generally, probes that contain repeat sequences as well as unique sequences require an excess (up to 50×) of competitor DNA to suppress repeat sequences from hybridizing. Failure to compete out repeat sequences results in a high specific background. (d) Post-hybridization washes were not stringent enough. Check the temperature of the washes and the concentration of formamide and increase if necessary. (e) Slides were allowed to dry during the detection and the antibodies/affinity molecules have been "trapped" on the specimen causing a widespread aspecific background. All the detection steps should be done quickly in order for the slides not to dry before completion of the washes.

References

1. Lawrence, J. B., Singer, R. H., and McNeil, J. A. (1990) Interphase and metaphase resolution of different distances within the human dystrophin gene, *Science* **249**, 928–932.

2. Trask, B. J., Massa, H., Kenwrick, S., and Gitschier, J. (1991) Mapping of human chromosome Xq28 by two-color fluorescence in situ hybridization of DNA sequences to interphase cell nuclei, *Am J Hum Genet* **48**, 1–15.

3. van den Engh, G., Sachs, R., and Trask, B. J. (1992) Estimating genomic distance from

DNA sequence location in cell nuclei by a random walk model, *Science* **257**, 1410–1412.

4. Weier, H. U. (2001) DNA fiber mapping techniques for the assembly of high-resolution physical maps, *J Histochem Cytochem* **49**, 939–948.

5. Collins, F. S. (1995) Positional cloning moves from perditional to traditional, *Nat Genet* **9**, 347–350.

6. Speicher, M. R., and Carter, N. P. (2005) The new cytogenetics: blurring the boundaries with molecular biology, *Nat Rev Genet* **6**, 782–792.

7. Volpi, E. V., and Bridger, J. M. (2008) FISH glossary: an overview of the fluorescence in situ hybridization technique, *Biotechniques* **45**, 385–386, 388, 390 passim.

8. Kulnane, L. S., Lehman, E. J., Hock, B. J., Tsuchiya, K. D., and Lamb, B. T. (2002) Rapid and efficient detection of transgene homozygosity by FISH of mouse fibroblasts, *Mamm Genome* **13**, 223–226.

9. Matsui, S., Sait, S., Jones, C. A., Nowak, N., and Gross, K. W. (2002) Rapid localization of transgenes in mouse chromosomes with a combined Spectral Karyotyping/FISH technique, *Mamm Genome* **13**, 680–685.

10. Nakanishi, T., Kuroiwa, A., Yamada, S., Isotani, A., Yamashita, A., Tairaka, A., Hayashi, T., Takagi, T., Ikawa, M., Matsuda, Y., and Okabe, M. (2002) FISH analysis of 142 EGFP transgene integration sites into the mouse genome, *Genomics* **80**, 564–574.

11. Cronkhite, J. T., Norlander, C., Furth, J. K., Levan, G., Garbers, D. L., and Hammer, R. E. (2005) Male and female germline specific expression of an EGFP reporter gene in a unique strain of transgenic rats, *Dev Biol* **284**, 171–183.

12. Goto, K., Yasuda, M., Sugawara, A., Kuramochi, T., Itoh, T., Azuma, N., and Ito, M. (2006) Small eye phenotypes observed in a human tau gene transgenic rat, *Curr Eye Res* **31**, 107–110.

13. Liska, F., Levan, G., Helou, K., Sladka, M., Pravenec, M., Zidek, V., Landa, V., and Kren, V. (2002) Chromosome assignment of Cd36 transgenes in two rat SHR lines by FISH and linkage mapping of transgenic insert in the SHR-TG19 line, *Folia Biol (Praha)* **48**, 139–144.

14. Hamta, A., Adamovic, T., Samuelson, E., Helou, K., Behboudi, A., and Levan, G. (2006) Chromosome ideograms of the laboratory rat (*Rattus norvegicus*) based on high-resolution banding, and anchoring of the cytogenetic map to the DNA sequence by

FISH in sample chromosomes, *Cytogenet Genome Res* **115**, 158–168.

15. Gomez-Fabre, P. M., Helou, K., and Stahl, F. (2002) Predictions based on the rat-mouse comparative map provide mapping information on over 6000 new rat genes, *Mamm Genome* **13**, 189–193.

16. Helou, K., Wallenius, V., Qiu, Y., Ohman, F., Stahl, F., Klinga-Levan, K., Kindblom, L. G., Mandahl, N., Jansson, J. O., and Levan, G. (1999) Amplification and overexpression of the hepatocyte growth factor receptor (HGFR/MET) in rat DMBA sarcomas, *Oncogene* **18**, 3226–3234.

17. Zullo, S., and Upender, M. (1995) Rat karyotyping by fluorescence in situ hybridization (FISH): localization of oncogene c-raf to 4q42, retinoblastoma antioncogene to 15q12, and mitochondrial D-loop-like sequences to the Y chromosome, *Genomics* **25**, 753–756.

18. Essers, J., de Stoppelaar, J. M., and Hoebee, B. (1995) A new rat repetitive DNA family shows preferential localization on chromosome 3, 12 and Y after fluorescence in situ hybridization and contains a subfamily which is Y chromosome specific, *Cytogenet Cell Genet* **69**, 246–252.

19. Helou, K., Walentinsson, A., Levan, G., and Stahl, F. (2001) Between rat and mouse zoo-FISH reveals 49 chromosomal segments that have been conserved in evolution, *Mamm Genome* **12**, 765–771.

20. McFadyen, D. A., and Locke, J. (2000) High-resolution FISH mapping of the rat alpha2u-globulin multigene family, *Mamm Genome* **11**, 292–299.

21. de Stoppelaar, J. M., Faessen, P., Zwart, E., Hozeman, L., Hodemaekers, H., Mohn, G. R., and Hoebee, B. (2000) Isolation of DNA probes specific for rat chromosomal regions 19p, 19q and 4q and their application for the analysis of diethylstilbestrol-induced aneuploidy in binucleated rat fibroblasts, *Mutagenesis* **15**, 165–175.

22. Adamovic, T., Trosso, F., Roshani, L., Andersson, L., Petersen, G., Rajaei, S., Helou, K., and Levan, G. (2005) Oncogene amplification in the proximal part of chromosome 6 in rat endometrial adenocarcinoma as revealed by combined BAC/PAC FISH, chromosome painting, zoo-FISH, and allelotyping, *Genes Chromosomes Cancer* **44**, 139–153.

23. Helou, K., Walentinsson, A., Kost-Alimova, M., and Levan, G. (2001) Hgfr/Met oncogene acts as target for gene amplification in DMBA-induced rat sarcomas: free chromatin fluorescence in situ hybridization analysis of amplicon arrays in homogeneously staining

regions, *Genes Chromosomes Cancer* **30**, 416–420.

24. Nordlander, C., Karlsson, S., Karlsson, A., Sjoling, A., Winnes, M., Klinga-Levan, K., and Behboudi, A. (2007) Analysis of chromosome 10 aberrations in rat endometrial cancer-evidence for a tumor suppressor locus distal to Tp53, *Int J Cancer* **120**, 1472–1481.

25. Samuelson, E., Nordlander, C., Levan, G., and Behboudi, A. (2008) Amplification studies of MET and Cdk6 in a rat endometrial tumor model and their correlation to human type I endometrial carcinoma tumors, *Adv Exp Med Biol* **617**, 511–517.

26. Buwe, A., Steinlein, C., Koehler, M. R., Bar-Am, I., Katzin, N., and Schmid, M. (2003) Multicolor spectral karyotyping of rat chromosomes. *Cytogenet Genome Res* **103**, 163–168.

27. Dugan, L. C., Pattee, M. S., Williams, J., Eklund, M., Sorensen, K., Bedford, J. S., and Christian, A. T. (2005) Polymerase chain reaction-based suppression of repetitive sequences in whole chromosome painting probes for FISH, *Chromosome Res* **13**, 27–32

28. Hoebee, B., and de Stoppelaar, J. M. (1996) The isolation of rat chromosome probes and their application in cytogenetic tests, *Mutat Res* **372**, 205–210.

29. Schrock, E., Zschieschang, P., O'Brien, P., Helmrich, A., Hardt, T., Matthaei, A., and Stout-Weider, K. (2006) Spectral karyotyping of human, mouse, rat and ape chromosomes – applications for genetic diagnostics and research, *Cytogenet Genome Res* **114**, 199–221.

30. Scherthan, H., Cremer, T., Arnason, U., Weier, H. U., Lima-de-Faria, A., and Fronicke, L. (1994) Comparative chromosome painting discloses homologous segments in distantly related mammals, *Nat Genet* **6**, 342–347.

31. Grutzner, F., Himmelbauer, H., Paulsen, M., Ropers, H. H., and Haaf, T. (1999) Comparative mapping of mouse and rat chromosomes by fluorescence in situ hybridization, *Genomics* **55**, 306–313.

32. Guilly, M. N., Fouchet, P., de Chamisso, P., Schmitz, A., and Dutrillaux, B. (1999) Comparative karyotype of rat and mouse using bidirectional chromosome painting, *Chromosome Res* **7**, 213–221.

Chapter 33

Fluorescence In Situ Hybridization on Early Porcine Embryos

Helen A. Foster, Roger G. Sturmey, Paula J. Stokes, Henry J. Leese, Joanna M. Bridger, and Darren K. Griffin

Abstract

Insight into the normal and abnormal function of an interphase nucleus can be revealed by using fluorescence in situ hybridization (FISH) to determine chromosome copy number and/or the nuclear position of loci or chromosome territories. FISH has been used extensively in studies of mouse and human early embryos, however, translation of such methods to domestic species have been hindered by the presence of high levels of intracytoplasmic lipid in these embryos which can impede the efficiency of FISH. This chapter describes in detail a FISH protocol for overcoming this problem. Following extensive technical development, the protocol was derived and optimized for IVF porcine embryos to enable investigation of whole chromosome and subchromosomal regions by FISH during these early stages of development. Porcine embryos can be generated in-vitro using semen samples from commercial companies and oocytes retrieved from discarded abattoir material. According to our method, porcine embryos are lyzed and immobilized on slides using Hydrochloric acid and "Tween 20" detergent, prior to pretreatment with RNase A and pepsin before FISH. The method described has been optimized for subsequent analysis of FISH in two dimensions since organic solvents, which are necessary to remove the lipid, have the effect of flattening the nuclear structure. The work in this chapter has focussed on the pig; however, such methods could be applied to bovine, ovine, and canine embryos, all of which are rich in lipid.

Key words: Fluorescence in situ hybridisation, Porcine embryos, Genome organisation, Lipid

1. Introduction

The normal functioning, or "nuclear health" of an interphase nucleus, including appropriate regulation of gene expression is related both to chromosome copy number and the spatio–temporal organization of the chromatin and associated proteins (termed "nuclear organization" or "genome organization") (Foster and Bridger 2005). A critical stage of development where gene regulation

Joanna M. Bridger and Emanuela V. Volpi (eds.), *Fluorescence in situ Hybridization (FISH): Protocols and Applications*, Methods in Molecular Biology, vol. 659,
DOI 10.1007/978-1-60761-789-1_33, © Springer Science+Business Media, LLC 2010

needs to be tightly controlled temporally is in the preimplantation stages of early embryo development when master regulator genes are being switched on and off in a highly coordinated manner. By implication, genome organization may similarly be tightly regulated at this developmental timepoint, both temporally and in three-dimensional space. Study of genome organization is considered to be fundamental to our understanding of the earliest stages of development; moreover, analysis of chromosome copy number in preimplantation embryo nuclei can provide insight into the origin of mosaicism, a common phenomenon leading to pregnancy complications and specific disease phenotypes such as Prader-Willi syndrome (1). While access to the early embryos of oviparous animals is relatively straightforward, in mammals it is complicated by the fact that fertilization and subsequent development usually occur internally. Embryos from in vitro-fertilization (IVF) provide the opportunity to circumvent this problem, and some studies have been able to assess genome organization in preimplantation human development (2–4). However, due to ethical constraints surrounding the use of human IVF embryos for research, such work has thus far been limited to the analysis of FISH preparations previously designed to determine chromosome copy number. The loci examined are therefore commonly limited to centromeres used for detection of aneuplopidy and requiring three-dimensional extrapolations from FISH procedures performed on flattened (2D) nuclei. In model organisms, some studies have been able to determine directly the 3D organization of the interphase nucleus preimplantation embryos of mammalian model organisms such as mouse (5) and cattle (6), however, such studies are in their infancy.

In many areas of scientific enquiry, the pig has been reported as an important model organism for human disease and genomics (7). Unlike the mouse, it is similar to humans in size and physiology (7); for this reason, it is the primary candidate for xenotransplantation studies (8). Many traits studied in detail for agricultural reasons (e.g., fatness, disease resistance, and fertility) are common issues in human health and since pig meat is the most commonly eaten in the World (9), material from most porcine cell types is relatively easy to obtain from abattoirs. From both a chromosomal and sequence-based standpoint, pigs are much more closely related to humans than mice (10) and, for all the above reasons, the sequencing of the porcine genome is near completion at the time of writing:

http://www.pre.ensembl.org/Sus_Scrofa/Info/Index
http://www.animalgenome.org/pigs/, http://www.ncbi. nlm.nih.gov/projects/genome/guide/pig/, http://www. sanger.ac.uk/Projects/S_scrofa/, http://www.projects.roslin. ac.uk/pigmap/, http://pigenome.nabc.go.kr/ (11).

Although the study of chromosome copy number and genome organization of preimplantation mammalian embryos is important, there remain certain technical issues, particularly when working with porcine embryos. With this in mind, the purpose of this chapter is to describe how IVF may be used to produce porcine preimplantation embryos, the nuclei from which may be fixed to glass slides in order to perform FISH.

2. Materials

2.1. In Vitro Porcine Embryo Production

Unless otherwise stated, all chemicals were sourced from Sigma-Aldrich UK

1. An 18.5-gauge needle.
2. A 10 mL disposable syringe.
3. Pig ovaries acquired from an abattoir (see Note 1).
4. Holding Medium – Tissue Culture Medium TCM199, supplemented with 5.0 mM NaHCO$_3$, 15.0 mM Hepes (Na salt/free acid) (Merck Biosciences Ltd), 0.05 g Kanamycin Sulphate/L 0.4 g Bovine Serum Albumin/L (Fraction V), and 0.04 g Heparin/L. This should be prewarmed to 39°C.
5. Oocyte maturation medium (TCM199 plus 0.1% (w/v) Poly (Vinyl alcohol), average mol wt 30,000–70,000 containing 0.5 μg porcine FSH mL/L, 0.5 μg porcine LH mL/L, 0.57 mmol cysteine, and 10 ng epidermal growth factor (EGF) mL/L).
6. Mineral oil, embryo culture tested.
7. 0.1% (w/v) hyaluronidase in 0.9% NaCl.
8. In vitro fertilization medium (IVF), medium – modified Tris-buffered medium (mTBM) (12); supplemented with 2.5 mM caffeine and 0.4% (w/v) BSA.
9. CO$_2$ Incubator.
10. Dissecting stereomicroscope.
11. Porcine Spermatozoa (see Note 2).
12. Bench top centrifuge.
13. 45%: 90% Percoll gradient (GE Healthcare Bio-Sciences AB (Uppsala)) – (see Note 3).
14. NCSU23 + 0.4% Bovine Serum Albumin (essentially fatty acid free) medium for embryo culture (13).
15. Bench-top Vortex.

2.2. Fixation and Preparation of Early Porcine Embryos for FISH

1. Hydrochloric Acid (HCl).
2. Tween-20.
3. Phosphate-buffered saline tablets.
4. Pulled glass pipettes – (see Note 4).
5. Poly-L-lysine slides.

6. Dissecting microscope (Olympus SZX 7 stereo microscope system) with a 39°C heated stage.

7. Diamond pen.

8. Ethanol series (70%, 90%, and 100%).

9. Pepsin.

10. RNAse A.

11. Sodium saline citrate (SSC).

12. Spreading solution (0.1% Tween 20, 0.01 N HCl).

2.3. Fluorescence In Situ Hybridization of Porcine Chromosomes in Nuclei of Early Porcine Embryos

1. Flow sorted porcine chromosome paints – (see Note 5).

2. Porcine genomic DNA – (see Note 6).

3. Herring sperm DNA.

4. 3 M Sodium Acetate.

5. High quality ethanol.

6. Hybridization mixture (50% formamide, 10% dextran sulphate, 2× Sodium Saline Citrate (recipe), and 1% Tween-20).

7. Water baths.

8. Hot block.

9. Glass coverslips 18 × 18 mm.

10. Rubber cement.

11. Humidified container.

12. Formamide.

13. 2× SSC (0.15 M Sodium citrate, 0.1 M NaCl).

14. Tween 20.

15. Bovine serum albumin (BSA).

16. Distilled H_2O.

17. Strepavidin-cyanine 3.

18. Vectashield containing 6-diamidino-2-phenylindole (DAPI) (Vector Laboratories, Bethesda).

3. Methods

Porcine embryos contain large amounts of lipid, mainly in the form of triglyceride (14) meaning that FISH on this material can be particularly challenging if not addressed. It is important therefore to remove sufficient amounts of the lipid to avoid impairing the experimental regime. We have pursued a number of methods to improve the visualization of FISH signals in porcine embryos. The method presented in this chapter is adapted from Rooney and Czepulkowski (15), which gives the best results, but has the effect of flattening the nucleus.

3.1. In Vitro Embryo Porcine Embryo Production

1. Aspirate oocytes from nonatretic antral follicles that are 3–6 mm in diameter using an 18.5-gauge needle attached to a disposable syringe containing 2 mL of prewarmed Holding Medium (see Fig. 1). The needle should puncture the follicle and the contents are removed by filling the syringe.

2. Pass the contents of the aspirated fluid through a 70 μm cell strainer (Falcon), emptied into a 9 cm Petri-dish and oocyte-cumulus complexes selected. Select OCCs (see Note 7), with an intact, evenly granulated ooplasm and a minimum of two layers of cumulus cells. Wash the OCCs twice in Holding Medium without heparin and a further three times in oocyte maturation medium (16).

3. Culture groups of 50 OCCs in 100 μL droplets of oocyte maturation medium previously covered in mineral oil and preequilibrated in air with 5% CO_2 in air for 40–44 h.

4. Prior to IVF, add 2 μL 0.1% (w/v) hyaluronidase to the droplets and incubate at 30°C; this will loosen of the expanded cumulus oophorus.

5. Wash the OCCs in IVF medium three times.

6. Place groups of 35 OCCs in 50 μL droplets of IVF medium, previously overlaid with mineral oil and preequilibrated in a 5% CO_2 incubator at 39°C.

7. Prepare the 45%: 90% Percoll gradient (see Note 3).

8. Thaw frozen spermatozoa in a 45°C water bath for 10 s. Motile spermatozoa are isolated via centrifugation at $1,000 \times g$ in a 45%: 90% Percoll gradient for 30 min (17).

Fig. 1. Abattoir-derived porcine ovary, removed from the reproductive tract.

9. To remove all traces of Percoll, the motile spermatozoa are resuspended in 4 mL pregassed mTBM and centrifuged at $500 \times g$ for 10 min.

10. Resuspend the pellet in mTBM and add 1.5×10^6 sperm to each of the IVF droplets containing the oocytes. Incubate the gametes for 6 h to allow fertilization to occur.

11. Collect presumptive zygotes and place in 1 mL of preequilibrated NCSU23 medium (18), supplemented with 0.4% (w/v) BSA, and vortex for 2 min to detach the spermatozoa and cumulus cells.

12. Wash the putative zygotes twice in preequilibrated NCSU23 + 0.4% (w/v) BSA, prior to culturing in 20 μL droplets of the same medium for 144 h (blastocyst) or 168 h (expanded blastocyst).

3.2. Preparation of Embryos for FISH

The following protocol was developed for porcine embryos involving using acid/Tween 20 prior to pretreatment with RNase A and pepsin for subsequent FISH procedures. It is based on previously published work (2, 4, 15, 19–22).

1. Wash the porcine embryos in 1× PBS prior to treatment.

2. Transfer the embryos using a finely drawn glass Pasteur pipette primed with spreading solution to a drop of spreading solution onto a poly-L-lysine coated slide (see Note 8). Allow the embryos to lyse completely and the lipid contents disperse over a relatively large area of the slide, thus freeing the nuclei from the cytoplasm. Lipid droplets from the lysed embryos clearly appear as black flecks within the spreading solution. This should be observed using a dissecting microscope (e.g., we used an Olympus SZX 7 stereo microscope system using 10× eyepieces and a 40× objective) with a 39°C heated stage. Add fresh spreading solution gently until the embryos are completely lysed (see Note 8).

3. Once embryo lysis occurs, immediately air-dry the embryos at 39°C on the heated microscope stage. The area of embryo lysis should be marked on the slide using a diamond pen.

4. Incubate the slides in PBS for 5 min, then through a 70%, 90%, and 100% ethanol series (5 min each).

5. The slides and adhered embryos can be stored for up to 2 weeks at room temperature in a sealed box containing silica crystals to provide a desiccated environment, or stored at –20°C for up to 1 year.

3.3. FISH on Early Porcine Embryos

The following pretreatment was adapted from van Minnen and van Kesteren, (1999) (23).

1. Incubate slides in 100 μg RNase A, 2× SSC for 1 h at 37°C, and rinse three times in 1× PBS.

2. Incubate the slides in 0.005% pepsin, 0.01 M HCl at 37°C for 15 min.

3. Rinse the slides three times in 1× PBS, dehydrate through a 70%, 90%, and 100% ethanol series, and air-dry.

4. Place 300 ng biotin-16-dUTP-labeled chromosome paint, 50 μg sheared porcine genomic DNA, and 3 μg herring sperm into an 1.5 mL Eppendorf tube and add one-tenth volume of 3 M sodium acetate and 2.5 volumes of 100% ethanol.

5. Incubate at –80°C for at least 1 h.

6. Centrifuge at 15,000× *g* for 30 min in a microfuge and remove supernatant.

7. Wash pellet in ice-cold 70% ethanol and centrifuge for a further 15 min at 15,000× *g*.

8. Remove supernatant and repeat step 7 one more time.

9. Dry the pellet on a hot-block set at 37°C (see Note 9).

10. Dissolve the probe in hybridization mixture at 50°C for a minimum of 2 h before performing FISH.

11. Denature the probe for 3.5 min at 75°C before application to the embryo slides.

12. Seal the probe on the slide with an 18×18 mm coverslip and rubber cement.

13. Place the sealed slide on a hot-block set at 75°C for 3 min 15 s.

14. Incubate the slides at 37°C overnight in a humidified container.

3.4. Posthybridization Washes and Visualization of Probe

1. Remove the coverslips carefully and wash the slides three times for 5 min each in 50% formamide, 2× SSC, pH 7.0, at 45°C.

2. Wash the slides with 0.1× SSC prewarmed to 60°C but placed in a 45°C water bath, three times, for 5 min each.

3. Transfer the slides to 4× SSC, 0.05% Tween 20 for 15 min at room temperature.

4. Apply the blocking solution to the slide – 150 μl 4× SSC, 0.05% Tween 20 containing 3% bovine serum albumin (w/v), and incubate for 20 min at room temperature.

5. Apply 150 μL of 120 mg/mL streptavidin cyanine 3, 4× SSC, 3% BSA, and 0.05% Tween 20 to each slide and incubate at 37°C for 30 min in darkness.

6. Wash the slides three times in 4× SSC, 0.05% Tween 20 in darkness at 42°C for 5 min each, before a brief wash in fresh deionized water.

7. Air-dry the samples and mount in Vectashield antifade mountant containing 2 μg DAPI as a counterstain.

8. View and image on an epifluorescence microscope. (Fig. 2).

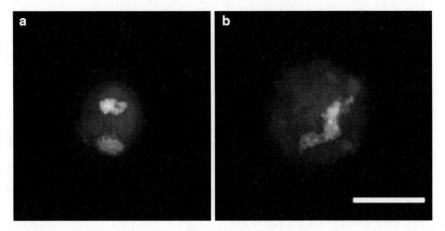

Fig. 2. Nuclei derived from the fixed in vitro constructed porcine embryos (**a**). A nucleus from a 2-cell embryo and (**b**). A nucleus from a blastocyst. DAPI staining is shown in *blue* and porcine chromosome X paint in *red* and porcine chromosome Y in *green*. Scale bar = 10 μm.

4. Notes

1. Many slaughterhouses or abbatoirs will allow collection of sows' ovaries not required by the meat trade to be used for scientific research. Ovaries should be transported back to the laboratory under temperature-controlled conditions. It is sufficient to use a thermos-style flask (for example Dilvac Dewar, UK) containing PBS prewarmed to 39°C supplemented with 400 μl of antibiotic–antimicotic solution (10,000 U penicillin G per mL, 10,000 μg streptomycin sulphate per mL; Invitrogen Life Technologies).

2. "Cryopreserved spermatozoa from a boar of proven fertility can be purchased from commercial genetic companies." Examples would include (but not exclusively) Genus Breeding, ACMC, JSR.

3. The Percoll™ gradient (45%/90%) wash is performed on sperm prior to IVF to remove dead sperm and extenders. The 90% Percoll™ is made by adding 0.6 mL of "Percoll additives" to 4.5 mL of Percoll. Percoll additives comprise 209 mmol $CaCl_2 \cdot 2H_2O/L$, 40 mmol $MgCl_2 \cdot 6H_2O/L$, 806 mmol NaCl/L, 31.1 mM KCl, 2.88 mM $NaH_2PO_4 \cdot 2H_2O$, 10 mg/mL Gentamycin. 42.1 mM Hepes Free Acid, 49.8 mM Hepes Na salt, and 0.64 mM Na–Lactate made up in sterile water. The 45% Percoll is made by adding 2 mL of "SPTL" to 2 mL of 90% Percoll. SPTL comprises 100 mM NaCl, 3.78 mM KCl, 0.38 mM NaH_2PO_4, 10 mg Gentamycin mL, 5.28 mM Hepes free acid, 6.24 mM Hepes Na Salt, 0.2 mM Na Lactate, 1.67 mM $CaCl_2 \cdot 2H_2O$, 0.32 mM $MgCl_2 \cdot 7H_2O$,

and 0.5% (v/v) phenol red solution, made up in water. To make the gradient, layer 2 mL of 45% Percoll on top of 2 mL of 90% Percoll in a 12 mL centrifuge tube and warm to 39°C. Thawed spermatozoa should be layered on top of the gradient and centrifuge-washed as detailed above.

4. Construction of hand-pulled Pasteur pipettes: The neck of the Pasteur pipette is gently heated over a burner until the glass begins to soften. It is then removed from the flame and pulled/stretched producing a narrow diameter pipette for aspirating oocytes and embryos.

5. The flow-sorted porcine chromosome material in our hands was obtained from Prof Malcolm Ferguson-Smith, Department of Clinical and Veterinary Medicine, University of Cambridge. The chromosomes were amplified by DOP-PCR to create to template stocks, primary and secondary. Labeled painting probes were obtained from the secondary template. At the time of writing, porcine chromosome paints are also available from Cambio (Cambridge).

6. To make porcine genomic DNA for suppression of repetitive sequences, we used pig liver that had been freshly removed from a culled pig and frozen at −20°C, and standard kit-based extraction protocols.

7. OCCs are oocytes surrounded by cumulus cells. The OCCs with two layers of cells should be collected as they are optimal for IVF.

8. The spreading solution is used to dissolve the embryo's zona pellucida and cytoplasm. However, due to the high lipid content of porcine embryos (14), blastomeres appear dark, therefore, it can be difficult to detect nuclei from the embryos at this stage.

9. Do not over-dry the pellet – dry it until it is just transparent.

References

1. Hanel ML, Wevrick R (2001). The role of genomic imprinting in human developmental disorders: lessons from Prader-Willi syndrome. *Clin Genet*, **59**, 156–164

2. McKenzie LJ, Carson SA, Marcelli S, Rooney E, Cisneros P, Torskey S, Buster J, Simpson JL, Bischoff FZ. (2004). Nuclear chromosomal localization in human preimplantation embryos: correlation with aneuploidy and embryo morphology. *Hum Reprod*, **19**, 2231–2237.

3. Diblík J, Macek M Sr, Magli MC, Krejčí R, Gianaroli L. (2007) Chromosome topology in normal and aneuploid blastomeres from human embryos. Prenat Diagn, **27**, 1091–1099.

4. Finch KA, Fonseka G, Ioannou D, Hickson N, Barclay Z, Chatzimeletiou K, Mantzouratou A, Handyside A, Delhanty J, Griffin DK. (2008). Nuclear organisation in totipotent human nuclei and its relationship to chromosomal abnormality. *J Cell Sci*, **121**, 655–663.

5. Lanctôt C, Kaspar C, Cremer T. (2007). Positioning of the mouse Hox gene clusters in the nuclei of developing embryos and differentiating embryoid bodies. *Exp Cell Res*, **313**, 1449–1459.

6. Koehler D, Zakhartchenko V, Froenicke L, Stone G, Stanyon R, Wolf E, Cremer T, Brero A. (2009). Changes of higher order chromatin

arrangements during major genome activation in bovine preimplantation embryos. *Exp Cell Res*, **315**, 2053–2063.

7. Lunney JK. (2007) Advances in swine biomedical model genomics. *Int J Biolol Sci*, **3**, 179.

8. Weiss RA. (1998) Transgenic pigs and virus adaptation. *Nature*, **391**; 327–327.

9. Guan TY, Holley RA (2003) Pathogen Survival in Swine Manure Environments and Transmission of Human Enteric Illness-A Review Sponsoring organizations: Manitoba Livestock Manure Management Initiative and Manitoba Rural Adaptation Council. In.: *Am Soc Agronom*, 383–392.

10. Rettenberger G, Klett C, Zechner U, Kunz J, Vogel W, Hameister H. (1995) Visualization of the conservation of synteny between humans and pigs by heterologous chromosomal painting. *Genomics*, **26**, 372–378.

11. Lim D, Cho YM, Lee KT, Kang Y, Sung S, Nam J, Park EW, Oh SJ, Im SK, Kim H. (2009). The Pig Genome Database (PiGenome): an integrated database for pig genome research. *Mamm Genome*, **20**, 60–66.

12. Abeydeera LR, Day BN. (1997) Fertilization and subsequent development in vitro of pig oocytes inseminated in a modified tris-buffered medium with frozen-thawed ejaculated spermatozoa. *Biol Reprod*, **57**, 729–734.

13. Petters RM, Reed ML. (1991) Addition of taurine or hypotaurine to culture medium improves development of one and two cell pig embryos in vitro. *Theriogenology*, **35**, 253.

14. Sturmey RG, Leese HJ. (2003) Energy metabolism in pig oocytes and early embryos. *Reproduction*, **126**, 197–204.

15. Rooney DE, Czepulkowski BH. (1986) Human Cytogenetics: A Practical Approach. Oxford, IRL Press.

16. Abeydeera LR, Wang WH, Cantley TC, Rieke A, Murphy CN, Prather RS, Day BN. (2000) Development and viability of pig oocytes matured in a protein-free medium containing epidermal growth factor. *Theriogenology*, **54**, 787–797.

17. Jeong BS, Yang X. (2001) Cysteine, glutathione, and Percoll treatments improve porcine oocyte maturation and fertilization in vitro. *Mol Reprod Dev*, **59**, 330–335.

18. Petters RM, Johnson BH, Reed ML, Archibong AE. (1990) Glucose, glutamine and inorganic phosphate in early development of the pig embryo in vitro. *J Reprod Fertil*, **89**, 269–275.

19. Coonen E, Dumoulin JC, Ramaekers FC, Hopman AH. (1994) Optimal preparation of preimplantation embryo interphase nuclei for analysis by fluorescence in-situ hybridisation. *Hum Reprod*, **9**, 533–537.

20. Harper JC, Dawson K, Delhanty JD, Winston RM. (1995) The use of fluorescent in-situ hybridisation (FISH) for the analysis of in-vitro fertilization embryos: a diagnostic tool for the infertile couple. *Hum Reprod*, **10**, 3255–3258.

21. Xu K, Huang T, Liu T, Shi Z, Rosenwaks Z. (1998) Improving the fixation method for preimplantation genetic diagnosis by fluorescent in situ hybridisation. *J Assist Reprod Genet*, **15**, 570–574.

22. Daphnis DD, Delhanty JD, Jerkovic S, Geyer J, Craft I, Harper JC. (2005) Detailed FISH analysis of day 5 human embryos reveals the mechanisms leading to mosaic aneuploidy. *Hum Reprod*, **20**, 129–137.

23. van Minnen J, van Kesteren RE. (1999) Methods towards detection of protein synthesis in dendrites and axons. In: Modern Techniques in Neuroscience Research (Windhorst U, Johansson H, eds.), Chapter 3, pp. 57–88. Springer-Verlag, Heidelberg, Germany.

Chapter 34

FISH on 3D Preserved Bovine and Murine Preimplantation Embryos

Daniela Koehler, Valeri Zakhartchenko, Nina Ketterl, Eckhard Wolf, Thomas Cremer, and Alessandro Brero

Abstract

Fluorescence in situ hybridization (FISH) is a commonly used technique for the visualization of whole chromosomes or subchromosomal regions, such as chromosome arms, bands, centromeres, or single gene loci. FISH is routinely performed on chromosome spreads, as well as on three-dimensionally preserved cells or tissues (3D FISH). We have developed 3D FISH protocol for mammalian preimplantation embryos to investigate the nuclear organization of chromosome territories and subchromosomal regions during the first developmental stages. In contrast to cells, embryos have much more depth and their nuclei are therefore less accessible to probes used to visualize specific genomic regions by FISH. The present protocol was developed to establish a balance between sufficient embryo permeabilization and maximum preservation of nuclear morphology.

Key words: Fluorescence in situ hybridization, Embryo morphology, Preimplantation embryos, Nuclear architecture

1. Introduction

In interphase nuclei of higher eukaryotes, the genome is organized in a spatially highly ordered fashion, including single gene loci (1, 2) as well as complete chromosomes, which are organized as individual chromosome territories (CTs) (3–5) (see Fig. 1). So far, the three-dimensional nuclear organization has been investigated mainly in cultured cells (adherent and in suspension) and to a much lesser extent in somatic tissues, like cryosections and paraffin embedded sections (6, 7), yet very little is known about higher order chromatin arrangements during early development. While 3D FISH for cultured cells (8) as well as for tissue sections (7)

Joanna M. Bridger and Emanuela V. Volpi (eds.), *Fluorescence in situ Hybridization (FISH): Protocols and Applications*, Methods in Molecular Biology, vol. 659,
DOI 10.1007/978-1-60761-789-1_34, © Springer Science+Business Media, LLC 2010

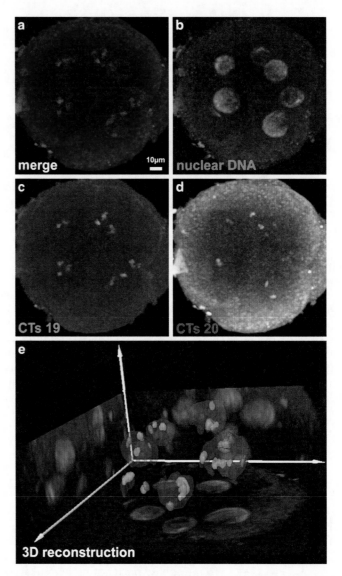

Fig. 1. Bovine in vitro fertilized preimplantation embryo (2 days after fertilization) containing seven nuclei. **(a)-(d):** Maximum intensity projections of nuclear DNA counterstained by DAPI **(b)**, **(a)**(blue) and of FISH-stained chromosome territories 19 **(c)**, **(a)**(green) and 20 **(c)**, **(a)** (red), respectively. Scale bar: 10 μm. **(e)**: 3D reconstruction using the same color code as in **(a)**. Note that nuclei at the bottom of the reconstruction could not be reconstructed properly with respect to their DAPI-stained nuclear DNA as they were further from the objective lense and therefore showed a weaker fluorescence.

is a well-established technique, applying such a 3D FISH protocol successfully on embryos turned out to be difficult due to the large amounts of proteins and RNAs in preimplantation stages.

The adaptation of a 3D FISH protocol for an embryonic specimen was challenging since originally it yielded poor hybridization efficiencies or failed to result in any specific hybridization signal, while harsh pretreatments resulted in reasonable hybridization efficiency but poor preservation of embryonic and nuclear morphology. In order to optimize a 3D FISH protocol for studies on embryos, we performed a series of experiments where we changed the fixation and/or permeabilization parameters systematically. The protocol described in this chapter was successfully applied to both bovine and murine embryos. We have been able to use whole chromosome painting probes as well as probes for subchromosomal regions. Probes were prepared according to a standard protocol (9), except that higher concentrations of probe had to be employed. Detection of FISH probes in embryos was carried out as published previously (8).

2. Materials

2.1. Fixation Procedure

1. Phosphate-buffered saline (1× PBS): 140 mM NaCl; 2.7 mM KCl, 6.5 mM Na_2HPO_4, and 1.5 mM KH_2PO_4, pH 7.2.

2. 3.7% formaldehyde (FA) in PBS: 1 mL of 37% formaldehyde (Sigma-Aldrich) is dissolved in 9 mL of 1× PBS.

3. 0.05 N and 0.1 N hydrochloride acid (HCl): 0.5 or 1 mL of 1 N HCl (Merck) is dissolved in 9.5 or 9 ml dH_2O, respectively.

4. 10% bovine serum albumin (BSA): 10 g of BSA (ICN Biomedicals) are dissolved in 100 mL of dH_2O.

5. PBS washing solution: 0.05% Triton X-100/0.1% BSA/PBS: 500 mL 1× PBS are mixed with 5 mL of 10% BSA and 250 μL Triton X-100.

6. Permeabilization solution 1: 0.5% Triton X-100/0.1% BSA/ PBS: 100 mL 1× PBS are mixed with 1 mL of 10% BSA and 500 μL Triton X-100.

7. Permeabilization solution 2: 0.02% RNase A (200 μg/ mL)/0.5% Triton X-100/0.1% BSA/PBS 100 mL 1× PBS are mixed with 1 mL of 10% BSA and 500 μL Triton X-100. Shortly before use, add 10 μL of 2% RNase A (Roche).

8. 20× SSC (sodium saline citrate) stock solution: 3 M NaCl, 300 mM Na citrate in 1,000 mL dH_2O. Stock solution is diluted to according to the working solutions required, i.e., 2× SSC (10 mL 20× SSC and 90 mL dH_2O).

9. SSC washing solution: 0.1% Triton X-100/0.1% BSA/2× SSC: 500 mL 2× SSC are mixed with 5 mL 10% BSA and 500 μL Triton X-100.

10. 50% formamide (v/v) (Merck) in 2× SSC containing 0.1% Triton X-100, pH 7.0.

2.2. Probe Preparation and Hybridization

1. Labeled chromosome or gene loci probes.
2. C_0t-1 DNA if necessary.
3. 100% ethanol for probe precipitation.
4. Hybridization mixture: 50% formamide, 10% dextran sulphate, 2× SSC.

2.3. Detection of FISH Signals and Embedding

1. 2× SSC including 0.1% BSA.
2. 0.1× SSC including 0.1% BSA.
3. 4× SSC /0.1% BSA and 0.2% Tween 20.
4. Antibody solutions: dilute antibody in 4× SSC/2% BSA/0.2% Tween.
5. Counterstain solution: TOPRO or DAPI in 4× SSC/0.2% Tween.
6. Glycerol solutions: 20, 40, or 60% glycerol in 4× SSCT.
7. 4-well cell culture plates (Nunc GmbH & Co.).
8. 12-well cell culture plates (Greiner bio-one).
9. Microtest plates: Greiner bio-one or alternatively: Petridishes, i.e., Greiner bio-one.
10. Glass slides with metal ring (Brunel Microscopes Ltd).

2.4. Equipment and Microscopes

1. Transferpettor (Brand GmbH).
2. Stereomicroscope (Zeiss).
3. Confocal laser scanning mircroscope (Leica).

---

3. Methods

The handling of embryos is different from the handling of other fixed specimens, since embryos do not readily attach to any surface, at least not without affecting their morphology as is the case if embryos are, e.g., "glued" to a slide or cover glass by air-drying the specimen. This necessitates a special handling of embryos during the FISH procedure. Embryos are transferred from one solution to the next by the use of a special pipette with very thin glass capillaries, a so-called "Transferpettor" (volume: 2–5 μL; Brand). All incubations are performed in 500–1,000 μL of the respective solution in 12-well cell culture plates, or if smaller volumes are necessary (i.e., glycerol incubations) in only 3–10 μL of solution in 60-well microtiter-testplates. Alternatively, all working steps might be carried out in drops (30–50 μL) of the respective solution as embryos can be tracked more easily, while in 12-well dishes embryos might eventually become lost by adhering to the walls of the well, for example. If embryos are incubated in small drops, one should be aware of a dilution effect as 5–10 μL of liquid (depending on the volume of the Transferpettor) are always cotransferred with the embryos thereby diluting the "target-solution". Therefore, it is advisable to always preincubate embryos in an initial drop of the same respective solution. The drop method is especially favorable with the paler and smaller mouse embryos since they are more difficult to track compared to bovine embryos, which are bigger and denser in contrast.

3.1. Fixation
of Embryos

1. Wash embryos briefly in PBS at 37°C.

2. Fixation of embryos is carried out in 3.7% formaldehyde/PBS for 10 min at room temperature (RT).

3. Perform a very short incubation (maximum 1 min) in 0.1 N HCl while embryos are monitored by a stereomicroscope until removal of the zona pellucida is obvious. If embryos are very sensitive to this treatment, the use of 0.05 N HCl is recommended.

4. Wash embryos twice for 10 min each in 0.05% Triton X-100/0.1% BSA/PBS.

5. Permeabilization is performed utilizing 0.5% Triton X-100/0.1% BSA/PBS for 1 h.

6. Perform another short incubation in 0.1 N HCl for approximately 1 min.

7. Wash embryos twice for 10 min each in 0.05% Triton X-100/0.1% BSA/PBS.

8. Wash embryos twice for 10 min in 0.1% Triton-X100/0.1% BSA/2× SSC.

9. Embryos are then incubated at least over night (better for 2 days) in 50% formamide/2× SSC at RT.

10. Wash embryos twice for 10 min in 0.1% Triton-X 100/0.1% BSA/2× SSC.

11. Embryos should then be equilibrated for 10 min in 0.05% TritonX-100/0.1% BSA/PBS.

12. A second permeabilization step is performed utilizing 0.5% Triton X-100/0.1% BSA/PBS including 0.02% RNase A for 1 h.

13. Wash embryos in 0.05% TritonX-100/0.1% BSA/PBS.

14. Briefly incubate embryos in 0.1 N HCl for max. 2 min.

15. Wash embryos for 10 min in 0.05% Triton X-100/0.1% BSA/PBS to remove HCl.

16. Wash embryos twice for 10 min in 0.1% Triton-X 100/0.1% BSA/2× SSC.

17. Prior to hybridization, embryos are incubated for at least 2 days in 50% formamide/2× SSC at RT.

3.2. Probe Precipitation

Labeled chromosome-specific painting probes or locus-specific probes derived from Degenerate Oligo Primer-PCR or nick translation (9) are used together with an adequate amount of unlabeled C_0t-1 DNA (9). The amount of probe should be at least twice as concentrated as for 3D FISH on cultured cells (8, 9).

1. Then add 2.5× volumes of 100% pure ethanol.
2. Incubate the solution for at least 20 min at −20°C.
3. Remove the supernatant and air-dry the pellet.
4. The pellet is dissolved in 6 µL of hybridization mixture that consists of 50% formamide, 10% dextran sulphate, and 2× SSC.

3.3. Denaturation of Specimen and Hybridization

1. Place 6 µL of hybridization mixture in the middle of glass slide containing an aluminium ring with a diameter of approximately 19 mm (Brunel Microscopes Ltd.).
2. Transfer 15 embryos into this hybridization mixture on the glass surface.
3. Carefully cover embryos with mineral oil. The oil is first added at the inner border of the metal ring and slightly shifted to cover the inner part of the area so as to prevent the embryos from moving to the metal ring (if they stick to the ring, they are barely visible and therefore difficult to remove).
4. For equilibration, place slides with embryos in a metal box in a water bath at 37°C.
5. Denaturation of embryo DNA is carried out for 3 min at 76°C on a hot block.
6. Hybridization is carried out for at least 3 days at 37°C in humid atmosphere, e.g., by placing slides with embryos in a metal box in a water bath.

3.4. Detection of FISH signals

1. Remove embryos carefully from the hybridization mixture and wash twice for 10 min in 2× SSC/0.1% BSA.
2. Perform two stringent washing steps for 10 min each in 0.1× SSC/0.1% BSA.
3. Incubate in 4× SSCT/0.1% BSA.
4. For blocking nonspecific antibody binding sites, incubate embryos in 4% BSA/4× SSCT for 10 min.
5. The first antibody layer (antibodies diluted in 2% BSA/4× SSCT) is incubated over night at 4°C.
6. Wash embryos twice in 4× SSCT/0.1% BSA at RT.
7. The second antibody layer (appropriate antibodies diluted in 2% BSA/4× SSCT) is incubated for 90 min at RT.
8. Wash embryos twice in 4× SSCT/0.1% BSA at RT.
9. To counterstain genomic DNA of embryonic nuclei, incubate embryos for 30 min in 0.1 µg/ml DAPI/4× SSCT.

10. Embryos are then subjected to an increasing glycerol dilution series, by incubating embryos for 5 min in 20, 40, and 60% glycerol/4× SSCT. These steps are performed in a 5 µL volume only in a 60-well microtiter-testplate. Embryos become pale and barely visible in glycerol solutions. It is easier therefore to rescue them from a small volume. Embryos are briefly rinsed in a first drop of the respective glycerol solution before the 5 min incubation step, to compensate for the dilution effect, due to the small volume in which they are incubated.

11. Incubate embryos briefly in Vectashield antifade medium (Vector Laboratories) supplied with 0.1 µg/ml DAPI before being embedded.

3.5. Embedding of Specimen

1. Prepare a dilution of Poly-L-Lysine in H_2O (final concentration: 1 mg/mL) and place approximately 20 µL in each well of a µ-slide (Ibidi).

2. Incubate Poly-L-Lysine on the slides for 60 min and drain afterwards.

3. Place embryos on the slides with the minimum amount of liquid as possible.

4. Remove excess antifade solution to enable direct contact of embryos to the surface of the slide. Embryos should only be covered by a very small amount of Vectashield.

5. Embryos are allowed to attach over night at 4°C onto the surface of the slide. Place them in a humid chamber to avoid drying.

6. The next day cover embryos carefully with an antifade solution containing 0.1 µg/ml DAPI.

3.6. Microscopy

1. Record the position of the embryos on the slides by using a stereomicroscope so that they can be easily found at the confocal laser scanning microscope (CLSM).

2. Embryos are imaged at a CLSM using a 63×/1.4 NA oil immersion objective lens for high resolution images.

3. Overview images can be taken using a 10×/0.3 Dry Phase 1 air objective lens.

4. Notes

1. The present protocol is originally developed for bovine IVF preimplantation embryos. However, we could also successfully apply to bovine nuclear transfer (NT) embryos as well as to mouse embryos. Since NT embryos are in general more sensitive than IVF embryos, they are treated in 0.05 N HCl instead of 0.1 N HCl to remove the zona pellucida.

2. Embryos are loose within the solution. Note that the bigger the volume, the more difficult it is to trace/track embryos. On the other hand, the smaller the volumes of liquids the bigger the dilution effect is when embryos are transferred. The protocol as exemplified where some steps are performed in drops, some in larger wells worked optimally in our hands, however, small modifications thereof are not likely to have any effect on the outcome of the experiment.

3. Note that solutions with a low surface tension tend to dissolve and not to build compact drops. Therefore, embryos in such dissolving solutions are at risk to dry out. Such solutions should rather be used in larger wells.

4. If you prefer working in 12-well plates, try to avoid that embryos touch the border of the well. It is hard to track embryos in those regions because of the suboptimal illumination and refraction.

5. After removal of the zona pellucida, embryos become more sensitive to the various treatments. HCl incubations should therefore be monitored with the help of a stereomicroscope to prevent morphological damage. Moreover, embryos without the zona pellucida stick more easily to any kind of surface. To minimize this stickiness, BSA and detergents like Triton X-100 or Tween 20 are added to the solutions. Following modifications of the protocol were successfully applied: (a) The incubation step in formamide during fixation procedure was reduced to an overnight incubation instead of 2 days (see step 9 of subheading 3.1). (b) The overnight antibody incubation (see step 5 of subheading 3.5) could be reduced to 2 h.

6. For embedding, any kind of slides/coverslips that allow a direct visualization of embryos through the bottom of the slides can be used. We recommend μ-slides (Ibidi) but were also successful using Lab Tek coverglass multiwall chambers (Nunc GmbH & Co. KG).

7. Note that with a conventional 63×/1.4 NA oil-immersion objective lens not all nuclei of an embryo can be recorded, due to the limited depth of field. Nuclei located in the lower part (the half located towards the slide) of the embryos can be appropriately focused while nuclei in the upper part of the embryo cannot.

Acknowledgments

We want to thank especially Irina Solovei, Marion Cremer, and Stefan Müller for helpful discussions on the project. This work was supported by the Deutsche Forschungsgemeinschaft (DFG) CR 59 and ZA 425.

References

1. Kupper, K., A. Kolbl, D. Biener, S. Dittrich, J. von Hase, T. Thormeyer, H. Fiegler, N. P. Carter, M. R. Speicher, T. Cremer and M. Cremer (2007). Radial chromatin positioning is shaped by local gene density, not by gene expression. *Chromosoma* **116**, 285–306.

2. Lanctot, C., T. Cheutin, M. Cremer, G. Cavalli and T. Cremer (2007). Dynamic genome architecture in the nuclear space: regulation of gene expression in three dimensions. *Nat Rev Genet* **8**, 104–115

3. Albiez, H., M. Cremer, C. Tiberi, L. Vecchio, L. Schermelleh, S. Dittrich, K. Kupper, B. Joffe, T. Thormeyer, J. von Hase, S. Yang, K. Rohr, H. Leonhardt, I. Solovei, C. Cremer, S. Fakan and T. Cremer (2006). Chromatin domains and the interchromatin compartment form structurally defined and functionally interacting nuclear networks. *Chromosome Res* **14**, 707-33.

4. Cremer, T., M. Cremer, S. Dietzel, S. Muller, I. Solovei and S. Fakan (2006). Chromosome territories–a functional nuclear landscape. *Curr Opin Cell Biol* **18**, 307–316.

5. Meaburn, K. J. and T. Misteli (2007). Cell biology: chromosome territories. *Nature* **445**, 379–781.

6. Solovei, I., J. Walter, M. Cremer, F. Habermann, L. Schermelleh and T. Cremer (2001). FISH on three-dimensionally preserved nuclei. In: J. Squire, B. Beatty and S. Mai. FISH: A Practical Approach. Oxford University Press, Oxford: 119–157.

7. Solovei, I., F. Grasser and C. Lanctot. FISH on Histological Sections. *Cold Spring Harb Protoc.*; 2007; doi:10.1101/pdb. prot4729.

8. Cremer, M., S. Muller, D. Kohler, A. Brero and I. Solovei (2007). Cell Preparation and Multicolor FISH in 3D Preserved Cultured Mammalian Cells. *Cold Spring Harb Protoc.*; 2007; doi:10.1101/pdb.prot4723.

9. Müller, S., M. Neusser, D. Köhler and M. Cremer. Preparation of Complex DNA Probe Sets for 3D FISH with up to Six Different Fluorochromes. *Cold Spring Harb Protoc.*; 2007; doi:10.1101/pdb. prot4730.

INDEX

Joanna M. Bridger and Emanuela V. Volpi (eds.), *Fluorescence in situ Hybridization (FISH):
Protocols and Applications*, Methods in Molecular Biology, vol. 659,
DOI 10.1007/978-1-60761-789-1, © Springer Science+Business Media, LLC 2010